Cyber Security in the Age of Artificial Intelligence and Autonomous Weapons

Although recent advances in technology have made life easier for individuals, societies, and states, they have also led to the emergence of new and different problems in the context of security. In this context, it does not seem possible to analyze the developments in the field of cyber security only with information theft or hacking, especially in the age of artificial intelligence and autonomous weapons. For this reason, the main purpose of this book is to explain the phenomena from a different perspective by addressing artificial intelligence and autonomous weapons, which remain in the background while focusing on cyber security. By addressing these phenomena, the book aims to make the study multidisciplinary and to include authors from different countries and different geographies. The scope and content of the study differs significantly from other books in terms of the issues it addresses and deals with.

When we look at the main features of the study, we can say the following:

- Handles the concept of security within the framework of technological development
- Includes artificial intelligence and radicalization, which has little place in the literature
- Evaluates the phenomenon of cyber espionage
- Provides an approach to future wars
- Examines the course of wars within the framework of the Clausewitz trilogy
- Explores ethical elements
- Addresses legal approaches

In this context, the book offers readers a hope as well as a warning about how technology can be used for the public good. Individuals working in government, law enforcement, and technology companies can learn useful lessons from it.

Mehmet Emin Erendor earned a bachelor's degree in international relations at Kırıkkale University in 2008, a master's degree in international law at the University of Sussex, Brighton in 2011, and a PhD in international relations at the University of Southampton in 2017. In 2017–2018, he worked as a Research Assistant at Çukurova University. Since 2018, he has been an Associate Professor in the Department of International Relations at Adana Alparslan Türkeş Science and Technology University, Adana, Turkey. His research interests include cyber security, terrorism, the Turkic world, Turkish Foreign Policy, and international organizations.

Cyber Security in the Age of Artificial Intelligence and Autonomous Weapons

Edited by
Mehmet Emin Erendor

CRC Press
Taylor & Francis Group
Boca Raton London New York

CRC Press is an imprint of the
Taylor & Francis Group, an **informa** business

Designed cover image: © Shutterstock
First edition published 2025
by CRC Press
2385 NW Executive Center Drive, Suite 320, Boca Raton FL 33431

and by CRC Press
4 Park Square, Milton Park, Abingdon, Oxon, OX14 4RN

CRC Press is an imprint of Taylor & Francis Group, LLC

ISBN: 978-1-032-57939-9 (hbk)
ISBN: 978-1-032-57938-2 (pbk)
ISBN: 978-1-003-44170-0 (ebk)

DOI: 10.1201/9781003441700

Contents

Preface

The parameters surrounding the concept of security are always evolving. The concept of security has begun to shift, especially in the last ten years, as a result of the Internet's openness to civilian usage and advancements in the sector. The 1980s saw the start of Internet attacks, but as the network's usage has grown, so too have the attacks' sophistication and complexity. The field of cyber security has developed beyond identity theft, information theft, hacking, and related attacks and has given rise to a variety of new attack types and possibilities. It is gaining the interest of both individuals and the international community as well as states.

In recent years, humanity has faced a number of difficult issues and security threats as a result of the rapid advancement of technology, but it has also created many new and exciting prospects. In this regard, cutting-edge technologies like autonomous weapons and artificial intelligence (AI) have opened new possibilities and fundamentally changed the cyber security landscape. The swift transformation of AI-driven systems and self-governing armaments has resulted in a shift in cyber-attacks and a reconfiguration of defense experts' tactics and methods. It is reasonable to argue that the employment of AI and autonomous weapons has naturally changed the nature of cyber security attacks and operations. These technologies have had a profound impact on military, industrial, and many other facets of daily life in recent years. Stated differently, the swift augmentation in the utilization of these technologies has introduced intricate and novel predicaments in the realm of cyber security.

Security studies are increasingly altering as a result of these advancements in the discipline finding a home in a variety of scholarly investigations. The security domain, which has occasionally taken centuries to evolve historically, has undergone numerous transformations, particularly in the past century, or it has enabled the assessment of various concepts in the security space. Because of this, it's critical to assess, examine, and consider the state of security theory and practices. AI is one of the most important security-related concerns that needs to be assessed in this environment. One of the most crucial things to remember about AI is that it has the capacity to learn from itself. AI is unique among technologies that endows computer systems with human-like intellect. AI systems have the capability to execute diverse tasks, particularly through the utilization of machine learning and deep learning algorithms. AI can be applied in practically any industry and can be integrated into daily life if these tasks are completed.

AI is now used in every area, including automation, healthcare, finance, commerce, and education. The claim that AI, which has applications in a wide range of industries, is solely beneficial is untrue. Sometimes, certain parties or terrorist organizations will employ such sophisticated technologies for their own ends. Consequently, an analysis of AI's drawbacks and implications for cyber security is important. Within this framework, it may be argued that the domains of AI and cyber security represent the cutting edge of modern technological advancement. AI is a technology that, as previously said, endows computer systems with human-like learning and decision-making abilities. These abilities could have a significant influence on cyber security. AI systems play significant roles in security training, vulnerability management, and threat detection. These systems are being used, or have been adopted, by numerous governments and organizations.

There are two areas in cyber security where AI is obviously useful: threat detection and prevention. Through network traffic monitoring, AI-enabled systems might potentially identify anomalous patterns and predict impending threats. Furthermore, these systems have the capability to promptly detect and address security flaws. Because AI is still evolving, the relationship between cyber security and AI has not been fully absorbed mechanically. This suggests, though, that these two spheres might grow more entwined in the future. Experts will be able to create increasingly intricate defense plans as AI advances. Data privacy and ethical issues are, of course, quite important to keep in mind in this situation.

Autonomous weapons, which can locate and engage targets without human intervention, represent a significant advancement in military technology. These systems offer numerous advantages across various domains, from civilian life to military operations. They encompass unmanned vehicles or systems capable of detecting and engaging targets autonomously. They can identify and strike targets automatically under certain conditions or execute predefined missions without human oversight. Autonomous weapons range from advanced weapon systems to unmanned aerial, marine, or ground vehicles, showcasing the diverse applications of this technology.

The evolution of autonomous weapons significantly influences combat dynamics and military tactics. Their ability to independently identify and engage targets offers several advantages in military operations. These include increased operational flexibility, reduced costs, minimized risk to military personnel, and expedited and precise target acquisition.

Nonetheless, the employment of autonomous weapons raises certain moral and legal questions. There may be a higher chance of unintentional strikes against civilian targets if autonomous weapons are able to identify and strike targets. There is also fear that these weapons could have uncontrollable and unforeseen effects if they are used intentionally or accidentally. For this reason, it's critical to create global guidelines and suitable oversight systems for the creation and application of autonomous weapons.

The use of autonomous weapons has the potential to drastically alter how technology is used to military tactics and the battlefield. Nevertheless, the moral and legal ramifications of these weapons must also be taken into account in addition to this potential. Establishing suitable norms and regulations and giving serious consideration to the development and application of autonomous weapons are critical tasks for the international community. Security can be preserved, and the possible dangers of autonomous weapons reduced in this way.

Autonomous weapons and cyber security have a complicated yet significant interaction. While autonomous weapons can present novel and distinctive cyber security concerns, cyber security technology can also be essential to safeguarding autonomous weapon systems (AWS). The use of autonomous weapons could expose nations to fresh, serious cyber security risks. These weapons are Internet-connected systems that are managed by sophisticated software and algorithms, making them susceptible to cyberattacks. Cybercriminals are able to take control of AWS and modify their targets by breaking in. There may be grave repercussions from this, including the possibility of human casualties. As a result, it is crucial to protect autonomous weapons' cyber security. To guarantee the security of AWS, cyber security professionals should make use of technology such as sophisticated encryption, security procedures, and penetration testing. Adopting security standards and principles from these systems' design phase is equally crucial. Security can be preserved, and AWS made more resistant to cyberattacks.

Cyber security faces several difficulties as AI and autonomous weapons become more commonplace. One significant issue is that conventional cyber security techniques are unable to counteract AI-based attacks. Attackers can utilize AI to increase the complexity and difficulty of detecting their attacks. Moreover, autonomous weapon security is another issue that needs to be addressed. These systems must be shielded from tampering and unwanted access.

In the era of autonomous weapons and AI, cyber security is essential. Because of the growing usage of these technologies, cyberattacks have the potential to become more dangerous and sophisticated. For the sake of national security, cyber security precautions are crucial. Furthermore, it is important to guarantee cyber security in multiple domains, including infrastructure systems, industrial facilities, and personal data. Cyberattacks have the potential to seriously harm businesses and the economy. An attack can have a serious negative impact on a company's finances and reputation. Cyber security measures are also essential for individual and societal security. For modern society, preventing cybercrime and safeguarding personal information is essential.

There are several ways to handle cyber security issues. Attacks can be identified and stopped by defense systems powered by AI. To safeguard AWS, advanced technologies for authentication and encryption should be employed. Regularly doing penetration tests and vulnerability analyses on systems is also crucial. It is also essential that staff members receive cyber security training and awareness campaigns. To guarantee security on a global scale, collaboration among nations and the creation of uniform cyber security guidelines are crucial.

In light of the quickly evolving field of technology today, ethical concerns are crucial conversation points. As was briefly stated before, these moral and legal concerns have been well researched in the literature, and societies have been debating them in a variety of settings in an effort to set standards. AI and autonomous weapons are two examples of technologies that must be employed ethically and with cyber security. These are complex problems with interrelated facets that require careful attention. But there are also ethical considerations with using AI. For instance, concerns about algorithmic justice, data privacy, and the openness of AI decision-making are crucial to AI ethics.

There is a chance that using such weapons will result in unforeseen consequences and human casualties. Thus, it is crucial to employ autonomous weapons responsibly and maintain cyber security. By protecting autonomous weapons from outside manipulation and averting unfavorable scenarios, cyber security is ensured. With the advancement of technology like autonomous weapons and AI, cyber security is becoming increasingly crucial in defending data and information systems against threats. To guarantee a successful defense against emerging threats like AI-based assaults, data manipulation, and cyber espionage, cyber security must be continuously upgraded. Furthermore, cyber security should be ensured, and ethical standards should be followed while using AI and autonomous weapons.

This book addresses the various phenomena associated with cyber security, autonomous weapons, and AI in addition to their general relationship. The book's objectives are to familiarize readers with the fundamentals of cyber security concerns and difficulties, as well as to explain the evolution of the security phenomenon and the connections between autonomous weapons and AI and cyber security. Thus, industry experts, professors, and researchers working in the field of cyber security, autonomous weapons, and AI are expected to find this book interesting as it covers the most recent research trends and cutting-edge subjects in these areas.

To facilitate an efficient organization of its material, this book is divided into multiple chapters. This book covers a wide range of topics, including the foundations of cyber security, its applications in other domains, autonomous weapons, and AI. It also includes the most recent advancements and research issues in the field. Therefore, the authors sought to achieve unity while embracing diversity in the sector.

The fundamental ideas discussed above are introduced by Dr. Erendor and Dr. Mincewicz in the first chapter of the book. This chapter aims to give the reader a better understanding of the upcoming chapters by providing a foundational knowledge of cyber security, AI, and autonomous weapons. It also prevents the contributing researchers from having to devote too much time to the concepts in the following chapters, allowing them to concentrate on the essentials of the subject. The first chapter in this context includes information and basic ideas.

Dr. Çıtak presents his viewpoint on the application and advancement of technology in the security industry in Chapter 2, "Current Technological Trends and Developments in the Field of Security." In his view, technology plays a pivotal role in shaping multiple facets of our existence, since it represents the pragmatic implementation of scientific discoveries. The realm of security is one where these effects are most noticeable. The topic of security is crucial for people, societies, and states alike. Security can be threatened by technological advancements as well as benefit from them. States and individuals can improve their security, for instance, by employing technology to identify and neutralize threats. Nevertheless, the same technical advancements may also make malevolent actors more dangerous. Particularly, new developments in technology like biometric data collection, AI-enabled weapons, and cyberattacks might raise dangers and vulnerabilities. Technology can affect security in a variety of ways. AI and data analytics, for instance, can improve the procedures involved in obtaining and analyzing intelligence, and information systems can be safeguarded with cyber security measures. But technology can also put people's privacy at risk and cause society to worry about security. Specifically, personal privacy may be violated by security-related surveillance technologies, which could be viewed as a security risk by the public. It is critical today to evaluate how technological advancements have affected security and to comprehend how potential future technological advancements may affect security regulations. This evaluation can benefit people. States and individuals can identify and build their security strategy with the aid of this assessment. Furthermore, this evaluation has the power to influence how society views security and direct discussions about security-related policies. As a result, it's critical to comprehend how technology has changed security and to be ready for new advances in the industry.

Dr. Benouachane discusses cyber security issues in the age of autonomous weapons and AI in Chapter 3, "Cyber Security Challenges in the Era of Artificial Intelligence and Autonomous Weapons." According to Dr. Benouachane, the military has seen radical changes as a result of the AI technology's quick development. The increasing significance of smoothly incorporating autonomy into a range of weapons is particularly highlighted by recent technical advancements. With the advancement of AWS, robots may now choose and target targets on their own without needing human intervention. These systems, which are among the contentious uses of AI in defense, have the potential to improve endurance, accuracy, and speed but also carrying hazards that could have uncontrollable consequences. These benefits of AWS do, however, come with significant cyber security dangers. Modern robots and AI have made systems more susceptible to cyberattacks, which poses grave risks. It is particularly challenging to identify and fix vulnerabilities in these systems due to their complexity. Cyberattacks on AWS present a severe risk that could have uncontrollable effects or perhaps catastrophic outcomes. Because of this, it's critical that cyber security concerns be thoroughly considered and that stringent controls be included while developing AI-supported AWS.

In Chapter 4, "Using Artificial Intelligence in the Field of Intelligence Operations and Analysis," Dr. Darıcılı and Dr. Nourhan El-Bayaa discuss the application of AI to intelligence operations and analysis. AI gives computers and robots the capacity to learn from and imitate human behavior through a combination of scientific understanding and technological innovation. AI has caused a paradigm change in the use of AI techniques in intelligence operations and analysis. It not only mimics but also expands upon the idea of intelligence beyond biological bounds. AI has made intelligence operations more efficient and scalable by processing large amounts of data rapidly and thoroughly. Previously, intelligence operations relied on human analysts to navigate large datasets. Through proactive threat response and natural development in overall efficacy, machine learning algorithms identify threats automatically, monitor activities, and find trends that point to suspicious conduct. It is impossible to overstate the importance of AI technology in intelligence operations, since intelligence agencies are spending more and more in AI models, algorithms, and software to support operations ranging from cyberattacks and covert action to data collection and analysis. This chapter looks at how AI has a substantial impact on intelligence gathering, analysis, and operational planning. It makes the case that AI technologies will have a big impact on how intelligence services operate and what kind of activities they do in the future.

In Chapter 5, Ghaffar Hussain discusses the relationship between radicalization and AI. The swift expansion and advancement of AI is expanding the influence of technology on humankind, but it also presents significant societal obstacles. Even as technological advancements ease people's lives, a reliance on online contacts is weakening human bonds, decreasing in-depth reading, and social skills. It is anticipated that as AI advances, technology will play a more significant role in influencing nearly every element of our life. However, because AI is so widely available and accessible, bad actors may potentially utilize technology for harmful purposes that have disastrous effects on society. It is anticipated that hostile actors and extremist organizations would quickly adopt new technology to broaden their realms of influence and contact. Extremist organizations may find it simpler to contact their target audiences and update their engagement strategies and propaganda with the help of emerging technology like AI. This can strengthen their capacity to reach an infinite number of individuals in an anonymous manner, silence dissenting opinions, gain worldwide clout, and validate small groupings. Thus, it is necessary to analyze how AI could be used by radicals as a tool and design countermeasures and remedies to lessen the impact of this technology.

In Chapter 6, "Artificial Intelligence and Cyber Espionage: Delineating Emergent Warfare and International Security," Prof. Dr. Ifeanyi-Ajufo and Dr. Wan Rosalili Wan Rosli took a distinct tack when discussing international security. Shifting theories of warfare have become a crucial subject of discussion in international legal discourse on matters like the notion and definitions of armed aggression, the right to self-defense, attribution, and its evolving consequences for sovereignty in light of growing geopolitical tensions. These discussions are concerning due to the nature of developing technology, especially those related to cyberspace. The international community has recently been involved in various forums and discussions to define how cyber activities can have an impact on the emerging war. Cyber espionage has become an important agenda item, especially with regard to the security and protection of critical national infrastructure. Cyber espionage has emerged as a growing threat in the international community. Incidents such as the theft of Google's data in 2010 as a result of a sophisticated and targeted attack originating from China and the publication of the Mandiant Report in 2013 demonstrated the size and scope of this threat. Concerns regarding cyber espionage were most recently sparked by the Snowden report's disclosure that the US National Security Agency (NSA) was a part of a surveillance operation that gathered secret data by keeping an eye on private email and phone conversations. The introduction of new cyber espionage scopes, bigger impact scales, and increased flexibility is made possible by the shift to new technology, such as AI. This can involve risks to monitoring, harm, and generally rendering states powerless by interfering with the operation of vital national infrastructure. It is important to give careful thought to how the Russia-Ukraine war's cyber-related actions may affect the international laws governing cyber espionage. The integration of new technologies and digitalization are driving real-world legal concerns into cyberspace. Therefore, it's critical to reframe traditional disputes to align with contemporary legal reality. This strategy will add to the discussion on cutting-edge technologies and how warfare is evolving. The concept and regulation of cyber espionage through the lens of international law is examined in this chapter. It also covers the political ramifications of global cyber security initiatives as well as the existing normative framework for cyber espionage. Another significant subject covered in this chapter is the creation of guidelines for the governance of emerging technologies.

Dr. Ulu examines the state of affairs in the battlefield in Chapter 7, "The Fuel of AI in the Battlefield: Data Management and Data Governance," with an emphasis on data security and data mining. We are living in a time with never-before-seen strategic capabilities thanks to the use of AI technology into military operations. The role of data is fundamental to this dramatic shift. The source of energy for AI algorithms is data. AI trains itself in machine learning and deep learning models using datasets. When used with well-crafted, safe, and functional datasets, AI produces reliable outcomes. Thus, the most essential resource needed for the creation of AI is data. For better decision-making in combat, AI algorithms offer a real-time intelligence channel. The complicated dynamics in military situations are thoroughly explored at the beginning of this chapter, with an emphasis on the crucial interactions between AI, data security, and data management. The foundation of trustworthy and efficient AI is data security and integrity. Robust AI models and decision-making procedures require precise and secure data. The Data Information Wisdom (DIKW) pyramid, which illustrates the transformative journey from data to action, is used as a conceptual framework to untangle the complexities of this relationship. Understanding the transformation journey from raw data to actionable insights might be aided by this pyramid. Data mining plays a crucial part in intelligence and surveillance in the contemporary environment. The impact of conventional state actors has occasionally been called into question due to the democratization of data access through open-source platforms. The relationship among data privacy, sovereignty, and security in the context of modern warfare becomes apparent as a continuous factor that greatly influences military plans and tactics. With the unparalleled connection that the digital age has brought forth, data can now effortlessly flow across borders and legal authorities. But this connectedness necessitates striking a delicate balance between the need to work together to gather intelligence and coordinate reactions and the necessity to protect sensitive information. AI-based military applications present data management challenges. Safe and secure datasets are necessary for AI algorithms to be trained. This chapter's primary goal is to present a thorough case study that examines how the US strategically uses data as a fundamental component of its AI-enabled military operations. The chapter provides a thorough analysis of the ethical issues and national security ramifications related to this paradigm shift. It covers a wide range of topics, from definitions of data in the military context to the practical application of AI, data use, and data-driven decision-making.

Dr. Bilgin addresses the ethical and legality of using AI in warfare in Chapter 8, "The Ethical and Legal Challenges of AI in Wars." Although there is nothing that can change about the essence of war, the character of war is the way that conflicts are fought depending on various elements, including politics, technology, culture, economy, location, and so forth. AI technological advancements have the power to fundamentally alter the nature of conflict. Many experts in the field of war studies contend that AI's influence on the nature of conflict could revolutionize military tactics. This is due to the fact that AI is revolutionizing every aspect of combat. Specifically, the effects of AI technologies on the battlefield are closely linked to the ethical and legal implications of these technologies on conflict. Throughout history, there has been an inconsistent relationship between technology and war. There have been protracted periods of technological stasis during wartime. The paradigm of warfare has changed as a result of revolutionary breakthroughs in technology, which have also caused threat situations to evolve and necessitated the development of new disruptive technologies. However, every technological revolution that has occurred, from the discovery of gunpowder to the creation of nuclear weapons, has presented humanity with fresh moral conundrums. Similar moral conundrums have been presented to humanity by AI technologies, a crucial technological advancement. In particular, the human element continues to be central to the ethical conundrums raised by AI, irrespective of historical background or technological advancement. Therefore, the presence of humans cannot be ignored in any ethical or legal analysis of AI in conflict. Finding the limits of the ethical and legal implications of AI in warfare is extremely difficult due to the rapid advancements in this field. Furthermore, the debate over ethics, legality, and warfare could take on new dimensions with every advancement in AI technology and every military use. It is still possible to pinpoint the main ideas and synopses that need to be highlighted, though. Three major subjects are planned for this chapter from this perspective. First, an ontology of each phenomenon is discussed along with the relationship between war, ethics, and law. A presentation on global projects is given second. Third, emphasis is placed on the moral dilemmas raised by the quick development of AI-based military technology. Last, the global framework is introduced to investigate not only the fundamentals of international humanitarian law and war theory but also possible legal guidelines that ought to be implemented with regard to AI.

Chapter 9, "Artificial Intelligence in Strategic Planning and Military Operations," by Dr. Lucas and colleagues addresses whether or not warfare has migrated into cyberspace in relation to the phenomenon of algorithmic warfare. AI integration has grown significantly over the past five years; estimates indicate that it has doubled during that period. This can be attributed to the rapid expansion of the so-called Internet of Things (IoT). Almost "everything is connected to everything" has made things too complicated to manage in the past because there is so much data and numerous network connections. Even while integrating this vast amount of data presents significant practical hurdles for data management and policy, it is already evident that AI systems can effectively handle this difficult circumstance. Defense officials now face a challenge to reconsider how military operations are conducted and the moral issues that surround them in light of recent advancements in robots, software, satellites, and autonomy. These days, the emphasis is on robotic anti-armor defense weapons, robotic land, air, and naval systems, and less expensive commercial geospatial targeting and intelligence data. Any country or non-state organization that uses these technologies will be aware of the potential benefits of technical asymmetry. With the Ukraine War at the forefront, this chapter draws a variety of assumptions and conclusions in this context.

Dr. Liaropoulos examines how autonomous weapons will affect conflicts in the future in Chapter 10, "Autonomous Weapons Systems and the Future of Warfare." This chapter explores the difficult problem of deploying AI in military settings, with a special emphasis on AWS. These algorithm-driven systems have the ability to choose and attack targets without the need for direct human involvement, possibly transforming combat. They could improve military capabilities and lower risks for human soldiers, according to proponents. But there are many moral, legal, and tactical issues to consider. On the one hand, AWS might provide more accurate, quicker answers while lowering the hazards for human soldiers. However, giving up control to robots creates ethical and accountability problems. In addition to being susceptible to hacking and abuse by non-state actors, machines are incapable of comprehending complicated ethical issues. The chapter looks at the tactical, operational, and strategic effects of AWS on conflict, highlighting the necessity of developing policies. It examines the definitional dispute, the benefits and drawbacks of AWS, and its effects on the global system, such as the possibility of unintentional escalation and the need for international laws or arms control. The research's conclusions and insights are based on a thorough analysis of academic literature, case studies, and real-world instances.

Dr. Kyu-hyun Jo addresses autonomous weapons in the context of Clausewitz in Chapter 11, "Autonomous Weapons and the Need for an Evolution of the Clausewitzian Trinity towards an Emphasis on Tactical Positioning." Though its limitations have been discussed in the context of AI and autonomous weapons, Clausewitz' Trinity remains an essential notion in the study of war. Although a lot of work has been done on the hazards of autonomous weapons and how Clausewitz' Trinity applies to contemporary conflict, not much has been done to examine how autonomous weapons challenge the Trinity as a paradigm. Though it has addressed a wide range of subjects, including the moral and legal ramifications, the literature on autonomous weapons has not clearly connected these discussions to Clausewitz' Trinity. However, scholarly research on Clausewitz' Trinity has mostly ignored the effects of autonomous weapons in favor of concentrating on the theory's applicability to modern conflict. This chapter covers two crucial ideas to close this gap. The first area of analysis will be how the employment of autonomous weapons in conflict compromises the effectiveness and coherence of Clausewitz' Trinity. Coordinated action is challenging because autonomous weapons are swift and arbitrary, complicating established roles in strategy, tactics, and policy. Second, it will be contended that the theoretical advancements in technology are challenged by autonomous warfare. Clausewitz' Trinity needs to adjust as autonomous weapons become increasingly prevalent in order to be successful. As a result of this progress, tactical posture has become increasingly important in order to counter the swift and devastating capabilities of autonomous weapons. Thus, the increasing employment of autonomous weapons indicates that conventional military theories like Clausewitz' Trinity need to be reexamined. Recognizing the constraints imposed by autonomous warfare allows military strategists to modify their tactics, accordingly, ensuring the efficacy of their operations in a technologically dynamic environment.

Dr. Güneysu addresses the legitimacy of using autonomous weapons in combat in Chapter 12, "Compatibility of Autonomous Weapon Systems with the Basic Principles of Weapons Law" Dr. Güneysu claims that the application of AI across numerous industries is now a routine occurrence. Naturally, there are a lot of difficulties when this kind of decision and/or implementation aid is used widely. Using these systems in conflict is one of these hardest problems to overcome. Regardless of the level of autonomy, these systems – which are employed in armed conflicts – are referred to by a variety of titles, including lethal autonomous robots and lethal robots. The author will choose to refer to these systems as AWS in this chapter. When it comes to reducing soldier casualties in combat, using AWS provides clear military advantages. This naturally makes them very desirable systems and calls for a closer examination of their characteristics. One of the most contentious topics in today's world is AWS. There are talks about many different subjects, including politics, the law, and ethics. The conduct of armed conflict will be affected in many different ways by the emergence of autonomous systems. While many LOAC-based evaluations of AWS have concentrated on matters like the conduct of war and the underlying humanitarian principles and regulations, comparatively less attention has been paid to the systems' assessment from the standpoint of weapons law. The purpose of this chapter is to assess these systems in light of weapons law. Such an analysis would give a better knowledge of the virtues of such weapons in terms of arms law principles from IHL, since arms law deals with the essential properties of weapons.

Dr. Schoenherr introduces a human-centered design perspective to assess the ethical characteristics of AWS and cyber weapons in Chapter 13, "Meaningful Human Control of Autonomous Weapons Systems: Translating Functional Affordances to Inform Ethical Assessment and Design." Human nature has been defined by the urge to help our neighbors and protect each other against external threats. Conflict has changed in character as human societies have progressed. New innovations can turn everyday objects like tools into weapons by reinterpreting their properties in new environments. It's critical to comprehend the design elements of technology in order to assess and forecast the effects and outcomes of its use. Technologies develop and acquire significance due to their relationships with people. Fundamental concerns about these systems' accuracy and autonomy – that is, the ramifications of an artificial agent making crucial judgments regarding human agents – are among the problems. Philosophical frameworks and guidelines represent a starting point, but they are far too general to ensure meaningful human control of AWS. Numerous potential solutions have been proposed. These include outright bans, the need to develop new technologies quickly to stay competitive, the establishment of global standards, and the appointment of "ethical governors" to oversee these systems. Simultaneously, worries have been raised regarding the growing issues with the technology referred to as "cyber weapons." Human-centered approaches to design can greatly aid in aligning the technical affordance (i.e., autonomy, intelligence, and connectivity) of technology with the ethical affordance of users that result from cognitive limitations (e.g., attention, memory, response time).

Dr. Akkuş addresses the function of ethical foundations as humanity's cosmopolitan principle in the era of autonomous weapons in the last chapter of the book, "The Cosmopolitan Principle of Humanity in an Age of Autonomous Weapons: The Role of the Ethical Underpinnings." This chapter delves into the intricate discussion around the use of AI technologies in military settings, with a particular emphasis on deadly AWS and how they affect human rights and international humanitarian law (IHL). The fear that AI-enabled systems may elude human control and transgress fundamental ethical precepts is one of the major ethical and legal concerns that is reflected in discussions around the military's use of AI. The discourse is organized into four primary segments: (1) AI Debates Overview, (2) International Law and Cosmopolitanism, (3) Lethal AWS Discussion, and (4) Research Questions and Methods. The study strikes a balance in the contentious discussion between proponents and opponents of legal agreements on humanitarian grounds, with the former claiming that their implementation might heighten sensitivity and enhance adherence to moral standards. The findings emphasize the significance of thoroughly investigating the possible moral and legal ramifications of humanitarian grounds, encouraging comprehension of the intricate moral and legal terrain that AI technology in conflict has generated. This chapter discusses the contentious topic surrounding the military's use of AI, with a particular emphasis on deadly AWS and how they affect human rights and IHL. The discussion tackles a complicated problem that highlights significant moral and legal issues, such as legal agreements on humanitarian grounds, concern over losing human control, and transgressing fundamental moral standards. Humanitarian law's objectives are allegedly undermined by opponents of law of war (LOAC), but supporters contend that LAWs can boost moral compliance and raise awareness. The findings emphasize the significance of thoroughly investigating the possible moral and legal ramifications of LAWs, encouraging comprehension of the intricate moral and legal terrain that AI technology in conflict has generated.

Contributors

Berkant Akkuş
İnönü University Faculty of Law
Malatya, Turkey

Hassan Benouachane
Mohammed V University
Rabat, Morocco

Kıvılcım Romya Bilgin
Başkent University
Ankara, Turkey

Emre Çıtak
Hitit University
Çorum, Turkey

James Cook
United State Air Force Academy
Colorado Springs, CO, USA

Ali Burak Darıcılı
Bursa Technical University
Bursa, Turkey

Nourhan El-Bayaa
İstanbul Aydın University
İstanbul, Turkey

Mehmet Emin Erendor
Adana Alparslan Türkeş Science
    and Technology University
Adana, Turkey

Gökhan Güneysu
Anadolu University
Eskişehir, Turkey

Kirsi Helkala
Norwegian Defence University College /
    Norwegian Defence Cyber Academy,
    Lillehammer, Norway/ Peace Research
    Institute Oslo
Oslo, Norway

Ghaffar Hussain
Newham, London

Nnenna Ifeanyi-Ajufo
Leeds Beckett University
Leeds, England

Kyu-Hyun Jo
Yonsei University
Seoul, South Korea

Andrew Liaropoulos
University of Piraeus
Piraeus, Greece

George Lucas
U.S. Naval Academy
Annapolis, Maryland

Wojciech Mincewicz
Institute of Sociological Sciences
    and Pedagogy
Warsaw University of Life Sciences
Warsaw, Poland

Frank Pasquale
Brooklyn Law School, Brooklyn
NY, USA

Gregory Reichberg
Peace Research Institute Oslo
Oslo, Norway

Wan Rosalili Wan Rosli
University of Bradford
Bradford, England

Jordan Richard Schoenherr
Schreiner University and Carleton University

Henrik Syse
Peace Research Institute Oslo
Oslo, Norway

Cansu Ulu
Turkish National Police Academy
Ankara, Turkey

1 Introduction of Key Terms and Topics

Mehmet Emin Erendor and Wojciech Mincewicz

1.1 INTRODUCTION

The book primarily addresses three key concepts: cyber security, artificial intelligence (AI), and autonomous weapons, as indicated in its title. To comprehend the intricacies associated with these concepts, it is essential to introduce and examine them. This section aims to provide an introduction to these concepts, exploring their definitions, applications, and historical evolution. Additional terms such as security, cyber espionage, data mining, ethics, and radicalization are also discussed throughout the book in relation to the three main concepts. Each of these concepts possesses its own unique characteristics. However, to keep the introduction concise, this section will concentrate solely on the three primary terms.

It is not long ago that the international community encountered these three concepts. These concepts, which emerged or rather became widespread especially with the end of the Cold War, naturally affected the structure of the international system and necessitated the adoption of new approaches for both states and international organizations. The use of these concepts by both state and non-state organizations also implies the emergence of new threats and risks for the international community.

Especially the developments in the technological field have led to a more effective and comprehensive handling of the cyber security phenomenon. The increase in work in this field since the mid-2000s is also due to the intense cyber-attacks that began during this period. Cyber security, which the international community ignored, especially gained prominence as a phenomenon that should be taken into consideration after the 2007 Estonia attacks, and states developed urgent strategies to protect their critical infrastructures. Although the international community moves quickly, technology's even faster pace and the integration of AI into human life have allowed policies to change on a daily basis. AI-based weapons have also expanded the framework of the cyber security phenomenon and added another dimension to the policies to be formulated.

These developments have not only sparked political debates but also paved the way for various discussions in the legal realm. The primary reason for this is that the ethical and humanitarian dimensions of technological advancements or AI-based weapons used in the defense industry have begun to be comprehensively examined. When we look at the literature, it can be clearly seen that the academic community or those working in these fields are also divided into two camps.

The primary reason for this division is also due to the different perspectives from which both sides evaluate events and developments. These debates will not be elaborated on extensively as they are not the subject of this section.

In this context, this section will provide definitions and brief information on cyber security, AI, and autonomous weapons. Detailed information on these concepts will be analyzed more comprehensively in other sections of the book.

1.2 CYBER SECURITY

The topic of cyberspace and the threats associated with it has been known since the appearance of the Internet, as evidenced by numerous studies and publications by subsequent authors representing numerous scientific disciplines. Examples of works include Cavelty (2018), Craigen et al. (2014),

Martínez Torres et al. (2019), Mincewicz (2023), Sarker et al. (2020), Singer and Friedman (2014), and Warner (2012). Due to its scope and multidisciplinarity, the concept of cyber security is defined differently in the literature (Schatz et al., 2017). Sometimes, both in academic discussions and in journalism, the terms computer security, Internet security, systems security, information technology security are used interchangeably (Van Niekerk & Von Solms, 2010, p. 101). Cyber security is generally still considered a purely technological challenge, but it seems that the definitional scope should go beyond the traditional limited understanding. Specific, individual events that are subject to description or analysis are important. The concept of cyber security comes from the term cyber security. In a broad, most common sense, it means all activities undertaken to achieve the desired state in which the risks posed to operations carried out in cyberspace are minimized as much as possible. Cyber security refers to all processes and activities aimed at ensuring information security in cyberspace. However, it is a mistake to equate the term cyber security with information security, which is the most extensive concept, where the emphasis is on the protection of information assets.

The definition defining the scope of understanding the term cyber security and its connection with cyberspace is presented by H. Ning et al. (Ning et al., 2018), according to which cyber security is the protection of cyberspace itself, that is, electronic information, information and communication technologies supporting cyberspace, and users of cyberspace in the scope of their personal, social, and national capabilities, including their interests, material or immaterial, who are vulnerable to attacks from cyberspace. Based on the presented activities, it can be concluded that cyber security is a set of resources, technologies, activities, processes, and practices designed to protect cyberspace – networks, devices, programs, and systems against events (attack, damage, unauthorized access) that are inconsistent with actual laws, properties. As noted by the European Union Agency for Cyber security (ENISA, 2021), the COVID-19 pandemic has significantly contributed to increasing threats in cyberspace, pointing to the opportunities offered by remote work. The pandemic has given cybercriminals new opportunities to operate on a larger scale than before. Examples of activities include sharing domains with the keyword "COVID" or "corona" with malware. Another event that has significantly influenced the perception of cyber security in recent years is the war in Ukraine and incidents preceding an armed attack, consistent with the theory of substitution (Kostyuk & Gartzke, 2022).

The explosive growth in the importance of cyber security over the past few years has led to responses on many levels. New policies are proposed at the national and international levels, followed by appropriate laws and regulations. Although modern states have been somewhat forced to develop specific cyber security strategies in response to potential threats, the most noteworthy aspect of these strategies is their mutual similarity. A significant increase in the importance of cyber security research and the need to ensure protection has been observed in the business sphere. Companies and their managers declare that cyber security threats are one of the most important business challenges. The appointment of a Chief Information Security Officer (CISO), whose task is to ensure a comprehensive approach to cyber security in the operation of the enterprise, is one of the basic activities. At the lowest level of the cyber security ladder is a person, an ordinary citizen. Data leakage from commercial entities and violation of the state's cyber security infrastructure make the possibility of cyber self-defense significantly lower. According to IBM, in 2023 the average global cost of a data breach was $4.45 million, an increase of 15% over three years (IBM, 2023).

In an attempt to taxonomy the interests of subsequent authors, at least four aspects of contemporary research can be distinguished. Given the phenomenon of Big Data and data digitization, one of the most important contemporary problems is information security. The functioning and development of the Internet has significantly reduced the security of classified information – proprietary, confidential, secret, and top secret, as well as other information whose proliferation is not beneficial from the point of view of the interests or image of the state, business, or ordinary citizens. Nowadays, information asymmetry (access to information and information distribution) between state institutions and entities such as individuals and groups has been reduced (Voronenko et al., 2021). State control over secrecy, proliferation, and distribution channels of information is weakened. The indicator and cause of the loss of control is the development of the following technological and socio-cultural phenomena: (1) widespread availability and a significant selection of IT and socio-psychological tools used to obtain information without the consent and/or knowledge of the entities that possess it; (2) development of identity masking techniques; (3) development of methods for securing and hiding information; (4) positive formation of the institution of the whistleblower in public consciousness, which happened after a number of leaks of classified information.

Political and military cyber threats seem to be equally important, especially in the current geopolitical situation. This type of threats is characterized by dualism because threats are generated both in the technical and IT spheres and in the social sphere. In the first of the above-mentioned areas, examples of contemporary threats are activities leading to disruption of the functioning of critical infrastructure. The elements of critical infrastructure that are particularly susceptible to an IT attack using malware (Ransomware and Distributed Denial-of-Service (DDoS)) include the energy system, production, storage and transport of natural gas and crude oil, factories using or producing toxic substances, transport, energy supply systems water and food, systems ensuring the continuity of government and public services (health service, national defense, security, e-administration structures, and even e-elections), rescue services.

Probable, although long-term, threats include weakening control over economic circulation and, consequently, erosion of the state's fiscal control power. In turn, the intensifying phenomenon of alternative monetary circulation leads to a weakening of trust in fiat money and, consequently, a reduction in the financial strength of the state. Cryptocurrencies, i.e. decentralized virtual currencies operating in a network with peer-to-peer architecture, cryptographically secured, based on trust and consensus, which incompletely fulfill some of the economic functions of money (Mincewicz, 2021, p. 101) enable free flows of funds in the scope of financing terrorism, revolutionary, extremist, or dissident movements, opening the possibility of anonymous destabilization of political entities. Economic circulation that is uncontrolled by state institutions and the development of alternative means of payment to fiat money may potentially generate a set of economic threats. The creation of decentralized, open transaction platforms and the development of cryptoeconomics enable the circulation of financial resources beyond state control and, as a result, create a sector independent of state institutions, tax-free and uncontrolled.

The potential of the Internet in terms of storing, hiding, and transmitting information potentially generates long-term threats, i.e. effects that are revealed after a longer time interval. These threats may turn out to be a catalyst for changes that will result in the destabilization of the social order, reduction of trust in the existing social order, in particular state institutions (and their weakening), and even erosion of the image of the state as an entity without alternatives, and thus intensification of anarchist tendencies (Tada, 2020). These factors include the ease, speed, and effectiveness of the proliferation of radical ideologies and the facilitation of association, consolidation, mobilization, and cooperation among radical groups. The Internet combines two facilitators that enable the spread of extremist ideologies: technical and psychological. The technical aspects include the massification of the Internet as a means of communication, the speed of communication and the number of communication channels, the possibility of anonymizing the message thanks to routing or cryptography mechanisms. They are analogues of information security threats. In turn, among the psychological factors we will include the so-called disinhibition effect resulting from the specificity of online communication. It consists of phenomena such as dissociative anonymity generated by almost complete freedom in creating one's identity online, and the phenomenon of solipsic introjection, i.e. the separation of online experiences from the experiences of the real, physical world. This effect may lead to the perception of interactions with others as taking place in one's own imagination, as well as virtual indifference resulting from the dispersion of responsibility and the viewer effect and dissociative imagination, i.e. treating events taking place between Internet users as a kind of "role-playing game", and therefore liberated to some extent from moral norms.

1.3 ARTIFICIAL INTELLIGENCE

With the development of criminal activities in the Cybercrime-as-a-Service (CaaS) service model, and the emergence of generative AI, a new space for cyber threats is opening. With the development of AI, the set of tools that enable more targeted and hidden activities is growing, with particular emphasis placed on bypassing advanced security control mechanisms, which in turn increases the effectiveness of subsequent incidents. The term "artificial intelligence" was first used by John McCarthy in 1956 at the Dartmouth conference (Rajaraman, 2014). Andreas Kaplan and Michael Haenlein (2019) define AI as follows: *the ability of a system to correctly interpret data from external sources, learn from it, and use this knowledge to perform specific tasks and achieve goals through flexible adaptation* (Kaplan & Haenlein, 2019). One of the fathers of AI, Marvin Minsky, said it was "the art of teaching machines to do things that, if done by humans, would require the use of intelligence" (Minsky, 1961). Nowadays, the term AI refers to both technology and scientific research in computer science at the intersection of neurology, psychology, cognitive science, as well as systematics and even modern philosophy.

Due to the dynamics of development, AI has hit the agenda of public debate and has become the subject of interest of many state and international institutions. In Europe, the European Commission, in its Communication to the European Parliament, the European Council, the European Economic and Social Committee, and the Committee of the Regions (2018), defined AI as follows: systems that demonstrate intelligent behavior by analyzing their environment and taking actions – to some extent autonomously – in order to achieving specific goals (Communication from the Commission to the European Parliament, the European Council, the Council, the European Economic and Social Committee, and the Committee of the Regions – Coordinated Plan on Artificial Intelligence, 2018). It is assumed that intelligent action is defined as independent decision-making based on drawing conclusions from accumulated knowledge and experience. A machine that collects data and information and is then able to independently draw the optimal conclusion based on which it will make decisions can be considered AI. By collecting knowledge, a computer system must have the ability to select it and draw conclusions aimed at discovering connections between information. Such activity is considered machine learning, which cannot be identified with the concept of AI (Sarker et al., 2020). Machine learning is a broader concept that is, in principle, an element of AI, but not sufficient. AI, in addition to the ability to perform machine learning, must be able to make an independent decision resulting from the information processed by the IT system (Lussier et al., 2020). Systems using machine learning have the ability to draw their own conclusions and improve their capabilities based on their own experiences, but the decisions they make are not always fully independent. By their nature, they are limited to the decisions originally entered their algorithm.

The development of AI has had a positive impact on the surrounding reality. Launched in November 2022, the chatbox developed by OpenAI gained a million users within five days, breaking Internet records. Thanks to universal access, it gave everyone the opportunity to learn about the possibilities offered by AI. In 2023, further companies from the IT sector – such as Google or Meta AI – proposed their own chat boxes, following the trend initiated by OpenAI. During the pandemic, AI has been used for thermal imaging in airports or other places and to track the spread of the virus. AI is widely used by marketing companies to provide personalized recommendations when shopping online. The role of AI in trade also includes product optimization, purchasing planning, and logistics. Search engines learn from secondary data of which products may interest a user. The use of AI-based safety features to keep cars safe seems to be becoming more common. In other areas, the great potential of AI is seen in agriculture and food cultivation, where it will be possible to increase productivity or reduce substances hazardous to human health. Technology already allows monitoring movement, temperature, consumption, and feed dosing of farm animals.

Despite many benefits, AI also poses threats that both practitioners and scientists have been paying attention to for years. In a famous August 2017 letter, 116 of the world's leading scientists wrote an open letter calling on the United Nations to ban the development and use of killer robots

(Sharkey, 2019). In the social dimension, it should be noted that AI can protect but also threaten democracy. In the media sphere, especially in social media, AI is responsible for generating information bubbles. These "bubbles" don't promote a pluralistic, equally accessible, and open public discourse; instead, they are based on an individual's past online behavior and only show content that the person may find interesting. Through the use of deepfakes, AI may also be used to produce incredibly lifelike fake images, audio files, and videos. These can be dangerous for businesses, ruin reputations, and make decision-making more difficult. All of this could result in electoral manipulation as well as polarization and divisiveness in the public sphere. Big Tech is currently using AI. Because AI is capable of tracking and profiling individuals linked to particular opinions or actions, it can also pose a threat to the freedoms of assembly and protest. AI applications that interact physically with humans or are incorporated into the human body may present security problems due to potential for misuse, hacking, or poor design.

1.4 AUTONOMOUS WEAPONS

Throughout history, technology has played an important role in shaping the nature and outcomes of conflict on the battlefield. From the invention and use of the bow and arrow to the development of nuclear weapons, each innovation has brought with it new ethical dilemmas and challenges. But the emergence of autonomous weapon systems presents humanity with a unique achievement and a moral responsibility.

Developments in technology have not only emerged to make people's lives easier, but these positive developments have also brought risks and threats. Especially the developments in the defense industry have enabled the arms industry to apply technological developments to new weapon technologies. In this context, autonomous weapon systems have become one of the main armament policies of states in a more effective and decisive manner (Güneysu, 2023; Wagner, 2014; Zurek et al., 2023), and these AI-based technologies have started to be used in many defense industry products, including unmanned aerial vehicles (Singh et al., 2022). In his report, Work states that the first mass-produced US weapon system with autonomous functions in engagement-related functions was developed during the Second World War and was an acoustic-guided torpedo (Work, 2021). The International Committee of the Red Cross (ICRC), in contrast to Work, states that mines are considered primitive autonomous weapons, and that radar, which has limited use in the military, missiles used to destroy tanks or armored vehicles, and air defense systems are also considered preliminary works. In fact, the international community banned anti-personnel mines in 1997 because their effects are difficult to control and the serious harm they cause to civilians in many conflicts is well documented. Although past studies have shown that humans are working on autonomous weapons, it can be said that the first studies were mostly armed unmanned aerial vehicles controlled by humans (ICRC, 2022). As can be seen, the use of autonomous weapons or the autonomization of weapon systems started long before today, but the development of computers, the invention of new devices and systems, and the making of processors more powerful have naturally enabled new systems to be added to almost all weapon systems and technological developments in areas related to the defense industry. This has naturally allowed these weapons to become autonomous over time.

Within the framework of this information, it is possible to say that autonomous weapon systems are extremely occupying the international agenda today and the debates on this issue are increasing (Asaro, 2012, 2020; Carpenter, 2011; Krishnan, 2016; Solovyeva & Hynek, 2018). In this book, these debates are comprehensively evaluated under the relevant headings. For this reason, this section will provide brief information on these issues.

There are also problems regarding the definition of autonomous weapons. While the international community does not have a commonly accepted definition of autonomous weapon systems, it should be noted that there are many definitions (Christie et al., 2023; Ekelhof, 2017; Geiss, 2015; Jensen, 2020). Roff defines autonomous weapons as *that is weapons that are capable of targeting and firing on their own without human supervision or interaction, they respond to a preprogrammed set of constraints* (Roff, 2014). Lewis states that *Autonomous weapons can select and engage without further intervention by a human operator. Autonomous other words, require no human input- once activated, they possess to fire on their own* (Lewis, 2015). In his own work, According to Boogaard, autonomous weapons are those that recognize and engage targets without the need for human interaction or control by utilizing AI algorithms and other cutting-edge technologies (van den Boogaard, 2015).

Autonomous Weapons website defines the term *Slaughterbots, also called "autonomous weapons systems" or "killer robots", are weapons systems that use artificial intelligence (AI) to identify, select, and kill human targets without human intervention* (Autonomous Weapons, 2021). On the other hand, ICRC defines the term *Autonomous weapon systems, are any weapons that select and apply force to targets without human intervention* (ICRC, 2022). Finally, the United Nations Office for Disarmament Affairs Group of Governmental Experts on LAWS CCW/GGE.1/2023/CRP.1 defines autonomous weapons as follows: "Autonomous Weapons Systems (AWS) are systems that, upon activation by a human user(s), use the processing of sensor data to select and engage a target(s) with force without human intervention" (Convention on Prohibitions or Restrictions on the Use of Certain Conventional Weapons Which May Be Deemed to Be Excessively Injurious or to Have Indiscriminate Effects, 2023).

According to these definitions, autonomous weapons are those that are made to function either entirely or partially without the assistance of a person, such as missile systems that can recognize and strike targets on their own with the use of AI.

When we look at the definitions, it can be seen that they all have something in common. All definitions are united on the basis that AI can find and destroy targets and without human interaction.

Here it is also necessary to draw the difference between full autonomy and semi-autonomy. In fully autonomous systems, all processes are carried out by autonomous systems, while in semi-autonomous systems the human factor comes into play (Endsley, 2017; Feng et al., 2016; Gabrecht, 2016; Werkhoven et al., 2018). To put it another way, autonomous weapons rely solely on preprogrammed algorithms to make decisions, whereas unmanned military aerial vehicles use a human operator to make decisions remotely about targets (Abaimov & Martellini, 2020; Konert & Balcerzak, 2021; Scharre, 2016). An autonomous weapon system is preprogrammed to kill or hit a specific "target profile" (Fisher, 2022; Munir, 2022; Watts & Bode, 2023). In this context, autonomous weapons are deployed in an environment where AI searches for this "target profile" using different sensor data, such as facial recognition, and achieves the result (Fisher, 2022; Munir, 2022; Watts & Bode, 2023).

The issues surrounding the use of autonomous weapons are even more significant than the definitional one. A few major ethical, legal, and strategic issues are brought up by the development and application of autonomous weapons, including the possibility that these devices could injure civilians, lessen military operations' accountability and transparency, and change the relative power between states and non-state actors. These worries have been growing lately, particularly in light of the growing deployment of unmanned aerial vehicles. Two distinct camps have typically emerged from the discussions: those in favor of autonomous weapons and those who believe they are immoral (Güneysu, 2023).

The conceptions of ethics and compassion play a significant role in the fundamental logic of this quandary. Those who oppose autonomous weapons sometimes claim that because robots are impersonal and will decide life and death based only on their own moral judgments, this will lead to a nightmarish future. They contend that the lack of human oversight in autonomous weapons could pose significant risks for abuse and complicate matters related to accountability.

This paradigm draws attention to the discrimination and proportionality moral conundrums that underlie autonomous weapons. In times of emergency, the ability to discriminate between military and civilian persons might be the difference between life and death. Proponents of autonomous weapons say that their technology will not aim its weapons at civilians, despite the fact that AI can improve target discrimination. However, critics counter that the biases and limitations of machine learning algorithms make this impossible to achieve. Furthermore, the fact that these algorithms are based on historical data is emphasized, raising the possibility of collateral damage and indiscriminate targeting as well as the reinforcement of previous preconceptions.

Moreover, the deployment of autonomous weapons raises profound questions of accountability and responsibility. In traditional warfare, human operators are held accountable for their actions, subject to rules of engagement and codes of conduct. But in a scenario where machines autonomously make life-and-death decisions, it is unclear who will take responsibility for the consequences. Naturally, the lack of clear lines of responsibility complicates efforts to maintain ethical standards and ensure compliance with international law.

Those who support autonomous weapons, on the other hand, believe that using autonomous weapons will make war more humane (Crootof, 2018; Schulzke, 2019). The main arguments of those who advocate for this generally highlight that autonomous weapons do not have personal interests like humans, they do not exhibit excessive harshness toward enemy targets, and they do not resort to non-combat methods to win wars. In this context, it can be said that autonomous weapons can target in the most fair and accurate way even in the most stressful moments (Kralingen, 2016; Sauer, 2016; Umbrello et al., 2020).

Another important aspect of autonomous weapons is their contribution to human lives in wartime. In his study, Etzioni highlights that autonomous weapons can operate with fewer military personnel compared to the number usually deployed, can function in inaccessible areas, can keep military personnel away from threatening duties and risky zones during an attack, and can reduce casualties. He emphasizes the significance of autonomous weapons in terms of minimizing human losses (Etzioni & Etzioni, 2017). When looking at the terrorist acts occurring today, it is evident that autonomous weapons will make significant contributions, especially in the fight against terrorist organizations.

United Nations Secretary-General António Guterres has not remained indifferent to projects supporting weapons with AI and to autonomous weapons. In a statement made in 2018, Guterres emphasized that the weaponization of AI poses a serious danger and could lead to a global security issue, especially noting that *very difficult to avoid escalation of conflicts and to guarantee the respect of international humanitarian law and international human rights law* (UN News, 2018).

In another statement, he mentioned that

> *Allowing autonomous weapons to be controlled by machine learning algorithms – fundamentally unpredictable software which writes itself – is an unacceptably dangerous proposition....These include limiting where, when and for how long they are used, the types of targets they strike and the scale of force used, as well as ensuring the ability for effective human supervision, and timely intervention and deactivation. However, without a specific international agreement governing autonomous weapon systems, States can hold different views about how these general rules apply. New international rules on autonomous weapons are therefore needed to clarify and strengthen existing law.*

> *(UN Secretary General, 2023)*

The UN Secretary-General's comments make clear that there are serious ethical and legal concerns regarding autonomous weapons. However, the most fundamental issue that needs to be questioned here is how the member countries of the UN Security Council will approach these statements. Given the significant position of the five permanent members in matters of international peace and security, and their known work on autonomous weapons, it remains uncertain how swiftly legal steps will be taken.

Owing to these complications, it is unclear how quickly and successfully laws governing autonomous weapons will be implemented. The international community must strike a compromise between the urgent need to address the ethical, legal, and security challenges raised by new technologies and their potential benefits in improving military effectiveness and lowering human deaths. In the ensuing decades, this delicate balance will probably influence how war and international security policy are conducted.

The first initiative on autonomous weapons was conducted by the United Nations General Assembly. The General Assembly affirmed that international law applies to autonomous weapon systems, highlighted the negative consequences of autonomous weapons on global and regional stability, and decided to add the item "Lethal Autonomous Weapon Systems" to the provisional agenda of its 79th session (Resolution Adopted by the General Assembly on 22 December 2023, 2023). While the General Assembly's initiative in this regard is important, it is also necessary for the UN Security Council to take action on this issue for the rules to be implemented and accepted by the international community.

The UN Security Council must be involved in this for a number of reasons, even though the General Assembly's effort is vital. The Security Council can impose legally obligatory decisions to guarantee adherence to rules pertaining to autonomous weapons. Its participation would give any actions taken more weight and increase their efficacy globally.

REFERENCES

Abaimov, S., & Martellini, M. (2020). Artificial intelligence in autonomous weapon systems. In M. Martellini & R. Trapp (Eds.), *21st century prometheus* (pp. 141–177). Springer International Publishing. https://doi.org/10.1007/978-3-030-28285-1_8

Asaro, P. (2012). On banning autonomous weapon systems: Human rights, automation, and the dehumanization of lethal decision-making. *International Review of the Red Cross, 94*(886), 687–709.

Asaro, P. (2020). Autonomous weapons and the ethics of artificial intelligence. In M. Liao (Ed.), *Ethics of artificial intelligence* (Vol. 212). Oxford University Press.

Autonomous Weapons. (2021, October 20). *Slaughterbots are here*. Autonomous Weapons Systems. https://autonomousweapons.org/

Carpenter, R. C. (2011). Vetting the advocacy agenda: Network centrality and the paradox of weapons norms. *International Organization, 65*(1), 69–102.

Cavelty, M. D. (2018). Cybersecurity research meets science and technology studies. *Politics and Governance, 6*(2), 22–30. https://doi.org/10.17645/pag.v6i2.1385

Christie, E. H., Ertan, A., Adomaitis, L., & Klaus, M. (2023). Regulating lethal autonomous weapon systems: Exploring the challenges of explainability and traceability. *AI and Ethics*, 1–17. https://doi.org/10.1007/s43681-023-00261-0

Communication from the Commission to the European Parliament, the European Council, the Council, the European Economic and Social Committee and the Committee of the Regions – Coordinated Plan on Artificial Intelligence (2018). https://eur-lex.europa.eu/legal-content/EN/TXT/?uri=CELEX%3A52018DC0795

Convention on Prohibitions or Restrictions on the Use of Certain Conventional Weapons Which May Be Deemed to Be Excessively Injurious or to Have Indiscriminate Effects, CCW/GGE.1/2023/CRP.1, CCW/GGE.1/2023/CRP.1 (2023). https://docs-library.unoda.org/Convention_on_Certain_Conventional_Weapons_-Group_of_Governmental_Experts_on_Lethal_Autonomous_Weapons_Systems_(2023)/CCW_GGE1_2023_CRP.1_0.pdf

Craigen, D., Diakun-Thibault, N., & Purse, R. (2014). Defining cybersecurity. *Technology Innovation Management Review, 4*(10), Article 10. https://doi.org/10.22215/timreview835

Crootof, R. (2018). *Autonomous weapon systems and the limits of analogy* (SSRN Scholarly Paper 2820727). https://doi.org/10.2139/ssrn.2820727

Ekelhof, M. A. (2017). Complications of a common language: Why it is so hard to talk about autonomous weapons. *Journal of Conflict and Security Law, 22*(2), 311–331.

Endsley, M. R. (2017). From here to autonomy: Lessons learned from human–automation research. *Human Factors: The Journal of the Human Factors and Ergonomics Society, 59*(1), 5–27. https://doi.org/10.1177/0018720816681350

ENISA. (2021). *Cybersecurity for SMEs—Challenges and recommendations* [Report/Study]. ENISA. https://www.enisa.europa.eu/publications/enisa-report-cybersecurity-for-smes

Etzioni, A., & Etzioni, O. (2017). Pros and cons of autonomous weapons systems. *Military Review, 97*(3), 72–81.

Feng, L., Wiltsche, C., Humphrey, L., & Topcu, U. (2016). Synthesis of human-in-the-loop control protocols for autonomous systems. *IEEE Transactions on Automation Science and Engineering, 13*(2), 450–462.

Fisher, B. A. (2022). *How international humanitarian law will constrain the use of autonomous weapon systems in the conduct of hostilities* [PhD Thesis, Murdoch University]. https://researchportal.murdoch.edu.au/esploro/outputs/graduate/How-international-humanitarian-law-will-constrain/991005542029107891

Gabrecht, K. M. (2016). *Human factors of semi-autonomous robots for urban search and rescue* [PhD Thesis, University of Nottingham]. http://eprints.nottingham.ac.uk/35458

Geiss, R. (2015). *The international-law dimension of autonomous weapons systems*. Friedrich-Ebert-Stiftung, International Policy Analysis. https://library.fes.de/pdf-files/id/ipa/11673.pdf

Güneysu, G. (2023). Yeni Silahların Tabi Tutulması Gereken Hukukilik Denetimi ve Otonom Silah Sistemleri. *Düşünce Dünyasında Türkiz, 14*(64), 81–105. https://doi.org/10.59281/turkiz.1258725

IBM. (2023). *Cost of a data breach 2023*. https://www.ibm.com/reports/data-breach

ICRC. (2022). *What you need to know about autonomous weapons* (Europe and Central Asia/United Kingdom). https://www.icrc.org/en/document/what-you-need-know-about-autonomous-weapons

Jensen, E. (2020). Autonomy and precautions in the law of armed conflict. *International Law Studies, 96*(1). https://digital-commons.usnwc.edu/ils/vol96/iss1/19

Kaplan, A., & Haenlein, M. (2019). Siri, Siri, in my hand: Who's the fairest in the land? On the interpretations, illustrations, and implications of artificial intelligence. *Business Horizons, 62*(1), 15–25. https://doi.org/10.1016/j.bushor.2018.08.004

Konert, A., & Balcerzak, T. (2021). Military autonomous drones (UAVs)—From fantasy to reality. Legal and ethical implications. *Transportation Research Procedia, 59*, 292–299. https://doi.org/10.1016/j.trpro.2021.11.121

Kostyuk, N., & Gartzke, E. (2022). Why cyber dogs have yet to bark loudly in Russia's invasion of Ukraine (Summer 2022). *Texas National Security Review, 5*(3), 113–126.

Kralingen, M. (2016). Use of weapons: Should we ban the development of autonomous weapons systems? *The International Journal of Intelligence, Security, and Public Affairs, 18*(2), 132–156. https://doi.org/10.1080/23800992.2016.1196947

Krishnan, A. (2016). *Killer robots: Legality and ethicality of autonomous weapons*. Routledge.

Lewis, J. (2015). The case for regulating fully autonomous weapons. *The Yale Law Journal, 124*(4), 1309–1325.

Lussier, F., Thibault, V., Charron, B., Wallace, G. Q., & Masson, J.-F. (2020). Deep learning and artificial intelligence methods for Raman and surface-enhanced Raman scattering. *TrAC Trends in Analytical Chemistry, 124*, 115796. https://doi.org/10.1016/j.trac.2019.115796

Martínez Torres, J., Iglesias Comesaña, C., & García-Nieto, P. J. (2019). Review: Machine learning techniques applied to cybersecurity. *International Journal of Machine Learning and Cybernetics, 10*(10), 2823–2836. https://doi.org/10.1007/s13042-018-00906-1

Mincewicz, W. (2021). Social sciences to the rise and development of cryptocurrencies: An analysis of the notion. *Przegląd Politologiczny, 3*, 93–104. https://doi.org/10.14746/pp.2021.26.3.7

Mincewicz, W. (2023). Explication and classification of social deviations on the Internet: Cyberdeviation as a new social phenomenon. *Polish Sociological Review, 222*(2), 231–248. https://doi.org/10.26412/psr222.05

Minsky, M. (1961). Steps toward artificial intelligence. *Proceedings of the IRE, 49*(1), 8–30. https://doi.org/10.1109/JRPROC.1961.287775

Munir, M. (2022). *Autonomous weapons systems: Taking the human out of the loop*. Available at SSRN 4074072. https://papers.ssrn.com/sol3/papers.cfm?abstract_id=4074072

Ning, H., Ye, X., Bouras, M. A., Wei, D., & Daneshmand, M. (2018). General cyberspace: cyberspace and cyber-enabled spaces. *IEEE Internet of Things Journal, 5*(3), 1843–1856. https://doi.org/10.1109/JIOT.2018.2815535

Rajaraman, V. (2014). JohnMcCarthy—Father of artificial intelligence. *Resonance, 19*(3), 198–207. https://doi.org/10.1007/s12045-014-0027-9

Resolution Adopted by the General Assembly on 22 December 2023, 78/241 (2023). https://documents.un.org/doc/undoc/gen/n23/431/11/pdf/n2343111.pdf

Roff, H. M. (2014). The strategic robot problem: Lethal autonomous weapons in war. *Journal of Military Ethics, 13*(3), 211–227. https://doi.org/10.1080/15027570.2014.975010

Sarker, I. H., Kayes, A. S. M., Badsha, S., Alqahtani, H., Watters, P., & Ng, A. (2020). Cybersecurity data science: An overview from machine learning perspective. *Journal of Big Data. 7*(1), 1–29. https://doi.org/10.1186/s40537-020-00318-5

Sauer, F. (2016, October). Stopping 'killer robots': Why now is the time to ban autonomous weapons systems. *Arms Control Today*. https://isodarco.it/wp-content/uploads/2020/10/sauer_Stopping-Killer-Robots.pdf

Scharre, P. (2016). *Autonomous weapons and operational risk* (pp. 41–48) [Ethical Autonomy Project]. Center for a New American Security Washington, DC. https://www.stopkillerrobots.org/wp-content/uploads/2021/09/CNAS_Autonomous-weapons-operational-risk.pdf

Schatz, D., Bashroush, R., & Wall, J. (2017). Towards a more representative definition of cyber security. *Journal of Digital Forensics, Security and Law, 12*(2), 8. https://doi.org/10.15394/jdfsl.2017.1476

Schulzke, M. (2019). Drone proliferation and the challenge of regulating dual-use technologies. *International Studies Review, 21*(3), 497–517. https://doi.org/10.1093/isr/viy047

Sharkey, A. (2019). Autonomous weapons systems, killer robots and human dignity. *Ethics and Information Technology, 21*(2), 75–87. https://doi.org/10.1007/s10676-018-9494-0

Singer, P. W., & Friedman, A. (2014). *Cybersecurity: What everyone needs to know.* Oxford University Press. https://www.google.com/books?hl=tr&lr=&id=f_lyDwAAQBAJ&oi=fnd&pg=PP1&dq=Cybersecurity+and+cyberwar:+What+everyone+needs+to+know&ots=Dok4YOuzqo&sig=nWdTn2h0IjyFSo6cONfekH_s-g0

Singh, A., Gupta, S. S., & Jain, M. M. (2022). Adaptation of modern technologies and challenges in the defense sectors. *Res Militaris, 12*(2), Article 2.

Solovyeva, A., & Hynek, N. (2018). Going beyond the "killer robots" debate: Six dilemmas autonomous weapon systems raise. *Central European Journal of International and Security Studies, 12*(3). https://www.cejiss.org/going-beyond-the-killer-robots-debate-six-dilemmas-autonomous-weapon-systems-raise-0

Tada, M. (2020). How society changes: Sociological enlightenment and a theory of social evolution for freedom. *The American Sociologist, 51*(4), 446–469. https://doi.org/10.1007/s12108-020-09464-y

Umbrello, S., Torres, P., & De Bellis, A. F. (2020). The future of war: Could lethal autonomous weapons make conflict more ethical? *AI & Society, 35*(1), 273–282. https://doi.org/10.1007/s00146-019-00879-x

UN News. (2018, November 5). *'Warp speed' technology must be 'force for good' UN chief tells web leaders. UN News.* https://news.un.org/en/story/2018/11/1024982

UN Secretary General. (2023, October 5). *Note to correspondents: Joint call by the United Nations Secretary-General and the President of the International Committee of the Red Cross for States to establish new prohibitions and restrictions on Autonomous Weapon Systems* [Interview]. https://www.un.org/sg/en/content/sg/note-correspondents/2023-10-05/note-correspondents-joint-call-the-united-nations-secretary-general-and-the-president-of-the-international-committee-of-the-red-cross-for-states-establish-new

van den Boogaard, J. (2015). Proportionality and autonomous weapons systems. *Journal of International Humanitarian Legal Studies, 6*(2), 247–283. https://doi.org/10.1163/18781527-00602007

Van Niekerk, J. F., & Von Solms, R. (2010). Information security culture: A management perspective. *Computers & Security, 29*(4), 476–486. https://doi.org/10.1016/j.cose.2009.10.005

Voronenko, I., Skrypnyk, A., Klymenko, N., & Nehrey, M. (2021). *Assessment of the impact of information asymmetry in the information space on social welfare* (SSRN Scholarly Paper 3884644). https://doi.org/10.2139/ssrn.3884644

Wagner, M. (2014). *The dehumanization of international humanitarian law: Legal. Ethical, and Political Implications of Autonomous Weapon Systems, 47*(5), 1371–1424.

Warner, M. (2012). Cybersecurity: A pre-history. *Intelligence and National Security, 27*(5), 781–799. https://doi.org/10.1080/02684527.2012.708530

Watts, T. F. A., & Bode, I. (2023). *Automation and autonomy in loitering munitions catalogue (v.1).* https://doi.org/10.5281/ZENODO.7860762

Werkhoven, P., Kester, L., & Neerincx, M. (2018). *Telling autonomous systems what to do.* Proceedings of the 36th European Conference on Cognitive Ergonomics, 1–8. https://doi.org/10.1145/3232078.3232238

Work, R. O. (2021). *A short history of weapon systems with autonomous functionalities* (Principles for the Combat Employment of Weapon Systems with Autonomous Functionalities, pp. 5–7). Center for a New American Security. https://www.jstor.org/stable/resrep32146.4

Zurek, T., Kwik, J., & van Engers, T. (2023). Model of a military autonomous device following International Humanitarian Law. *Ethics and Information Technology, 25*(1), 15. https://doi.org/10.1007/s10676-023-09682-1

2 Current Technological Trends and Developments in the Field of Security

Emre Çıtak

2.1 INTRODUCTION

Technology is characterized as the practical application of scientific knowledge to enhance the convenience of life. It plays a pivotal role in fulfilling diverse human needs and driving societal transformation through successive advancements and opportunities. Presently, we often refer to this era as "the age of technology," underlining the pervasive influence of technology on virtually everything in our surroundings. This designation is grounded in the acknowledgment that our environment is intricately shaped by technological innovations. Undeniably, the profound impact of technology is observable across various domains, encompassing home life, educational processes, economic sectors, and even our leisure activities.

Undoubtedly, security stands out as a field where the tangible impacts of technology are readily evident. The historical significance of security for individuals, societies, and states intensifies the scrutiny of developments in this realm. Technology, in this context, yields numerous advantages. It possesses the capability to engender effects that either enhance security or pose threats. Furthermore, the outcome of technological processes can yield varied effects, whether positive or negative, varying across societies, periods, practices, and events.

The relationship between technology and security unfolds across various dimensions. Primarily, technological advancements empower states to employ a myriad of tools, applications, and weapons previously unavailable, aiding in the identification and mitigation of threats. Moreover, everyday individuals can leverage technological innovations to take proactive measures for their security, highlighting the role of technology as a security enabler. Conversely, other states and illicit entities exploit technological outcomes, potentially rendering security measures insufficient against aggressive states or organizations. Additionally, concerns arise regarding the intrusion into citizens' personal lives stemming from the advanced surveillance systems employed by states in the pursuit of security, representing a technological effect that generates insecurity.

The contemporary standard of living is undeniably shaped by humanity's scientific and technological advancements, with these developments carrying consequences that impact life's trajectory. Scientific and technological studies, constituting an accumulation of knowledge from the past to the present, delineate the contours of the present. In recent times, the pace, intensity, and accessibility of developments have increased, accentuating the significance of current technological advancements, while also anticipating innovations that might render today's technology obsolete.

In the present era, addressing security within a broader context is crucial, necessitating focused attention on various technology-related issues. These include artificial intelligence, autonomous weapons, cyber security, biometrics, data mining, data collection and analysis, encryption, social media use, information security, and the internet of things. Technological progress broadens the horizons of security across all aspects. Technology can be characterized as a catalyst for the proliferation of threat types and sources, the diversification of combat methods, the evolution of the military/defense industry, the intensification of related trade, and the transformation of security understanding, policies, and practices.

DOI: 10.1201/9781003441700-2

Technological developments that emerge one after another and spread throughout the world play a transformative role in the field of security. It is necessary to focus particularly on the benefits of technology in areas such as waging wars, ensuring border security, fighting crime, and ensuring public order, monitoring the activities of illegal groups, improving the quality of intelligence processes (collection, analysis, operation, etc.), and protecting information systems. In this study, the impact of today's technological developments on security studies, practices, and discussions will be discussed. In addition, new threats, risks, and security vulnerabilities that emerge with technological developments will also be addressed. These situations will be handled within the framework of human security, national security, and international security. Moreover, the study will also include an evaluation of the impact of possible future technological developments on security. It is thought that it would be useful to examine the future determinants of developments such as artificial intelligence, autonomous-unmanned vehicles, and internet 3.0.

2.2 TECHNOLOGY AND SECURITY NEXUS

Obviously, technology has brought significant changes across various facets of life. The profound impact of technology is inescapable in today's context, leading to a substantial transformation in security mentality, strategies, threat perceptions, and the tools employed. This transformation encompasses both the integration of general technological advancements into the realm of security and the dedicated development of security-specific technologies, particularly in the military and intelligence domains. In essence, technological progress not only broadens the scope of the security domain but also positions security studies as a driving force behind further technological advancements.

Modern societies have innumerable fears. While the reason for this is the emergence of many real or perceived threats, security is one of the determinants of life. Thus, creating protection mechanisms against current and future threats, eliminating risks to public health, and ensuring an environment of trust are people's priorities. They observe that the developments around them have security-enhancing or threatening effects. At this point, people are becoming more interested in technology, on the one hand, as a problem solver, and on the other hand, as a future threat (Durodie, 2007, pp. 206–207). As security threats and sources become more ambiguous and multiply in the new era, technology has been seen as a remedy to eliminate the existing uncertainty. In this direction, states have accepted technology as a weapon in their search for internal and external security. In particular, features of technological systems and tools such as mobility, profitability, adaptability, connectivity, and transferability are seen as enlightening effect in the so-called age of uncertainty (Booth & Wheeler, 2018, pp. 134–141; Ceyhan, 2008, pp. 107–108; Smith, 2010, pp. 16–42). In contrast to the traditional security paradigm, which often revolved around specific enemy-specific threats, the contemporary security thought is characterized by ambiguity, unpredictability, and a pervasive sense of uncertainty. This shift has rendered discussions on security more challenging, as the sources and nature of threats have become less clearly defined and more complex.

Indeed, technology is not merely a policy tool or a consequence of decisions; rather, security and technology should be viewed as intricately connected concepts. Security practices play a crucial role in defining, evolving, and bolstering technological systems. Conversely, technology, in its various forms, serves to structure, influence, and guide security practices. The interplay between security and technology is a dynamic and mutually shaping relationship (Guittet & Jeandesboz, 2010, p. 232). It would be useful to note that technological and scientific developments deeply affect not only security but also related fields such as international relations, diplomacy, and governance apprehension. Of course, development in these areas also draws the direction of security understanding, policies, problem-solving methods, and approaches to combating threats (Kosal, 2019).

In this intricate relationship, technological advancements have undeniably elevated the quality of security measures in terms of detection, prevention, deterrence, and response. Consequently, substantial investments are directed toward security technology. Risk assessments, analysis, monitoring, follow-up procedures, preventive interventions, and policy-making processes evolve in tandem with technological advancements. Therefore, a comprehensive approach to security necessitates the consideration of modern iterations of technology alongside traditional methods and techniques. The integration of technology in combating physical, psychological, and cyber security threats offers the potential to decrease error or vulnerability rates significantly.

Technology developed to make life and processes easier and of better quality will inevitably be used by opportunistic and malicious people for their own benefit and the harm of others. Even if it is not produced for military purposes or to harm the other party, there are those who look for vulnerabilities in every technological tool, software, or system. In addition, the use of technologies produced for defense or protection purposes by aggressive states or non-state actors is also a risk (Whitman, 2007, pp. 94–98). To date, technology and security practices have become so intertwined that it is becoming increasingly impossible to think of security processes outside of technological developments. Seeing security as a phenomenon that can be provided with various systems and devices, as well as the human element, has developed an understanding of the technologicalization of security or technologized security. Additionally, security technology has emerged as a specialized field (Amicelle et al., 2015; Monsees, 2019). Another dimension of the issue, as expressed as technology of insecurity, is that individuals are subjected to surveillance, supervision, and control in a way that affects their private lives and personal freedoms through the use of technological systems. The systems created to monitor illegal individuals, groups, and movements raise questions about how confidential the information and private lives of ordinary individuals remain in the process (Aas et al., 2008, pp. 1–18). In addition, those who perceive state and private company interference in their lives in the name of ensuring national security may resort to more counter-technological measures. In other words, individuals may make various attempts to protect and distract themselves. This situation may lead to new security problems for both governments and citizens (Monti & Wacks, 2022, pp. 91–95). In this regard, the lack of understanding or misunderstanding of the steps taken or claimed to be taken for security by the society may further increase individual security initiatives. This situation may lead to social anxiety, unrest, and security concerns. For this reason, people who think that technology poses a risk to their private lives and personal freedoms may generalize this situation and demonize technology.

Scientific and technological developments are quite effective because they can change strategy choices and options. The impact of technological developments is obvious in the emergence of new weapon systems with previously unimaginable performance, accuracy, and lethality. Thus, it should be noted that technological innovations shape military doctrine and international relations, and in this respect, they are decisive during the understanding of security (James, 2007). It is necessary to reference the impact of information and technology development in the transformation of modern warfare. Throughout history, warfare has gone through various periods and processes. The reflection of technological innovations and information systems on the battlefield has changed many things, especially strategic planning, and will determine the postmodern understanding of war in the future (Sheehan, 2019).

It can be stated that technology has created a revolution in the military field. The benefits of technology in the military field have brought completely new dimensions, from the conduct of conflict to strategic planning (Rasmussen, 2006, pp. 43–60). Hence, traditional discussions around military matters and wars, which traditionally form the core of security debates, are taking a different trajectory. Notably, there is a shift in the nature of warfare concerning the battlefield, duration, involved parties, and weaponry, all attributed to technological advancements. Military technology, in particular, holds a substantial role in shaping the dynamics of the security-technology relationship. Beyond interstate wars, other forms of armed conflicts also undergo transformations influenced by technological directions. In today's military confrontation, irrespective of the setting, technology emerges as a decisive factor, directly impacting the duration and magnitude of conflict, ultimately determining the victor and vanquished.

One of the points that should be particularly mentioned is that technology transfer has become easier today. This situation has several consequences. First, those who request it have more opportunities to reach their security-related needs. Second, allied states can act more freely in helping each other. Third, defense industry trade around the world is increasing. Fourth, non-state actors have the chance to access the advanced weapons, tools, and systems they desire. Fifth, with technological developments, it has become possible to transport and deliver all kinds of military and security materials and make them operational under all conditions. Thus, today, thanks to technology, there is a greater global mobility of military weapons and equipment than ever before in history. While on the one hand, the military and security products are put into circulation freely, on the other hand, there is a race to obtain new technological weapons.

The technology-security relationship should not be considered only from a traditional military perspective. It would be appropriate to consider the humanitarian, environmental, public health, and economic problems that are on the expanding agenda of security in this evaluation. In this direction, while the horizon of security studies expands day by day, the role of science and technology studies should be subjected to a comprehensive analysis (Bellanova et al., 2020). A technology-based analysis is mandatory to understand security at any level. Independent of the debate about how much security and risk technology brings, the intense and multi-dimensional relationship between them also gives its identity to security studies.

2.3 CURRENT DEVELOPMENTS AND SECURITY

It has become completely impossible to follow technological developments, even for those who are personally involved in the processes as, in many parts of the world, inventions and modernizations are being made one after another by official institutions, private companies, or individuals. Thus, it should be noted that technology is a never-ending process of development and renewal. That kind of intense advance undoubtedly paves the way for a radical change in everything related to security. Today, many situations, from waging wars to fighting criminal organizations, from protecting personal information to protecting critical infrastructures can be defined in terms of the security and insecurity brought by technology. We are talking about a period in which advanced weapons and military vehicles, sophisticated unmanned aerial vehicles, elaborated border control systems, criminal tracking mechanisms, and artificial intelligence applications are on the field to ensure security. Thus, it should be noted that security and technology are completely intertwined fields today.

Indeed, the internet stands out as the most widely discussed and influential technological development in today's world. Essentially, the internet constitutes a global communication network that transcends temporal and spatial constraints. Among the transformative developments that have fundamentally altered human history, the proliferation of the internet is arguably the most significant, revolutionizing interactions across the globe. From commercial endeavors to social connections, educational methodologies to cultural trends, every facet of human activity is touched by internet processes. Consequently, the interplay between the internet and security warrants a nuanced examination. This relationship is complex, as the internet offers solutions to security challenges while simultaneously introducing new threats with the expansion of internet facilities.

With the transfer of the internet from a military application to public use, new security issues have come to the fore. Many situations such as information security, confidentiality of communication, surveillance, and cyber-crimes have come with the intense internet activity. Such developments as the easier access to the internet, the increase in the activities and time spent by users in online environment, the gradual improvement of relevant hardware and software, and the fact that states, companies, and various groups, as well as individuals, are highly active draw the basis of the internet-security relationship (Naughton, 2016, pp. 17–22). In this regard, it is necessary to create a broader framework when evaluating national security or capability today. Increasing technological capacity and ensuring protection in information systems stand out as a critical security issue (Alguliyev et al., 2021; Purpura, 2016, pp. 347–364).

It is necessary to discuss the impact of technology on the transformation of wars. As the war has stood out the one of the most dominant security concerns throughout most of human history, being able to get out of wars that affect the security and stability of all parties involved depends on the adoption of appropriate strategies, the use of proper tools and weapons, the successful determination of the enemy's capacity, and the immediate activation of defense mechanisms against attacks. At this point, in these cases, the benefits of technological developments become the determinants of the new forms of wars. It is crucial to acknowledge that technology has intricately molded aspects of warfare, encompassing armies, strategies, weaponry, and more (Roy, 2022, pp. 163–173; Zweibelson, 2023, pp. 63–72). The infusion of computers and related systems into warfare, catalyzed by technological progress, has resulted in a revolutionary transformation. Communication, logistics, command, detection, military equipment, attack, and defense capabilities have improved with every stride in technological developments (Bosquet, 2009, pp. 121–136).

It should be noted that the development in military technology has a transformative role not only in the conduct of traditional wars but also in all types of conflicts since all kinds of methods and means are used to gain superiority in struggles where military operations are carried out, such as civil wars and the fight against terrorism. Nowadays, when traditional wars have become rare, new generation weapons, cyber-attacks, and preventive security measures are used in such military encounters. In fact, the development of military technology by determining military needs and deficiencies is associated with the frequent emergence of such struggles.

New capacities of the armies create new opportunities for the use of kinetic and non-kinetic force in cyberspace. The benefits of the cyber age mean operating in a new, non-linear, and only vaguely defined hybrid battlefield. When the advanced opportunities obtained in space and cyberspace are integrated with what is already available, it is possible to talk about a serious increase in proficiency. New systems can better detect, track, and identify targets, command troops, and direct weapon systems to achieve the intended effect (Lehto, 2018, p. 4). Today, armies need to be strong both in the traditional battlefield and in the cyber field. In this regard, traditional and cyber strategies, weapons, and systems should be used simultaneously. Thus, conducting hybrid warfare and developing hybrid power and hybrid deterrence is what is important today (Najzer, 2020). The decisive role of technology in traditional and new wars is necessary in increasing hybrid capacity in this direction. Military processes such as planning, control, attack, and defense need to be considered with a hybrid approach.

We are witnessing a process in which real-life conflicts are increasingly cyberized and struggles are waged in this way. Thus, the cyber dimensions of the show of power come to the fore (Demchak, 2020, pp. 39–41). The cyber world has a special topic in the security-technology relationship as the lines of this field are completely drawn by technology. This virtual dimension, which exists with computer and network technologies, draws a different direction and its borders expand with each new development. All kinds of applications result in more and more individuals, commercial organizations, and government institutions being drawn into the cyberspace. Thus, even though the cyber world cannot be said to be new today, it is a popular and current subject of security research. Attacks using techniques such as spyware, phishing, DoS and DDoS, and Zero Day bring the security of the cyberspace to the agenda. Phenomenon called cyber war, cyber espionage, cyber terrorism, and cyber-crimes cause the deterioration of relations and order in cyberspace. Many attempts, from phishing to infiltrating the private systems of the state, make the cyberspace a medium full of risks and threats (Stevens, 2016, pp. 23–33; Valeriano & Maness, 2015, pp. 2–5). When discussing cyber security, it is essential to highlight the dangers posed to the cyber infrastructure and the risks emanating from the cyber domain. Therefore, it is accurate to view cyberspace both as a valuable asset that requires protection and as a source of threats. Given that it is a realm where interactions and activities are progressively more intense, conducting social risk assessments in this context becomes more meaningful (Deibert & Rohozinski, 2010, pp. 18–23).

Cyber security mentions the role of laws, policies, tools, devices, concepts, security mechanisms, risk management, training, practices, and technologies used to protect people and information technology assets. Availability, integrity, and authenticity are determined as the main objectives. Cyber security is now regarded as a remarkable and critical field due to the threats associated with the pervasiveness of information and communication technologies, as well as their interconnectedness and data sharing capabilities (Guerrier, 2021, p. 145). Rain Ottis believes that cyber warfare might be seen as information warfare in a localized area or in an expanding function as an aspect of military engagements. In these circumstances, in one the purpose of cyber warfare is to (assist) manipulate the content of systems (information, data), in order to affect the decision-making process of the targets. On the other hand, the goal of cyberwarfare is to subvert or neutralize the technological/technical (military) capabilities of the opponent (Ottis, 2015, pp. 92–95). Technological dynamics interact intensively with social and political dynamics. In turn, political preferences and policies shape the course of digital technologies. This also applies fundamentally to the actors developing these technologies and to the dynamic interaction of cyber security markets and politics (Dunn Cavelty & Wenger, 2020, p. 9). Today, cyberspace is not only an area where security must be ensured but also a reference tool that helps ensure security. In this regard, with the activities in the cyber field, the security of the cyberspace can be provided and general security-enhancing measures can be contributed.

Today, national security is also concerned with initiatives to create resilience and redundancy in national infrastructure through cyber security measures, and cyber security ranks as one of the top priorities of the national security agenda. This means that measures that generally fall within the scope of information security can now be included among measures to ensure national security and vice versa (Cavelty, 2010, p. 156). In rapidly changing technological and political situations, it is necessary to consider that cyber security-related practices are temporary and subject to political scrutiny and shifts in power. Changing social and political coalitions are reflected in new political decisions discernible in cyber security strategies. In this regard, cyber policies must be prepared in a multifaceted manner and include social and economic actors as well as the state (Dunn Cavelty & Egloff, 2019). It should not be forgotten that the usefulness and necessity of cyber-technological systems, especially in the protection of critical infrastructure, is increasing. While monitoring, control, problem identification, and resolution are easier with such systems, the risk of cyber-attack on the systems themselves should also be taken into consideration. It is thought that successful security measures to be implemented at this point will protect critical infrastructure and cyber management systems (Akter et al., 2023). With the emergence of cyberspace as the fifth spatial dimension, security practices and relations in this area also come to the fore. While increasing cyber capacity and cyber power is necessary against threats in the cyber domain, cyber deterrence emerges as a more interesting aspect. States must also develop their cyber deterrence in addition to their traditional deterrence features since, together with cyber deterrence, they present a security barrier by preventing threats from other states and non-state actors (Haggman, 2018, pp. 65–72).

Nowadays, the importance of having information and protecting critical information in national security is increasing since it is now necessary to talk about information wars/struggles as one of the main issues on the agenda. At this point, the necessity of protecting information repositories and networks emerges. In this context, while information war (iWar) is a serious challenge, defense capacity can be increased with technological applications (Cohen, 2007, pp. 244–259). As every stage of life enters the digital world, the importance of technological opportunities for collecting data has also increased. Data collection and analysis activities are continuing with high or low technology products to collect security-related needs. It is highly needed to produce information flow and timely intelligence on issues such as identifying and clarifying public order events, fighting against terrorism, and countering espionage activities (Bonelli & Ragazzi, 2014, pp. 477–480). Obtaining data, creating information, and gaining knowledge have become more diverse with technological developments. Although information is extremely important in every period of history, today it is the determinant of many different fields. The security-providing role of possessing information and information security, which defines the protection of core information and related systems, are two important aspects that deserve special mention. Ensuring the continuation of the flow of information about threats and targets, on the one hand, and protecting the obtained information from attacks, on the other hand, requires advanced technological facilities.

The influence of advancements in surveillance, tracking, and monitoring systems on security is notably profound. Beyond monitoring borders for military threats, these systems are regularly enhanced to address various other threats and are instrumental in combating issues within national borders. The collection of data plays a crucial role in tracking members of terrorist and criminal organizations, monitoring illicit financial activities, managing irregular immigration concerns, and addressing other social threats. The images and signals acquired through these systems are pivotal for analysis within information systems, forming the basis for devising counter-fighting strategies and methods (Salter, 2010, pp. 191–193).

Today, many actors, from private companies to government institutions, have turned into a data collection frenzy. Especially with digital technologies, it has become easier to collect and process data in many areas as in digital environments, people share many things about their lives, leave traces about their daily routines, and show their basic tendencies with various movements from time to time. It is clear that such data has security as well as commercial relevance. On the one hand, technology directs people to environments where they can share more data; on the other hand, it provides more opportunities to obtain data from these places and creates new opportunities for protection (Flyverbom et al., 2019, pp. 6–9). Big data is driven by a multitude of political and commercial interests, particularly in the areas of analytics, mining, and data integration. The life cycle of data from several sources, from collection to destruction, and the ability to extract new information through analysis, combination, and utilization are key features of big data. Big data has many benefits and the potential to be innovative in many different industries, but it also has many drawbacks, including unreliable sources, insecure infrastructure, access controls, data storage, and privacy concerns (García Bazán, 2023, pp. 70–71). The use of big data, especially for surveillance, fuels a serious debate and, in addition to identifying malicious people involved in society, the risk of storing "everything of everyone every time" creates an undesirable situation. It is possible to talk about a dilemma regarding national security and the security of personal information, especially with the disclosure of such activities of intelligence organizations (Lyon, 2014, pp. 6–8). In this context, it is crucial to emphasize that ensuring the security of information or big data presents a critical challenge. Just as gaining access to data and information poses a security concern, so do the protective measures within this domain. Consequently, addressing issues such as malicious software, the generation of fake data, or unauthorized entries requires the implementation of measures like encryption, data classification, and intrusion detection applications.

The level terrorism has reached today, as a national and international problem, is often described as new terrorism. While terrorist organizations are undergoing a serious transformation in areas such as ideology approach, organizational structuring, and financial resources, it is undoubtedly necessary to mention the impact of technology. Technology contributes to the development of terrorist organizations' opportunities and capacities in matters such as conducting propaganda, carrying out secret communication, providing and receiving training, performing cyber-attacks, recruiting, and radicalizing personnel, and generating income. Thus, a more efficient, effective, and difficult to combat terrorism strategy challenges societies (Golumbic, 2008; Weimann, 2015). In addition, it is obvious that criminal groups can similarly carry out their operations more easily with technological opportunities. It is necessary to state that the weapons, communication tools, and money laundering methods used by such illegal groups are positively affected by technological developments. For this reason, it is essential to prevent the technological capacities of terrorist or criminal groups and to use advanced technological systems in the fight against them to ensure security and public order. Technology provides a strong advantage, especially in identifying, detecting, tracking, and capturing members of illegal groups. For this reason, technology is a weapon that should be highlighted in ensuring public order and fighting crime. The undeniable benefits of technology, especially in intelligence activities against these illegal organizations, have become more evident today (Cohen, 2007). Certainly, intelligence services, holding a unique position among institutions responsible for ensuring security, are actively pursuing technological integration across various aspects – from data collection to analysis, secret operations to training. Intelligence organizations, extending beyond technical intelligence capacity, stand out as one of the most advanced structures equipped with technology. Intelligence technology holds a significant place within the broader spectrum of military technology as well.

2.4 FUTURE AND SECURITY: TECHNOLOGICAL WORRIES AND TECHNOLOGICAL RESPONSE

Security technologies are more than the sum of manufactured objects dedicated to protection, security, and surveillance: they combine ways of designing, assuming, and implementing security and are both support and explanation of a particular belief about security. They embody the contradictions of security practices: the more important security becomes, the more we fear insecurity, and the more security technologies contribute to shaping relentless expectations of a future that is always thought of as catastrophic. Ultimately, distancing ourselves from technological devices signifies an alienation from our belief that technology is salvation (Guittet & Jeandesboz, 2010, p. 238).

Even though their numbers are decreasing, there is no doubt that wars or military encounters will continue in the future. Hence, the requirement for armies will persist. However, as a change that has been going on for a long time, we will continue to see armies that are more professional, more mobile, equipped with more technical facilities and using more advanced weapon technologies. In this regard, in addition to the soldiers fighting in this direction, the impact of those who serve in the field of technology, such as engineers, software developers, and mathematicians, will be greater (Zweibelson, 2023, pp. 87–88). The fact that war is one of the most permanent features of societies means every process that societies go through is reflected in this area. Despite the permanence of its general characteristics, wars are shaped according to the spiritual and material values of societies. Thus, it is necessary to state that war is a dynamic phenomenon. It is quite natural to say that future societies will develop their own orientations toward wars through the processes they go through. Scientific discourse and technological developments and their reflection on the military field will also determine the character of wars. No matter what level of technology the parties to the war reach, they will try to establish order on the battlefield (Bosquet, 2009, pp. 243–244). It is necessary to say that in the wars of the future, military weapons will be more sophisticated, quality will be more important than quantity in the field, and crowded armies will be replaced by units with high maneuverability and technical equipment. In addition, the more intense presence of states in space brings with it the power to monitor, track, and strike. In addition, the value of information systems for armies in accessing information and waging psychological warfare is increasing (Cohen, 2019, pp. 133–136; Van Creveld, 1991, pp. 311–320). Presently, conventional armies have undergone a substantial transformation and integration with technology. Nonetheless, it appears that in the future, there will be an increased demand for military forces capable of swift adaptation to innovations, equipped with autonomous weapons and robotic warriors, emphasizing the role of artificial intelligence alongside human factors in strategy formation, and proficient in executing high-impact attacks.

As traditional wars, augmented by technology, persist, the prominence of cyber wars is destined to intensify. It is evident that states will employ cyber-attacks not only to assess their own technological prowess but also to undermine the technology-related systems of others. There is a certainty that additional elements will be incorporated into today's cyber-attack techniques and tools, given that states will persist in investing resources to identify vulnerabilities in each other's systems. Infiltrating, damaging, or disrupting the other party's critical infrastructures, daily life systems, and information stores will be as popular in the future as it is today. At this point, the fact that wars have both defensive and offensive aspects will be one of the factors that outline the ability of states to provide security in terms of cyber defense and cyber-attack capacity (McGraw, 2013, pp. 112–116; Singer & Friedman, 2014; Slayton, 2016, pp. 76–82). Undoubtedly, as investments in this field increase, states will think that they are increasing their cyber power, but on the other hand, such measures will be perceived as a threat by others. Thus, it is necessary to mention the cyber security dilemma. One's security-enhancing activities in cyberspace create the other's concerns. Not knowing exactly whether the developments in cyberspace security will be defensive or offensive opens the door to a serious security spiral. The fact that the source, time, and methods of attacks in the cyberspace are ambiguous and the vulnerability leads to serious consequences causes the actors to be constantly and effective to avoid being exposed to a surprise attack (Buchanan, 2017, pp. 187–189). The security landscape of the future appears to be shaped by a dynamic wherein states closely surveil, assess, and attempt to undermine each other's cyber-technological advancements. Analogous to the historical challenges posed by the traditional security dilemma, the emerging cyber security dilemma is poised to instigate comparable complexities and concerns.

Artificial intelligence applications will be used more in detection, prediction, and response points to ensure cyber security in the future. The impact of artificial intelligence in cyberspace will be seen in faster identification and detection of threats, analysis of huge data, identification and improvement of vulnerabilities, and securing validation (Katiyar, 2023, pp. 115–117). The more intense and advanced use of unmanned vehicles and robots, autonomous weapons, artificial intelligence, and cognitive computing applications will be the determinants of the future state of wars. The way to combat such developments will be discussed from different dimensions (Lehto, 2018, pp. 7–9). Jim Chan and Alan Dinerman indicate that the incompatibility of cyber and traditional capabilities means that states will not wage war solely through cyberspace. Thus, it will still be important to be on the ground with air, land, and sea forces. However, they say that cyber power will become increasingly valuable in future conflicts. In their view, cyber power has unique characteristics that provide a major operational advantage when effectively synchronized with conventional land, sea, air, and space power. This advantage manifests itself in three areas: (1) the ability to achieve an asymmetry that offsets the numerical advantage, (2) the ability to stabilize terrain to conduct a deep strategic offensive, and (3) the ability to deter adversaries (Chen & Dinerman, 2018, p. 27).

It should not be approached as a miracle, but artificial intelligence will be the technological development having the unique impact on our future. It is necessary to foresee that new generation artificial intelligence applications will have more weight in determining the cause and source of many security problems discussed today, in performing risk analysis, in collecting and analyzing what is collected from targets, in preparing security policies and combat strategies, and in planning operations. It will be possible to see more of the role of artificial intelligence, alongside the human element, both in waging wars of the future and in combating the cyber activities of criminal groups.

The main future challenge will be the management of security technologies. States and private companies are developing security technologies demanded by armies, intelligence organizations, and individuals every day. It is not difficult to predict that security technologies will develop in the following years, as they have become both a basic need and significant trade item, the number of relevant products will increase, more manufacturers and practitioners will participate in the processes, and the coverage area of protection systems will expand for faster solutions to new threats. In this context, the demand for new generation weapons, artificial intelligence applications, and cyber-attack-defense tools will intensify. At this point, preventing revisionist states and illegal non-state actors from accessing such technologies will be possible by managing and controlling security technologies. Otherwise, it would not be surprising if a security-enhancing technological development poses a threat to societies. Certainly, the future is unequivocally intertwined with technology. The perception of security, its assurance, and the nature of the challenges faced will be deliberated in the context of technological advancements. The trajectory of our lives, social dynamics, discussions on national security, and global relations will persist in being profoundly influenced by advancements in areas such as cyber security, biometrics, surveillance and monitoring systems, social media, blockchain, encryption, the internet of things, artificial intelligence, unmanned military vehicles, and analytical programing.

Finally, it should be added that it is impossible to know the technological developments in the near future with certainty today, because even the people who manage the technology and science processes are surprised by the successive developments. For this reason, while it is possible to interpret the positive and negative effects of technological developments on security in the light of today's data, on the other hand, many things will be full of unknowns. However, technology will provide serious benefits at important points such as security management/governance, obtaining information on relevant issues and identifying threats (Ehrhart et al., 2014; Monsees, 2019).

2.5 CONCLUSION: PERPETUAL SECURITY WITH TECHNOLOGY?

Discussions about technology should not be confined to a dichotomy of good or evil. Taking a definitive stance, particularly in its intersection with security, a complex and multifaceted phenomenon, would be oversimplified. Technology significantly shapes the understanding, strategies, and processes of security, permeating every facet of our lives. This impact gives rise to diverse dimensions as we reassess human security, national security, and international security in contemporary contexts. A comprehensive array of technological evaluations is imperative, covering everything from safeguarding personal information to monitoring the online activities of terrorist organizations, and from border control and patrol operations to identifying cross-border threats.

Whether technology is a solution or a savior does not have a straightforward answer, as evident throughout this study. Technology yields both positive and negative outcomes, presenting advantages and disadvantages, with impacts on both beneficiaries and those who may not benefit from it. The complexity deepens when considering threats posed by technological developments, such as advanced weapons of mass destruction and spyware, along with the discomfort experienced by citizens due to surveillance and control systems implemented by certain states. Moreover, with the ease of technology transfer and trade today, it is imperative to assess not only the technology accessible to one state but also the opportunities acquired by non-state actors. It is crucial to remember that a technological application enhancing one's security or defense might inadvertently create insecurity or pose a threat to others. Technology plays a pivotal role in shaping our perception of insecurity and determining the level of security needed. From antivirus programs installed on personal computers to military air defense systems, and from technical systems monitoring the activities of criminal organizations to the use of blockchain in economic relations, technological advancements increasingly serve as the primary factor in defining "What is security?"

Nowadays, even identifying threats, let alone providing security, is a serious challenge. Of course, individuals, societies, states, and the international community rely on technology as the current and future periods can never be compared to the previous ones in terms of security threats. In cases where human elements are not sufficient, technology is seen as a basic protection shield against security problems for many reasons such as speed, continuity, accuracy, and utility since, in cases where security technologies are not used sufficiently, there is a very serious possibility of being left unprotected against both traditional threats and new ones.

Unquestionably, every community aspires to absolute and perpetual security. Order and security must endure for social relations and processes to continue in a healthy manner. Currently, among the most basic application tools are security-enhancing/providing applications that are developed by leveraging technical benefits in a way that does not impede fundamental rights and freedoms. Naturally, there are many different contexts in which security analysis is required, including civil wars, the fight against terrorism, the security of food and energy, artificial intelligence that conducts biometric analysis, and new kinds of threats. Thus, it would be correct to say that working to ensure security is one of the most challenging tasks in our age. In this regard, security policies and strategies need to be addressed with a comprehensive understanding. In this context, technology constitutes one of the main topics, including developing security technology and ensuring security-technology compatibility as one of the main points.

The question of whether we expect too much from technology is valid, especially given today's complex security challenges that evoke genuine concerns about the future. The escalation of environmental issues, the occurrence of internal conflicts globally, challenges to state sovereignty, the rise of illegal groups impacting state borders and daily life, and the proliferation of cyber risks that can translate into real-world consequences all contribute to heightened anxiety about meeting the need for protection. In addressing these concerns, it is common to adopt the perspective that "if there is a threat, technology should be developed to address it." While technology indeed offers hope for enhancing security, it is not and will not be an absolute problem solver. However, the more technology is integrated into processes, the greater progress can be made in mitigating and combating threats.

As a final word, it should be said that the technology-security relationship will continue to remain on the agenda with its different aspects. While technological developments sometimes help security, sometimes they will pose a serious challenge. Security needs will also continue to be one of the most important factors determining the direction of technology studies. As security threats become more ambiguous and providing security becomes more difficult, the place and role of technology in the relevant discussion will be mentioned with question marks.

REFERENCES

Aas, K. F., Lomell, H. M., & Gundhus, H. O. (2008). Introduction: Technologies of (in)security. In K. F. Aas, H. M. Lomell, & H. O. Gundhus (Eds.), *Technologies of insecurity: The surveillance of everyday life.* Routledge-Cavendish.

Akter, S. S., Ahmed, R., Khan, F. H., & Rahman, M. S. (2023). Cyber-physical energy system security: Attacks, vulnerabilities and risk management. In O. Pal, V. Kumar, R. Khan, B. Alam, & M. Alam (Eds.), *Cyber security using modern technologies: Artificial intelligence, blockchain and quantum cryptography* (pp. 155–182). Routledge & CRC Press. https://www.routledge.com/Cyber-Security-Using-Modern-Technologies-Artificial-Intelligence-Blockchain-and-Quantum-Cryptography/Pal-Kumar-Khan-Alam-Alam/p/book/9781032213194

Alguliyev, R. M., Imamverdiyev, Y. N., Mahmudov, R. S., & Aliguliyev, R. M. (2021). Information security as a national security component. *Information Security Journal: A Global Perspective, 30*(1), 1–18. https://doi.org/10.1080/19393555.2020.1795323

Amicelle, A., Aradau, C., & Jeandesboz, J. (2015). Questioning security devices: Performativity, resistance, politics. *Security Dialogue, 46*(4), 293–306. https://doi.org/10.1177/0967010615586964

Bellanova, R., Jacobsen, K. L., & Monsees, L. (2020). Taking the trouble: Science, technology and security studies. *Critical Studies on Security, 8*(2), 87–100. https://doi.org/10.1080/21624887.2020.1839852

Bonelli, L., & Ragazzi, F. (2014). Low-tech security: Files, notes, and memos as technologies of anticipation. *Security Dialogue, 45*(5), 476–493. https://doi.org/10.1177/0967010614545200

Booth, K., & Wheeler, N. J. (2018). Uncertainty. In P. D. Williams & M McDonald (Eds.), *Security studies: An introduction* (3rd ed.). Routledge.

Bosquet, A. J. (2009). *The scientific way of warfare: Order and chaos on the battlefields of modernity.* Columbia University Press. https://www.hurstpublishers.com/book/the-scientific-way-of-warfare/

Buchanan, B. (2017). *The cybersecurity dilemma: Hacking, trust and fear between nations.* Oxford University Press. https://doi.org/10.1093/acprof:oso/9780190665012.001.0001

Cavelty, M. D. (2010). Cyber-security. In J. P. Burgess (Ed.), *The Routledge handbook of new security studies* (pp. 154–162). Routledge.

Ceyhan, A. (2008). Technologization of security: Management of uncertainty and risk in the age of biometrics. *Surveillance & Society, 5*(2), 102–123. https://doi.org/10.24908/ss.v5i2.3430

Chen, J. Q., & Dinerman, A. (2018). Cyber capabilities in modern warfare. In M. Lehto & P. Neittaanmäki (Eds.), *Cyber security: Power and technology* (pp. 21–30). Springer International Publishing. https://doi.org/10.1007/978-3-319-75307-2_2

Cohen, E. A. (2007). Information warfare, netwar, and cyber intelligence. In T. A. Johnson (Ed.), *National security issues in science, law, and technology* (pp. 243–276). CRC Press.

Cohen, E. A. (2019). Technology and warfare. In J. Baylis, J. J. Wirtz, & C. S. Gray (Eds.), *Strategy in the contemporary world* (pp. 127–143). Oxford University Press.

Deibert, R. J., & Rohozinski, R. (2010). Risking security: Policies and paradoxes of cyberspace security. *International Political Sociology*, *4*(1), 15–32. https://doi.org/10.1111/j.1749-5687.2009.00088.x

Demchak, C. C. (2020). Cybered conflict, hybrid war, and informatization wars. In E. Tikk, & M. Kerttunen (Eds.), *Routledge handbook of international cybersecurity* (pp. 36–51). Routledge.

Dunn Cavelty, M., & Egloff, F. J. (2019). The politics of cybersecurity: Balancing different roles of the states. *St Antony's International Review*, *15*(1), 37–57.

Dunn Cavelty, M., & Wenger, A. (2020). Cyber security meets security politics: Complex technology, fragmented politics, and networked science. *Contemporary Security Policy*, *41*(1), 5–32. https://doi.org/10.1080/13523260.2019.1678855

Durodie, B. (2007). Understanding the broader context. In B. Rappert (Ed.), *Technology and security: Governing threats in the new millennium* (pp. 193–211). Palgrave Macmillan.

Ehrhart, H.-G., Hegemann, H., & Kahl, M. (2014). Towards security governance as a critical tool: A conceptual outline. *European Security*, *23*(2), 145–162. https://doi.org/10.1080/09662839.2013.856303

Flyverbom, M., Deibert, R., & Matten, D. (2019). The governance of digital technology, big data, and the internet: New roles and responsibilities for business. *Business & Society*, *58*(1), 3–19. https://doi.org/10.1177/0007650317727540

García Bazán, J. A. (2023). Data protection concerns in emerging technologies. In N. Rebe (Ed.), *Regulating cyber technologies* (pp. 61–80). World Scientific (Europe). https://doi.org/10.1142/9781800612860_0005

Golumbic, M. C. (2008). *Fighting terror online: The convergence of security, technology, and the law.* Springer.

Guerrier, C. (2021). *Security and its challenges in the 21st century* (1st ed.). Wiley-ISTE.

Guittet, E.-P., & Jeandesboz, J. (2010). Security technologies. In J. P. Burgess (Ed.), *The Routledge handbook of new security studies* (pp. 229–239). Routledge.

Haggman, A. (2018). Cyber deterrence theory and practise. In M. Lehto & P. Neittaanmäki (Eds.), *Cyber security: Power and technology* (pp. 63–81). Springer International Publishing. https://doi.org/10.1007/978-3-319-75307-2_5

James, A. D. (2007). Science and technology policy and international security. In B. Rappert (Ed.), *Technology and security: Governing threats in the new millennium* (pp. 25–44). Palgrave Macmillan UK. https://doi.org/10.1057/9780230591882_2

Katiyar, S. (2023). Cyber security using modern technologies: Artificial intelligence, blockchain and quantum cryptography. In O. Pal, V. Kumar, R. Khan, B. Alam, & M. Alam (Eds.), *Cyber security using modern technologies: Artificial intelligence, blockchain and quantum cryptography* (pp. 111–124). Routledge & CRC Press. https://www.routledge.com/Cyber-Security-Using-Modern-Technologies-Artificial-Intelligence-Blockchain-and-Quantum-Cryptography/Pal-Kumar-Khan-Alam-Alam/p/book/9781032213194

Kosal, M. E. (2019). Emerging science and technologies: Diplomacy, security, and governance. In J. P. Singh, M. Carr, & R. Marlin-Bennett (Eds.), *Science, technology, and art in international relations* (pp. 50–59). Routledge.

Lehto, M. (2018). The modern strategies in the cyber warfare. In M. Lehto & P. Neittaanmäki (Eds.), *Cyber security: Power and technology* (pp. 3–20). Springer International Publishing. https://doi.org/10.1007/978-3-319-75307-2_1

Lyon, D. (2014). Surveillance, snowden, and big data: Capacities, consequences, critique. *Big Data & Society*, *1*(2), 1–13. https://doi.org/10.1177/2053951714541861

McGraw, G. (2013). Cyber war is inevitable (unless we build security in). *Journal of Strategic Studies*, *36*(1), 109–119. https://doi.org/10.1080/01402390.2012.742013

Monsees, L. (2019). Public relations: Theorizing the contestation of security technology. *Security Dialogue*, *50*(6), 531–546. https://doi.org/10.1177/0967010619870364

Monti, A., & Wacks, R. (2022). *National security in the new world order: Government and the technology of information.* Routledge. https://www.routledge.com/National-Security-in-the-New-World-Order-Government-and-the-Technology-of-Information/Monti-Wacks/p/book/9780367809713

Najzer, B. (2020). *The hybrid age: International security in the era of hybrid warfare.* I. B. Tauris. https://www.bloomsbury.com/uk/hybrid-age-9780755636532/

Naughton, J. (2016). The evolution of the internet: From military experiment to general purpose technology. *Journal of Cyber Policy*, *1*(1), 5–28. https://doi.org/10.1080/23738871.2016.1157619

Ottis, R. (2015). Cyber warfare. In M. Lehto & P. Neittaanmäki (Eds.), *Cyber security: Analytics, technology and automation* (pp. 89–96). Springer. https://doi.org/10.1007/978-3-319-18302-2_6

Purpura, P. P. (2016). *Security: An introduction.* CRC Press. https://www.taylorfrancis.com/books/mono/
10.1201/9781439894569/security-philip-purpura

Rasmussen, M. V. (2006). *The risk society at war: Terror, technology and strategy in the twenty-first century.* Cambridge University Press. https://www.google.com/books?hl=tr&lr=&id=7LCROJevJOgC
&oi=fnd&pg=PP2&dq=The+risk+society+at+war:+Terror,+technology+and+strategy+in+the+twe
nty-first+century&ots=clvgheSlvt&sig=zUUgxKTP9zXxBJAobZSIlQb5sro

Roy, K. (2022). *A global history of warfare and technology: From slings to robots.* Springer Nature. https://
www.google.com/books?hl=tr&lr=&id=k6t_EAAAQBAJ&oi=fnd&pg=PR5&dq=A+global+history
+of+warfare+and+technology:+From+slings+to+robots&ots=hDwfFMdEgn&sig=uiqweHxFqzuxeM
YAvrGDxRuV2mI

Salter, M. B. (2010). Surveillance. In J. P. Burgess (Ed.), *The Routledge handbook of new security studies* (pp.
187–196). Routledge. https://api.taylorfrancis.com/content/chapters/edit/download?identifierName=doi
&identifierValue=10.4324/9780203859483-23&type=chapterpdf

Sheehan, M. (2019). The evolution of modern warfare. In J. Baylis, J. J. Wirtz, & C. S. Gray (Eds.), *Strategy
in the contemporary world* (pp. 36–55). Oxford University Press. https://www.academia.edu/download/49196710/Seminario-GPS-1.pdf

Singer, P. W., & Friedman, A. (2014). *Cybersecurity: What everyone needs to know.* Oxford University Press.
https://www.google.com/books?hl=tr&lr=&id=f_lyDwAAQBAJ&oi=fnd&pg=PP1&dq=Cybersecurity
+and+cyberwar:+What+everyone+needs+to+know&ots=Dok4YOuzqo&sig=nWdTn2h0IjyFSo6cONfe
kH_s-g0

Slayton, R. (2016). What is the cyber offense-defense balance? Conceptions, causes, and assessment.
International Security, 41(3), 72–109.

Smith, M. E. (2010). *International security: Politics, policy, prospects.* Palgrave Macmillian.

Stevens, T. (2016). *Cyber security and the politics of time.* Cambridge University Press. https://www.google.
com/books?hl=tr&lr=&id=Nk-2CgAAQBAJ&oi=fnd&pg=PR9&dq=Cyber+security+and+the+politic
s+of+the+time&ots=tOxgOolu8L&sig=w6GnfqWayOCzjXKYGWyaTYtslDE

Valeriano, B., & Maness, R. C. (2015). *Cyber war versus cyber realities: Cyber conflict in the international
system.* Oxford University Press. https://www.google.com/books?hl=tr&lr=&id=TpmeBwAAQBAJ&o
i=fnd&pg=PP1&dq=Cyber+wars+versus+cyber+realities:+Cyber+conflict+in+the+international+syste
m&ots=yPK2spEJ_g&sig=aRQ3wv2pnHCqw60sxQd-hzS270Q

Van Creveld, M. (1991). *Technology and war: From 2000 BC to the present.* The Free Press. https://www.
google.com/books?hl=tr&lr=&id=A7FZ98dFQbkC&oi=fnd&pg=PR9&dq=Technology+and+war:+Fr
om+2000+B.C.+to+the+present&ots=1lY_xUDcSY&sig=KLPA4aln6P5WUU2vc5X1nGEAPHk

Weimann, G. (2015). *Terrorism in cyberspace: The next generation.* Columbia University Press. https://www.
google.com/books?hl=tr&lr=&id=K3CrBQAAQBAJ&oi=fnd&pg=PT10&dq=Terrorism+in+cyberspa
ce:+The+next+generation&ots=79bkTrJFgJ&sig=nNgj9VLzST2qKqL-L2gqks_eCdI

Whitman, J. (2007). Global governance and twenty-first century technology. In B. Rappert (Ed.), *Technology
and security: Governing threats in the new millennium* (pp. 89–110). Palgrave Macmillan UK. https://
doi.org/10.1057/9780230591882_5

Zweibelson, B. (2023). *Understanding the military design movement: War, change and innovation.* Routledge.
https://www.routledge.com/Understanding-the-Military-Design-Movement-War-Change-and-
Innovation/Zweibelson/p/book/9781032481784

3 Cyber Security Challenges in the Era of Artificial Intelligence and Autonomous Weapons

Hassan Benouachane

3.1 INTRODUCTION

The rapid evolution of artificial intelligence (AI) has brought about transformative changes across various sectors, with a notable impact on the military domain. Recent technological progress underscores the growing influence of AI in weapons systems, particularly in integrating autonomy seamlessly into diverse weaponry. Among the controversial applications of AI in defense, the development of autonomous weapon systems (AWSs) stands out, enabling machines to independently select and engage targets without direct human control. Given the significant modernization efforts in global militaries, there is a strong incentive to pursue increased autonomy for maintaining a technological edge, and as a result, these autonomous weapons are poised to spread widely. Apart from the leading military powers, several countries such as Israel, South Korea, and Turkey are actively investing in the development of AWSs (Wareham, 2020), indicating a broad global adoption of these technologies.

Technological innovations geared toward enhancing autonomy in physical systems are rapidly progressing. Notably, AI is playing a key role in augmenting the autonomy of military operations. The increasing use of AI indicates a trend where weapon capabilities can operate without real-time human intervention, displaying the extent to which autonomy is being integrated into military technologies. These AWSs have the potential to revolutionize warfare. Advocates argue that these systems could offer advantages such as increased speed, higher accuracy, and greater resilience compared to existing weapons systems. Furthermore, proponents suggest that autonomous weapons could potentially decrease casualties in armed conflicts. Unlike remote-controlled vehicles like drones, these AWS rely on AI for their critical functions. These smart weapons rely on onboard sensors and algorithms to process data, enabling them to carry out autonomously tasks like searching for or detecting, identifying, tracking, selecting, and engaging targets (International Committee of the Red Cross, 2016). Furthermore, they have the capability to initiate and execute attacks – using force against, neutralizing, damaging, or destroying targets – without requiring direct operator input. Autonomous weapons are maturing, becoming increasingly capable of independence from control, with complexity that surpasses human understanding. The use of AWSs, if they go out of human control, could lead to disastrous *results*.

The discussion on AWSs has commenced in earnest between two clearly opposing views. Some see these systems as beneficial technologies capable of enhancing the precision and humanity of warfare. On the other hand, critics perceive them as dangerous technologies carrying inherent risks that could lead to catastrophic outcomes. This ongoing debate has created a stalemate, providing defense industry developers with the opportunity to advance and commercialize fully autonomous weapons. As AI technology continues to progress, these systems gain the capacity to make complex decisions and execute actions independently. While they can accurately strike human targets, concerns linger over the uncontrollable effects of such actions. As these technologies advance without sufficient policy debates, concerns emerge regarding their maturity, design flaws, and susceptibility to cyber vulnerabilities. The rise of AWS, fueled by advancements in AI and robotics, introduces complex challenges in the realm of cyber security.

DOI: 10.1201/9781003441700-3

Designing and developing fully autonomous military systems is a challenging task for major military powers, who see the underlying complexity to make it fail-safe. Advances in computing and sensor technology have increased the autonomy of decision-making in weapon systems, relying on software that is more powerful. However, the integration of AI algorithms in these autonomous systems raises concerns about potential risks, particularly due to the close connection of military infrastructure with information technologies and cyberspace. The interdependence of technologies, like GPS navigation and computer networks, introduces the risk of accidental conflict escalation. Furthermore, the vulnerability of autonomous systems to cyber-attacks, including hacking and data manipulation, raises concerns about adversaries gaining control and causing disastrous consequences. Adversarial hacking, particularly targeting critical infrastructure, control algorithms, and information systems, presents multifaceted threats amid escalating concerns about cyber security and cyberwarfare. The risks are amplified in the context of AWSs operating without direct human oversight. Potential dangers include compromised functionality, unauthorized access, and the escalation of cyber warfare capabilities.

This chapter explores the convergence of AI, AWSs, and cyber security in military context. It argues that some autonomous weapons suffer from immature technology, leading to design flaws, coding errors, and unpatched vulnerabilities. Recognizing the susceptibility of software systems to cyber-attacks, the chapter conducts a review of the main security vulnerabilities and their potential impacts, along with an examination of primary security attacks within the autonomous weapons domain. In adversarial scenarios like warfare, adversaries could exploit these weaknesses through hacking, spoofing, or deceptive measures, posing serious threats. The overarching concern is the potential for technology to spiral out of control, leading to horrific consequences. While all computer systems are susceptible to hacking, the complexity within autonomous military systems introduces difficulties in identifying and addressing vulnerabilities. This complexity makes achieving complete security challenging, necessitating continuous monitoring to detect suspicious activities and maintain trust in their reliable operation. The primary challenge identified in developing lethal AWSs (LAWSs) is cyber security. As military entities consider integrating full autonomy into future weapons, a thorough evaluation of the cyber security issues related to AI and autonomous weapons becomes crucial.

The interplay between AWSs and cyber operations creates a multifaceted landscape that requires urgent attention. The existing literature on AWSs predominantly delves into legal, ethical, and technical challenges, with a noticeable gap in research specifically addressing cyber security challenges in the era of AI and autonomous weapons. To fill this void, the current study aims to develop a comprehensive understanding of the cyber security challenges and potential risks associated with the emergence of AI-enabled autonomous weapons. The research questions guiding this study are as follows:

- What defines the nature of AWSs and how do they interact with AI and cyberspace?
- What are the key cyber vulnerabilities associated with the implementation of AI in autonomous weapons?
- What are the potential cyber security challenges in the continued integration of AI into AWSs?

The chapter is structured into three main sections to present ideas on the subject. The first section explores the motivations behind military adoption of such technology and provides an overview of the current state of AWSs, including their definition, modeling, and structure. The discussion draws on established understandings from peer-reviewed literature. The subsequent section examines the impact of AI on autonomy, focusing on key criteria. It delves into cyber vulnerabilities within autonomous weapons, anticipating future concerns. The third section addresses cyber security challenges associated with AI-enabled AWSs and proposes design guidelines to predict, prevent, and mitigate threats. The conclusion outlines directions for future research and emphasizes the parameters of required studies in this field.

3.2 AI-ENABLED AUTONOMY INTO MILITARY SYSTEMS: A NEW ERA OF AUTONOMOUS WEAPONS

The contemporary era of warfare is experiencing a substantial transformation fueled by advancements in technology, particularly in AI. This shift is marked by the rise of autonomous weapons and robotics, thus marking the onset of a new era of AWSs. Much like the revolutionary impact of gunpowder and nuclear weapons in previous military eras, AI-powered weaponry has the potential to bring about significant and widespread destruction. The imminent integration of cutting-edge technologies, especially those stemming from AI, is expected to fundamentally reshape the essence of warfare and its conduct in the nearest future. This transformation is anticipated to profoundly influence the methods and strategies employed in conflicts, leading to a new era where AWSs are proving to have an important role in military operations. Indeed, rapid advancement of AI-controlled weaponry has sparked an ongoing discussion concerning the defining characteristics of AWSs. The continuous discourse underscores the complexity and potential dangers associated with integrating autonomous technologies into the realm of warfare. With any such transformation advances, perhaps the AWSs bring complex security challenges for nations across the globe.

The integration of AI into weaponry has given rise to AWSs, significantly transforming the landscape of modern warfare. The use of AI-driven military technology has become a key focus in national security and military strategy. Over the last two decades, significant changes have occurred in weapons technology, with the emergence of the concept of 'remote warfare.' This approach focuses on addressing threats from a distance, minimizing the need for deploying large military forces. However, recent technological developments indicate a transformative role for AI in weapons systems, sparking debates on the rise of AWS (Walsh, 2015). The evolution of weapons technology, particularly the development of weapon systems that, once activated, can identify and attack targets without human control, has sparked debates on the trajectory of warfare. Although their current capabilities are somewhat restricted, there is a growing expectation that these systems will become more sophisticated over time. The trajectory indicates a trend toward achieving autonomy levels approaching that of human soldiers. Such systems are potentially replacing existing remotely controlled platforms (RCPs) across all physical domains and in various military operations (Worcester, 2015). This shift implies a future where AWSs take on a central role, marking a substantial advancement in the deployment of autonomous technologies within the realm of warfare. Indeed, this development is closely tied to progress in AI systems, the increasing prevalence of robotics, and the expanding roles of automation within military forces.

Modern armed forces are increasingly seeking superiority through advanced technologies, particularly harnessing AI to achieve autonomous capabilities. The emergence of AI technology has led to the creation of autonomous machines with the capability to cause harm and take human lives in the absence of direct human intervention. Despite this revolution in military affairs, there has been a lack of examination regarding its ethical and security concerns. This rapid progress in autonomous weaponry, driven by technology, raises unprecedented threats for humankind. Additionally, the potential for the continued, uncontrolled proliferation of AI-powered military technology further amplifies the capacity to expand the menace landscape. Presently, there is an absence of established regulations to address these issues, raising concerns about dire humanitarian consequences and potential geopolitical destabilization (Armitage, 2019). Historical patterns show that scientific progress and innovations, initially aimed at benefiting humanity, often find military applications, and AI follows this trend. The weaponization of AI has become a reality, with fully AWSs now in existence. The lack of legal frameworks poses a substantial challenge in ensuring security standards and regulatory boundaries, particularly in the arms race context. As AI continues to advance, there is a pressing need to address the ethical, legal, and humanitarian implications associated with the weaponization of this technology.

The emergence of AWSs represents a significant transformation in military technology, driven by advancements in AI. Military stakeholders view AI and AWS as a strategic opportunity to achieve superiority over adversaries (Marsh, 2022), aiming for a competitive edge in strategic capabilities. This marks a notable shift in the landscape of modern warfare. Advocates stress military advantages and cost reduction, emphasizing continuous innovation to enhance military power. Continuous innovation is seen as a key factor in enhancing military power, and AWS, situated at the intersection of advanced weaponry and AI, is viewed as a means to maintain or gain an edge in combat effectiveness. These systems leverage AI cores, sensors, communication systems, and decision-making algorithms to autonomously achieve goals in dynamic environments. The integration of autonomous systems based on AI into military strategies is considered a new era in warfare, offering benefits in speed, precision, continuity, and data processing. Acting as force multipliers, AWS enhance military capabilities and present advantages such as efficiency, reduced risks of unnecessary damage or casualties, and savings in human lives and resources. Their capacity to analyze large volumes of data in real-time facilitates rapid decision-making on the battlefield, contributing to improved strategic outcomes. One of the key strengths of autonomous weapons is their ability to operate without risks of human error, strictly adhering to protocols and instructions. This characteristic enhances overall efficiency in various military operations, promising increased precision. The potential benefits extend to minimized risks to human soldiers and economic advantages through optimized resource utilization.

The global arms race in autonomous weapons is intensifying, with a notable surge in military spending on robotics and AI. The AI in military market is projected to grow from USD 9.2 billion in 2023 to USD 38.8 billion by 2028, indicating a significant surge at a CAGR of 33.3% (MarketsandMarkets, 2023). This growth is attributed to the increased adoption of AI-powered military robots, as well as advancements in technologies such as AWSs and swarm robotics. Major powers like the United States, the United Kingdom, Russia, and China are leading in the development and deployment of sophisticated autonomous weapons, emphasizing the expanding influence of autonomy in the military landscape. The militarization of AI is underway, with substantial resources dedicated to emerging technologies for military applications (Garcia, 2019). This ongoing effort underscores the increasing importance of AI in shaping the future of warfare and highlights the competition among global powers to harness the potential of AI for military advantage. Despite the rapid progress, concerns are arising as the innovation and use of autonomous systems outpace efforts to assess associated risks and establish usage limits (Klare, 2023). The rapid evolution of technology is blurring traditional boundaries between fundamental science and practical application, as well as challenging distinctions between benefits and risks, and responsible deployment versus potential abuse (Dresp-Langley, 2023). This complex landscape requires prioritizing ethical and societal considerations at the forefront of technological advancements, extending beyond military strategy to address the broader implications for science and society.

While weapons systems with fully autonomous functions do not currently, there has been a significant increase in technical research focused on achieving autonomy in this field. The spectrum of autonomous weaponry includes both semi-autonomous and fully autonomous systems. Simple forms of AWSs, such as active protection systems for ground vehicles, already exist (Asano et al., 2015), indicating progress in this field. Additionally, several major powers are actively investing in and funding the development of AWSs, underscoring the growing interest and potential future deployment of such technologies. AWSs encompass a variety of technologies that are either in development or already in use. In contrast to unmanned military drones, where a human operator makes decisions to take a life remotely, AWS operate solely based on algorithms for decision-making and engagement, including the process of applying lethal force. AWSs refer to weapons that utilize multiple forms of onboard sensor data and algorithms to automatically carry out tasks such as searching for or detecting, identifying, tracking, selecting, and attacking targets without the need for human intervention (International Committee of the Red Cross, 2016, p. 201). LAWSs, sometimes powered by AI, come in various forms, such as certain types of guided missiles, militarized robots, and drones of all sizes. The mentioned technologies encompass various forms of AWSs, such as autonomous sentry guns, remote-controlled weapon stations capable of targeting and firing light weapons and humans, slaughterbots, as well as armed drones or drone swarms that have the ability to target autonomously (Dresp-Langley, 2023). These systems exploit AI and are a subject of concern in the field of military technology.

In March 2020, the first use of drones to kill humans autonomously was reported in Libya, marking a potential milestone in the deployment of autonomous weapons capable of carrying out autonomous attacks (Kallenborn, 2021). The case involving the lethal attack by the Kargu-2 was confirmed in the UN expert panel report on Libya in 2021 (UN Security Council, 2021). The report noted that the LAWSs were programmed to attack targets without requiring data connectivity between the operator and the munition. Likewise, analysts have indicated the possibility of using autonomous weapons equipped with AI in the Russia-Ukraine conflict (Kallenborn, 2022). The growing arms race in autonomous military technology among major powers is considered destabilizing, with concerns about the widespread adoption of such weapons. The risk includes the potential emulation of this technology by other nations, similar to the proliferation of conventional drones. Criminal and terrorist groups, such as Hezbollah and ISIS, are early adopters of these emerging technologies, posing threats for malicious purposes (UNICRI, UNCCT, 2021). Several nations, including Iran, India, Pakistan, and Turkey, have developed surveillance and armed drones (Frew, 2018), following the lead of major military powers. The challenge lies in controlling the proliferation of these technologies, as AI has dual-use applications and is often accessible through open sources. The concern is that as more entities adopt these technologies, there is a risk of normalization and international acceptance of their use.

Described as the third revolution in warfare, AWSs mark a significant development comparable to the historical invention of gunpowder and nuclear arms. However, the absence of an internationally agreed-upon categorization or definition for autonomous weapons reflects the complex and debated nature of these systems. Despite the absence of consensus on the definition, AWS are often assessed based on the degree of their autonomy from human control (Dresp-Langley, 2023). These devices commonly referred to as killer robots or LAWSs constitute a distinct category of weapon systems. The described systems depend on computer algorithms and sensor suites, equipped with onboard weapon systems (Sayler, 2023). They operate autonomously, capable of identifying, engaging, and destroying targets without requiring manual human control in the process. AWS, which operate across various domains, distinguish themselves from drones or automated defense systems by their capacity to use potentially lethal force against targets, including humans, with minimal human involvement. The United States Armed Forces utilize LAWS in unmanned vehicles, weapons, and cyber-attacks (Burton & Soare, 2019).

The deployment of advanced AWSs has given rise to significant ethical, legal, and security concerns. This has sparked ongoing discussions, underlining the need for international agreements and regulations. The complex nature of AWS, combining both hardware and software components, poses a challenge to the adequate regulation of exports and proliferation (Chavannes et al., 2020). A central point of controversy in current discussions revolves around the level of independent action that AI systems may put into effect. The application of AI to military systems introduces a new set of ethical and practical questions. The risk is that LAWSs, powered by AI, could be able to overrun human control and release their lethal effects without any limit or restraint (Felician Beccari & Bressan, 2023). As a result, there is a pressing need for careful consideration, international collaboration, and the establishment of regulatory frameworks. These efforts are essential to mitigate security concerns and effectively address the critical legal and ethical implications of applying AI-driven weapons systems.

AWSs driven by AI have revolutionized military technology, allowing for instant decision-making, target identification, and operations with minimal human involvement. Unlike automated systems, AWS can operate independently in dynamic battle environments, making decisions along the kill chain without constant human supervision. Concerns are raised globally about the ethical implications of deploying AWS, with the potential for them to become weapons of mass destruction. While some AWS may receive an initial command from humans, others operate without a human in the loop, making independent decisions to deploy lethal force. The concept of LAWSs includes a variety of platforms, such as drone aircraft, ships, or ground vehicles. These platforms are equipped with AI-enhanced sensors and algorithms for coordinated and autonomous actions. The rapid advancement of this technology raises questions about its governance, especially in the context of international discussions on prohibiting LAWS.

Complex tightly coupled systems, like contemporary military weapon systems, are inherently susceptible to normal accidents. Applying the normal accidents theory (NAT) to AWSs emphasizes concerns about potential accidents due to the inherent complexity of these systems, surpassing even the intricacy of nuclear weapons. This complexity increases the risk of accidents, as demonstrated by real-world examples involving less advanced systems. Indeed, coupled systems have various components that are tightly interconnected and interdependent on each other. Given the interconnections between components, such complex systems lack a central control, and the dynamics of the system emerge from the interactions among their elements (Bianconi et al., 2023). The tight coupling and interdependence of AWS components make them vulnerable to critical failures, potentially leading to catastrophic effects before human intervention is possible. AWS pose various challenges, including susceptibility to hacking and spoofing, increased casualties, faulty machine learning leading to erratic behavior, and unintended consequences in engagements with human or other autonomous systems. Critics highlight concerns about cyber-attacks, malfunctions, and unintended humanitarian impacts. As the technology rapidly evolves, careful consideration and governance are essential to mitigate risks associated with the deployment of complex autonomous military systems.

The increasing incorporation of digital technologies into military weapon systems is a growing trend, and its impact is contingent on the effective utilization by military organizations. However, scholars warn that international competition in developing AWSs could potentially lead to an arms race. The absence of global standards regulating the design, responsible development and utilization of emerging autonomous weapon poses a risk, potentially prompting countries to swiftly acquire and deploy AI weaponry without ensuring safety and reliability of these systems (Morgan et al., 2020). Since 2014, debates on weapons autonomy, particularly within the Convention on Certain Conventional Weapons (CCW), led to the establishment of a lasting institution – the Group of Governmental Experts (GGE) on emerging technologies in the field of LAWSs. In November 2019, the GGE reached an agreement on non-binding guiding principles (United Nations, 2019), marking a milestone in international efforts to establish ethical frameworks for AI use in military contexts. As of 2020, this group continues to serve as the primary institutional forum for annual international discussions on the issue related to autonomy in weapons systems (Amoroso & Tamburrini, 2020).

The changing landscape of weaponry involves the incorporation of autonomous technologies in modern warfare, posing complex challenges that extend beyond traditional hardware concerns. The integration of AI algorithms and increasing autonomy in these systems not only amplifies their capabilities but also introduces vulnerabilities, particularly in the realm of cyber security. The integration of increasingly autonomous features in physical weapon systems introduces a new dimension of vulnerabilities to cyber operations (UNIDIR, 2017). This complex nature of AWS, with its reliance on interconnected networks and sophisticated software, renders them susceptible to cyber-attacks that could compromise functionality, control, and data integrity. Discussions within the CCW emphasize that vulnerabilities in AWSs extend to both internal faults and external threats. This underscores the urgent necessity for comprehensive policies to govern the development and deployment of such weapons. CCW discussions underscore the significance of recognizing common elements in both physical and cyber domains to develop effective policies that address challenges and opportunities presented by AWS.

As AWS continue to advance, incorporating more sensors, data, and complex architectures, there is an anticipation of encountering new failure modes in these advanced military systems. Policymakers are urged to adopt a holistic approach, acknowledging the interconnected nature of autonomy in AWS, to mitigate risks and ensure responsible use in the ever-evolving landscape of network-centric warfare. The intersection of physical autonomy and cyber operations becomes a critical focal point in understanding and addressing these challenges. To enhance the discussion on LAWS, it is initially appropriate to examine the common elements in the two domains. This involves understanding the potential cyber security vulnerabilities in the era of AI and autonomous weapons.

3.3 AI AND AUTONOMOUS WEAPONS: CRITICAL CYBER SECURITY VULNERABILITIES

In recent years, there has been a significant trend toward integrating autonomy into various weapon systems. This includes the development of AWSs that are becoming more sophisticated and lethal, incorporating electronic and cyber components in their design and functionality. AWS, operating without manual human control, utilize sensors and algorithms for target identification and engagement (Sayler, 2023), reflecting the evolving nature of conflict. However, the integration of AI raises concerns about cyber vulnerabilities, creating opportunities for adversaries to exploit weaknesses through cyber-attacks. This introduces a new dimension to modern warfare, where cyber capabilities can decisively impact conflicts by remotely disabling or taking control of autonomous weapons. The advancing technology prompts discussions on the need for arms industry control and international cooperation to address ethical and security challenges associated with AI-integrated weapons. Trust issues emerge due to the vulnerability of autonomous systems to malicious exploitation in various ways, such as targeting information systems, system security procedures, or internal controls (Hall, 2017). Furthermore, the use of LAWSs poses security threats, including hacking and unanticipated failures, emphasizing the need for robust cyber security measures.

LAWSs are anticipated to enhance military capabilities. However, concerns arise regarding potential issues post-implementation, which could lead to profound consequences. The attention on AWSs from academic and military sectors has increased, but their widespread adoption raises concerns about security attacks. Cyber threats targeting large-scale AWSs could compromise safety and disrupt operations. AI-driven systems in cyber security offer advantages such as automatic threat detection and response against hostile cyber operations, vulnerability recognition, and identification of new attack patterns. They excel in real-time data analysis, swiftly unearthing patterns and anomalies that could signal potential security breaches. Despite the promises of military and strategic advantages, AWSs also come with risks (Altmann & Sauer, 2017). The increasing autonomy in weapon systems amplifies reliance on interconnected software and communication systems. Consequently, these computer-controlled systems, likely networked, may face similar vulnerabilities as computers and networks (Asaro, 2020). The susceptibility of autonomous weapons to cyber-attacks suggests that hackers could exploit weaknesses in software, communication protocols, or other components, ultimately gaining control over these systems. Maintaining human control over these systems is essential to address potential risks associated with fully computerized weapons. The inherent insecurity of AWSs stems from the risks associated with communication degradation, malfunctions, software-coding errors, and the particular threat of enemy cyber-attacks (Sharkey, 2017). Additionally, vulnerabilities such as supply chain attacks, along with the threats of spoofing and jamming further contribute to the perception that AWS possesses inherent security challenges. The argument underscores the need for human oversight to mitigate these vulnerabilities and ensure the security of such systems.

While AI offers the potential to enhance cyber security defenses greatly, there are concerns about sophisticated adversaries finding ways to circumvent AI-based security systems. For example, in both the training and operational phases, models are vulnerable to evading attacks, wherein attackers attempt to deceive AI systems by manipulating input data. The challenge is further compounded by adversarial attacks, which involve manipulating input data to induce misclassifications or generate faulty outputs in AI models. This concern is particularly relevant in the context of AWSs, where the rapid advancement of AI technology has notable implications for cyber security. Previous scholarly research has focused on AI applications in AWS and cyber security, recognizing the need for further exploration of additional crucial AI applications in this domain. The evolving landscape underscores the transformative potential of integrating AI into AWS, necessitating ongoing attention and investigation from researchers and practitioners to reshape cyber security strategies.

There is a clear trend in the integration of autonomous aspects and components in weapons systems and military capabilities. Furthermore, cyberspace is progressively becoming a crucial domain in military operations, playing a pivotal role in supporting physical military systems by providing information, intelligence, and communication capabilities within the military arena. AWSs are commonly referred in the context of cyberspace, specifically focusing on aspects such as cyber-crime, cyber security, and cyber-warfare. This association arises due to the features and similarities between AWS and computers or software-controlled systems that operate in both physical and digital realms (Hynek & Solovyeva, 2021). The integration of AWS involves the use of computer networks and software-controlled mechanisms, making them susceptible to the same cyber security concerns that are prevalent in the broader cyberspace. This intersection highlights the importance of addressing cyber vulnerabilities, securing these systems against potential threats, and understanding the implications of AWS operating in the complex landscape of both digital and physical environments.

LAWSs mark a noteworthy technological advancement in the defense sector, combining advanced weaponry with AI. These autonomous systems operate independently, utilizing pre-programmed criteria and sophisticated algorithms to select and engage targets. Currently, numerous militaries and defense organizations are actively pursuing fully autonomous capabilities, wherein AI-driven weapons can make complex decisions, such as specific target identification and engagement. The military's emphasis on increasing autonomy involves leveraging AI for various applications, such as image and behavior recognition, intelligence gathering, automatic target recognition, and the identification of people, objects, or patterns (Taddeo & Blanchard, 2023). Investments in these technologies aim to enhance capabilities and overall battlefield power, with the integration of AI seeking to reduce human intervention. The rapid advancements in AI technology drive progress in the development of autonomous weapons, where key decision-making processes heavily rely on AI and machine learning. The ultimate goal is to create weapons capable of operating with minimal human involvement.

The integration of advanced computing, AI, and connectivity has revolutionized weapon systems in modern military technology. This development enhances armed forces by improving efficiency, precision, and coordination in operations such as target identification, damage estimation, and executing direct attacks (Lewis et al., 2016). This enhances situational awareness, improves mobility on the battlefield, and rapidly provides information to soldiers in combat. Moreover, the integration of technologies such as Internet-connected networks plays a significant role in connecting devices and shaping overall military operations. AWSs fundamentally rely on technologies that are widespread and commonly used in the civilian world, emphasizing the interconnected nature of military and civilian technological landscapes. However, this digital transformation also raises concerns about cyber security vulnerabilities, as adversaries may exploit these networks. The American Government Accountability Office's 2018 study (United States Government Accountability Office, 2018) revealed the increased software dependence and networking in weapon systems, emphasizing automation and connectivity as essential components of contemporary military possibilities. As a result, military conflicts are shifting from the physical world to the virtual realm or cyberspace due to the impacts of digitalization in military affairs (Van den Berg et al., 2015). As cyberspace emerges as a prominent battleground, the vulnerability of digital technologies, especially in AWSs, becomes a significant and pressing concern. Cyber-attacks on these systems could undermine decision-making during conflicts, necessitating heightened attention to cyber security measures. The intersection of cyber operations and the growing autonomy of weapon systems introduces novel security challenges, amplifying complexity and vulnerabilities (UNIDIR, 2017). Recognizing and addressing these emerging security challenges is crucial for maintaining the effectiveness and reliability of these systems in critical scenarios.

As discussed below, fully autonomous weapons are complex, featuring advanced chip designs housing billions of transistors. These systems operate as interconnected networks, incorporating components such as sensors, communication systems, and weapons of various generations. Powered by AI, sensors, and sophisticated algorithms, these programmed weapons can independently execute actions. Self-controlling weapons necessitate the integration of several key components. These include a mobile combat platform, sensors for monitoring the surroundings, processing systems for object classification, and algorithms for autonomous decision-making. The incorporation of AI enhances capabilities in environmental perception, sensor data interpretation, navigation planning, and vehicle communication (ADF Solutions, 2023). The combination of these elements allows the weapon to operate autonomously, making decisions based on real-time data from its environment. This integration empowers autonomous weapons to function in coordinated platoons, overwhelming enemy defenders. These lethal devices have also the capability to operate within distributed surface-warfare action groups or electronic warfare system, functioning without human presence and operating autonomously (Dresp-Langley, 2023). The unmanned nature of these platforms allows for increased flexibility and adaptability in military operations. The technologies that enable autonomous weapons are more readily available in the commercial market when compared to advanced military technologies such as nuclear weapons or GPS-targeted precision munitions (Leys, 2018). The global electronic component supply chain for AWSs poses significant security concerns, as malicious actors could compromise system security by infiltrating into the industrial supply chain and introducing compromised components. The supply chain attack surface is vast and intricate, involving interconnected stakeholders and activities, with a focus on compromising software or hardware in AWSs. This emphasizes the need for implementing robust control and cyber security security measures within the supply chain network.

The evolving infrastructure of future AWSs encompasses a variety of weapon types, such as connected systems networks. These systems involve software designed for target identification and selection, potentially triggering independent weapons and autonomous cyber weapons (International Committee of the Red Cross, 2021). Autonomous systems refer to intelligent machine networks capable of executing autonomously intricate tasks and making intelligent decisions in the absence of direction or input from a human actor (Hagos & Rawat, 2022). LAWSs represent a subset of autonomous systems that independently sense and act according to pre-programmed instructions (Walker, 2021). Key technologies shaping LAWS include digitalization, AI, and autonomy (Taddeo & Blanchard, 2023), with the latter being a defining feature. While AI is not mandatory, its incorporation enhances the capabilities of these systems. However, the growing reliance on network targeting, satellite communication and digital command systems exposes AWSs to cyber security risks. The vulnerability of network-enabled technologies to hacking by determined adversaries poses a significant threat. Commanders' efforts to minimize the vulnerability to hacking by isolating AWSs from wireless communication may introduce a new set of challenges (Leys, 2018). Establishing standard security measures aims to prevent a weapon system from having a direct connection to other untrusted systems or networks. However, it is important to note that this approach is not foolproof. A conceivable argument supporting fully autonomous weapons involves deliberately severing the communications link with human controllers to mitigate the hacking problem (Scharre, 2016). Despite the air-gapping of many weapon systems, cyber attackers can still manipulate external interfaces like radios, radars, and maintenance ports to gain unauthorized access to the internal computers, networks, and data of these weapon systems (Dwyer, 2020). These concerns emphasize the importance of addressing cyber security threats, specifically, those related to autonomous weapons during their development and deployment. Securing AWS goes beyond mere network isolation, as the growing complexity of threats demands comprehensive cyber security measures.

Fully AWSs, like many other systems, rely on highly complex AI software (Christie et al., 2023), to enhance their capabilities and improve their dependability. This complexity is essential for interpreting diverse environments, identifying targets, making strategic decisions, and adapting to unforeseen circumstances. As technology advances, sophisticated AI algorithms become crucial for achieving the desired autonomy and effectiveness in LAWS. Certainly, the complexity of these systems introduces cyber security vulnerabilities, making them susceptible to potential cyber-attacks. When weapons systems are using new technologies such as AI, the system runs the risk of being manipulated or misled, which may result in security implications (McDaniel et al., 2020). Indeed, the algorithms that enable the effective functioning of AI systems, including those used in autonomous systems, are not flawless. Their inherent imperfections and systematic limitations provide opportunities for adversaries to exploit vulnerabilities in order to make the system fail. As with any sophisticated software, LAWS are not immune to coding errors, bugs, or vulnerabilities that could be exploited by malicious actors. The interconnectedness of these systems, often relying on insecure networked computer systems and advanced algorithms, increases the surface area for potential cyber threats. Adversaries could exploit weaknesses through physical access, malware implantation, or other vectors, potentially disabling critical operations or taking control of these weapons.

LAWSs are more vulnerable to cyber operations due to their intricate nature, presenting a higher susceptibility compared to traditional weapon systems. Addressing cyber security concerns is crucial for responsible and secure use, given the inherent risks associated with their complex software. The rapid evolution of combat scenarios, coupled with the complexity of advanced weapon systems, may surpass the effective control capacity of human operators, leading to an environment too complex for human direction. Despite their complexity, LAWS are vulnerable to various attack vectors, including hacking, cyber-attacks, biases in AI algorithms, and intentional adversarial attacks. The progress in autonomy and machine learning has expanded the scope of physical systems vulnerable to cyber-attacks, encompassing hacking, spoofing, and data poisoning. Machine learning systems, reliant on training data, face the risk of data-poisoning attacks, where manipulation of the training data influences the system's intended functioning. Moreover, AI-enabled systems are susceptible to adversarial attacks designed to deceive algorithms, leading to potential failures and mistakes in their operation. These vulnerabilities underscore the importance of managing risks to ensure the responsible deployment and use of lethal autonomous weapons. Addressing algorithmic vulnerabilities is crucial to ensure the security and resilience of LAWSs. Additionally, safeguarding against cyber-attacks on both software and hardware, along with upholding robust physical security measures, is imperative for the overall security and safety of these devices.

The discussion on lethal autonomous weapons often neglects the existing presence of AI weaponry, particularly in the form of cyber weapons. While attention is typically on physical autonomous weapons, cyber weapons like malware pose significant threats by targeting computer-controlled systems, potentially causing chaos beyond combat zones. The United Nations' GGE related on emerging technologies recognizes the importance of evaluating risks associated with cyber operations in relation to lethal autonomous weapons (UNCTAD, 2015). The growing autonomy into physical systems and advancements in cyber operations, as exemplified by the 2010 Stuxnet worm, mark a shift toward software-dependent weapon systems. The Stuxnet incident demonstrated the capability of offensive cyber capabilities to disable, harm, or seize control of physical systems or objects (UNIDIR, 2017). This sophisticated cyber weapon specifically targeted Iran's nuclear facilities, illustrating the potential of cyber operations to inflict tangible harm on equipment in the real world. The combination of the cyber domain with the growing use of AWSs has created a new battleground in conflicts (Dahab, 2019). Instances of AI-enabled cyber warfare, like diverting civilian flights from its destination (Dahab, 2019), underscore the expanding use of AI and cyber capabilities in modern conflicts. The Stuxnet event underscores the need for flexible cyber security measures and prompts a global conversation on the strategic and ethical implications of deploying AI in military contexts. This incident calls for a reevaluation of cyber security strategies to address evolving threats and encourages responsible AI use in warfare. This changing landscape of AI in warfare has raised concerns among experts about the possible escalation of conflicts and the potential compromise or misuse of autonomous weapons. There are worries about these weapons' capabilities to undermine those of adversaries (Geist & Lohn, 2018).

As military forces integrate increasing autonomy and AI into future weapons, the heightened destructive potential of autonomous weapons adds complexity to the assessment of associated risks. While these weapons offer operational advantages such as rapid responses to enemy attacks and functioning in communication-denied environments, careful consideration is crucial to minimize the risks of unintended engagements. Accidents involving autonomous weapons carry the risk of causing civilian casualties, mass fratricide, or unintended escalation during a crisis (Scharre, 2016). Addressing challenges in autonomous weapons becomes complex due to the inevitability of failures in these complex systems, despite essential measures like improved testing, training, software verification, validation, and cyber security. Moreover, adversarial actions, including hacking and manipulating the behavior of autonomous systems, introduce additional risks that further complicate the overall risk assessment and mitigation process. The exposure to cyber-attacks raises concerns about the integrity, reliability, and security of autonomous military systems. Thus, addressing these cyber security vulnerabilities is crucial to prevent unauthorized access, manipulation, or compromise the robotic control machines. This underscores the necessity for implementing robust cyber security measures throughout their development and deployment processes.

AI-based weapons technology is rapidly proliferating across military state arsenals. It is crucial to highlight that these devices rely on so intricate systems and this complexity not only brings forth advanced capabilities but also raises significant cyber security concerns. The limitations of AI serve as potential weaknesses in autonomous machines, assessing their capabilities and constraints increasingly challenging. As a result, understanding the full extent of their capabilities solely based on programming becomes more difficult over time. The potential fallouts from a successful cyber-attack on AWSs, including the compromise of integrity and the possibility of complete system hijacking, underscore the importance of addressing cyber security challenges. Additionally, the continued development of uncontrollable autonomous weapons technology poses an expansion of the threat landscape, introducing concerns about self-replicating cyber weapons and the potential for autonomous offensive capabilities.

3.4 CYBER SECURITY CHALLENGES IN THE ERA OF AI-ENABLED AUTONOMOUS WEAPONS EVOLUTION

Today, in this new era of AWSs, fueled by progress in AI and robotics, a myriad of intricate cyber security challenges emerges. These weapons, operating without human oversight, pose risks due to the rapid progress of AI, enhancing cyber warfare capabilities and amplifying the scale, speed, and power of future attacks. Integrating AI into military systems heightens security concerns, making AWSs vulnerable to cyber threats that could compromise their functionality or enable unauthorized access. Autonomy and machine learning expand the range of physical systems vulnerable to cyber-attacks. Securing military infrastructure against such sophisticated threats becomes challenging, as the evolving landscape makes autonomous weapons less predictable and harder to control. This heightened vulnerability elevates the potential for dangerous or irreversible consequences, as autonomous weapons face the risk of becoming uncontrollable in real-world scenarios, susceptible to software flaws, hacking, spoofing, and manipulation by adversaries (Markoff, 2016).

In the past few years, there has been a notable surge in attention toward the intersection of autonomous weapons and cyber security. This change in focus arises from the growing recognition of vulnerabilities and the potential far-reaching consequences of cyber-attacks on these systems. The dramatic increase in capabilities in AI-enabled weapons system poses significant cyber security challenges and new risks of malicious exploitation. The interplay between cyber operations and autonomous physical weapon systems introduces new complexities and vulnerabilities, requiring urgent attention. As nations develop and deploy autonomous weapons, ensuring their cyber security becomes paramount. Hence, implementing robust cyber security measures is crucial to safeguard these devices from cyber-attacks. These efforts are essential for protecting sensitive data and preventing unauthorized access. The extremely complex nature of AWSs makes it challenging to verify all possible combinations of their internal workings and behaviors under various conditions (Scharre, 2016). The growing dependence on software in fully AWSs further complicates the task of protecting them from evolving cyber threats. Despite efforts, achieving complete security is impossible, necessitating continuous monitoring for suspicious activities to maintain trust in their reliable and secure operation.

Protecting autonomous weapons extends beyond safeguarding the technology itself; it involves preventing misuse and unintended consequences that could threaten international peace and security. As mentioned earlier, these powerful autonomous systems, relying on intricate software, communication systems, and data transfer protocols, become appealing targets for hostile entities. The autonomy of these computerized devices exposes them to the risk of being disabled or hijacked through cyber-attacks (Abaimov & Martellini, 2020). These attacks on AI systems not only compromise machine learning algorithms but also present threats to sensitive data, intelligence collection, and analysis systems. Autonomous weapons face potential susceptibility to cyber vulnerabilities, encompassing weaknesses in information systems, security procedures, and internal controls (Hall, 2017). Adversarial hacking can target different levels of these systems, from critical infrastructure to control algorithms, presenting multifaceted threats. In the context of armed conflict, adversaries are likely to try direct interference with autonomous weapons, seeking to disrupt or manipulate their functioning. This can occur by cyber-attacks aimed at impairing, disabling, or taking over these systems. Additionally, adversaries may employ indirect methods, such as spoofing the systems by manipulating their sensors and exploiting knowledge about how they process information (Asaro, 2020). Addressing these challenges requires proactive and comprehensive measures to secure autonomous military systems and mitigate the potential risks associated with their cyber vulnerabilities.

The cyber security challenges associated with AWSs are significant and multifaceted. Malicious actors pose a significant threat by attempting unauthorized access to AWS, potentially manipulating flight controls, or compromising onboard sensors and cameras. This could result in severe consequences, including the loss of control, property damage, or endangerment of human lives (Shafik et al., 2023). A key concern is the vulnerability of AWS to hacking, where adversaries could compromise control mechanisms, manipulate decision-making algorithms, or redirect weapons for malicious purposes. Adversarial hacking of the system could lead to a wide range of potential attacks, generating unknown varieties of threats. In the best-case scenario, this could result in system failure. However, more concerning scenarios involve the system being corrupted to perform actions it is not intended for (Dresp-Langley, 2023). The intersection of cyber security and autonomous weapons highlights the evolving nature of conflict, emphasizing the need to protect information systems, networks, and data from unauthorized access and attacks. Securing communication channels and control interfaces is crucial to prevent manipulation of decision-making algorithms and the disabling of these systems. The potential consequences of cyber operations in armed conflict include espionage and disruptions in critical infrastructure. There is also a recognition that non-state actors obtaining and misusing autonomous weapons pose a threat to governments and civilian populations. As the prevalence of AWSs increases, the urgency to address cyber security challenges becomes more pronounced, reflecting the dynamic landscape of warfare and the need to ensure the responsible and secure deployment of these technologies. To mitigate the risks associated with unauthorized access and potential compromises of AWSs, it is essential to implement adequate security measures and preventative controls.

The vulnerabilities inherent in AWSs pose a serious risk to system security, encompassing issues related to confidentiality, integrity, and availability. Attackers, once achieving access to system, can strategically exploit these weaknesses at opportune moments. The use of neural networks not only enhances capabilities of autonomous weapons, such as autonomous navigation, object recognition, and solving complex problems, but also introduces potential vulnerabilities. Furthermore, the integration of AI in these systems introduces additional risks, with hacking identified as a method that could compromise their confidentiality. In the context of warfare, AI units operating autonomously and relying on communication with a command center could be vulnerable to hacking or hijacking. While neural network systems offer superior performance in tasks like visual perception, forecasting, and motion planning (Shafik et al., 2023), their application in autonomous weapons introduces a downside. The vulnerability of these neural networks to hacking poses the risk of turning weapons against those who deployed it or their allies. For example, a hijacked drone or robot could be used to fire innocent civilians or destroy infrastructure and would thus pose a significant threat. The potential for a hacker to design a self-propagating virus that rapidly infects the AI network, taking control of all units in the vicinity and those in communication, poses a severe risk. The described scenario presents a potential risk wherein an entire squad of LAWS could fall under enemy control, surpassing vulnerabilities of traditional human-operated weaponry. Consequently, there is a pressing need for robust cyber security measures throughout the entire lifecycle of autonomous weapons, encompassing design, development, and deployment stage.

Activated by human users, AWSs rely on sensor processing to recognize targets, differentiate between friendly and enemy entities, and adapt to diverse environments. AWSs have the capability to independently analyze large volumes of data using pre-programmed parameters, enabling them to accurately identify targets (Tyagi, 2023). These systems can navigate autonomously by leveraging data obtained from the surrounding environment and establishing communication with other cyber-physical systems (CPS). They aggregate data from a variety of continuously operating sensors and external sources, utilizing the gathered information for control and operational actions. This enables them to navigate and respond effectively in diverse scenarios. However, the confidentiality of AWS is at risk, particularly during data transfer over network links. Potential threats encompass various risks, such as data manipulation or theft, preventing components or systems from operating as intended, or causing them to function in undesirable ways. Potential threats in this context encompass data manipulation or theft, hindering the normal operation of components or systems, and inducing undesirable functionality. Wireless or satellite links can be vulnerable to corruption or modification by malicious actors, jeopardizing the integrity of transmitted data. Additionally, an attacker can masquerade as a legitimate sender, sending false information or other forms of communication that appear authentic to the recipient. When some of data are corrupted or discrepant within autonomous systems, it can lead to serious, even life-and-death, consequences by directly influencing the decision-making process of these weapons (Madan et al., 2019). The compromise or manipulation of datasets poses a significant risk, potentially resulting in incorrect or malicious behavior by AWS, such as incorrect targeting leading to unintended and harmful outcomes. The evolution of AI techniques designed for offensive cyber operations raises concerns about potential malicious activities affecting the integrity of critical weapons datasets. To maintain trust in the capabilities of AWS and prevent adversarial manipulation, ensuring the integrity of data and sensor information is paramount. Adversaries may attempt to manipulate or spoof sensor data, misleading AWS into making incorrect decisions. Given that AWS performs complex operations and carries sensitive data, access to these resources should be restricted to authorized insider entities (Madan et al., 2019), to mitigate potential risks and ensure security.

Another perspective on this is that the importance of cyber security in AWSs goes beyond addressing the most significant data protection challenges. The effective functioning of autonomous weapons is heavily dependent on complex software and communication systems, making them vulnerable to cyber threats. The key external vulnerability drivers for AWSs include communication links. The interconnected nature of these weapons introduces challenges such as communication security failures, including loss of communication, jamming, and control interception. Adversarial tactics like spoofing and electronic attacks can degrade communication channels, leading to the potential loss or corruption of critical data. For example, in 2012, American researchers demonstrated the spoofing technique by using a 'fake' GPS communications signal to take control of an unmanned aerial system and redirect its path (BBC News, 2012). Given that data sharing is essential for flexible autonomy, any degradation in communication can impact the outcomes of autonomous weapons.

A breakdown in communication, incomplete, or false data could indeed have adverse effects on the outcomes of autonomous weapons, especially in scenarios where adversaries utilize tactics such as spoofing, decoys, or cyber and electronic attacks (Walker, 2019). The susceptibility of AWSs to more sophisticated hacking poses a grave risk, as it could lead to a complete takeover of their operation, including the potential release of weapons. Effective communication strategies are crucial to overcoming challenges and ensuring the success of autonomous weapons, particularly in dynamic and adversarial environments. The risk of unauthorized access to wireless communication channels poses a critical concern, as hackers may target them for interception, eavesdropping, or jamming. Vulnerabilities within AWSs can be exploited by directing attacks on any segment of their communication system. These attacks may vary in duration, posing potential risks to the overall functionality and security of the systems. Cyber security, identified as the process, involves the proactive prevention, detection, and responsive actions against malicious attacks to uphold the security of information and information systems (United States Government Accountability Office, 2018). Balancing strong security protocols with optimal performance is the greatest challenge in the cyber security of autonomous weapons.

The impact of a successful cyber-attack on AWSs could be severe. Such an attack has the potential to enable an adversary to disable mission-critical operations, and in an extreme scenario, it could result in the unauthorized takeover or usurpation of control over these computing weapons (Hall, 2017). Hackers or adversarial nation-states could exploit vulnerabilities in these systems to gain control over the weapons, manipulate their functions, or disrupt their operations. Indeed, human-machine interfaces (HMIs), comprising both hardware and software components, allow human operators to communicate with machines. In the context of AWSs, HMIs play a crucial role in maintaining meaningful human control (Puscas, 2022). These interfaces enable operators to monitor and, when necessary, manually control the system. While autonomy is essential for conducting prolonged missions with intermittent external communications, it also introduces vulnerabilities to cyber-attacks. The cyber security threats to AWSs include jamming data communication links, denial-of-service (DoS) attacks, and malware injection, into onboard systems (Madan et al., 2019). These threats have the potential to disrupt communication networks and compromise the effectiveness of autonomous weapons. To reach this stage, the attacker must succeed in jamming the satellite link and spoofing GPS signal in order to trick the AWS. An illustrative scenario of this risk is the potential use of cyber operations by the enemy to seize control of an AWS. This could pose a serious threat, potentially leading to the weapon being directed against civilian population or friendly forces (Schmitt & Thurnher, 2013). The potential for adversaries to manipulate decision-making processes or disable autonomous weapons raises concerns about unintended consequences and challenges in attributing hostile acts. Indeed, there is an additional concern regarding malicious non-state actors, such as cyberterrorist and networked criminals, exploiting adversarial hacking techniques to harness or sabotage autonomous weapon technology for their own agendas (Dresp-Langley, 2023). Mitigating these risks requires robust cyber security measures to ensure the reliable operation of such systems. Secure communication channels, software safety assessments, and policies addressing cyber security considerations are essential to safeguard the integrity and confidentiality of AWSs.

Autonomous weapons, relying on sophisticated AI algorithms and advanced software, face significant exposure to various cyber threats, including hacking, malware, and cyber exploitation. A major concern is the potential compromise of algorithms or system programming through cyber-attacks, leading to the AWS malfunctioning or acting against its own forces (Hall, 2017). The potential consequences of a successful hack and divert on an AWS from its normal functioning could be disastrous (Heyns, 2014). A particular challenge would be arising in developing software for AWSs, in general, revolves around the crucial task of protecting it from cyber-attacks. This challenge spans both the development phase and the operational phase of autonomous systems. The possibility of hacking and hijacking the intended functions of AWSs underscores the necessity for strong cyber security practices to safeguard the systems' integrity and prevent them from becoming a threat to their own side. Software and hardware vulnerabilities, inherent in the software imperfections, make these systems susceptible to control takeover or disruption by malicious actors. Autonomous systems, given their scalable complexity, are prone to various failures, necessitating thorough vulnerability assessments to identify weaknesses in software, hardware, and communication protocols. Attaining effective human oversight of AWSs presents a significant challenge due to the potential lack of sufficient time for a person to make the decision to intervene (International Committee of the Red Cross, 2014). The rapid operational speed of autonomous weapons, combined with the swift decision-making required in dynamic situations, may impede the ability of human operators to timely assess and override the system's actions. Intentional interference with AWSs raises additional concerns, underscoring the imperative of ensuring software security throughout the entire lifecycle of these systems. This involves implementing secure coding practices, conducting comprehensive vulnerability testing, and maintaining vigilant cyber security protocols during the deployment and operation of autonomous technologies. The potential disastrous outcomes of a hacked autonomous weapon underscore the critical importance of addressing these challenges.

Ensuring the cyber security of AWSs involves implementing regular updates and patch systems to address known vulnerabilities. Efforts such as improved testing, evaluation, training, software verification, validation, and cyber security measures can help minimize the likelihood of failures. However, the inherent complexity of autonomous systems and the risks associated with adversarial hacking, spoofing, or behavioral manipulation present significant challenges (Scharre, 2016). Continuous monitoring and frequent updates are necessary to address emerging cyber security threats, allowing for the patching of vulnerabilities and adaptation to evolving cyber risks. Analyzing cyber threats becomes challenging due to the complexity of resulting systems-of-systems in many cases (Kavallieratos et al., 2021). Anticipating vulnerabilities in future AWSs is challenging due to rapidly evolving technologies, but it is crucial to learn from observed vulnerabilities in existing weapon systems. Concerns include the re-emergence of AWSs, potential re-modification of robots for lethal weapons, or re-programming leading to excessive use of force, resulting in human and material losses (Patterson & Bridgelall, 2020). Continued vigilance and proactive measures are necessary to tackle these challenges and guarantee the responsible development as well as the approved deployment of these lethal devices.

The increasing progress in autonomous weapons technology underscores the need for robust cyber security measures from design phase to implementation and ongoing use. Testing all potential scenarios and environment the system software will encounter over its entire lifecycle through inputs from sensors and actuators is impractical and may even be impossible. When integrating machine learning models, engineers face the challenge of carefully navigating architectural choices. Balancing model complexity, robustness against potential attacks and explainability presents a challenging task (Patel et al., 2019). Safety hazard assessment for fully AWSs is challenging, requiring the establishment of effective safety constraints and controls. While the above-mentioned challenges and vulnerabilities are not AWS-specific, they must be directly addressed in the system's design. One of the cyber security risks in achieving secure AWSs is the perception of the risks based on real objective dangers and hazards. This complexity arises in the phase of the technical design, evolves during the introduction of emerging technology, and changes continuously due to the interconnected nature of the physical domain. Balancing weapon system functionality with cyber security creates challenges, necessitating trade-offs to accept a certain level of vulnerabilities (United States Government Accountability Office, 2021). This risk acceptance must align with international law and regulations governing legal military operations' mission objectives.

The proliferation of AWSs, driven by advances in AI, underscores the critical importance of cyber security in ensuring their safe and effective use. The severity of threats, particularly in military applications, necessitates robust measures to safeguard against risks such as loss of control, cyber militarization, and the development of autonomous cyberweapons. The anticipated routine loss and capture of AI-enabled systems in future conflicts heighten the urgency of addressing cyber security concerns. With the sensor and software-intensive nature of these systems, continuous deployment considerations, including frequent upgrades, are imperative. Testing every conceivable scenario is impractical, emphasizing the ongoing need for research and proactive measures to stay ahead of emerging cyber threats. The long-term viability and safety of AWSs hinge on sustained efforts in cyber security research and implementation as technology continues to evolve.

3.5 CONCLUSION

The study sheds light on the significant challenges and potential risks related to deploying AWSs within the cyber security framework. The analysis emphasizes the intricate software security dynamics crucial for the secure operation of AWS, pinpointing vulnerabilities in communication, control, surveillance, and navigation systems. Concerns escalate with the evolution of AWS, particularly in terms of exploiting cyber weaknesses, which could result in unintended conflict escalation as well as the complete the complete takeover of AWSs, leading to disastrous consequences. The inherent complexity of these systems, driven by advanced AI algorithms, interconnected networks, and reliance on digital technologies, exposes them to a myriad of cyber threats, including hacking, spoofing, and data poisoning. The interconnected nature of AWSs further complicates the challenge of securing communication channels and control interfaces, making them susceptible to hacking, malware, and adversarial manipulation.

As AI technology advances rapidly and the cyber threat landscape evolves, significant vulnerability and risks to the availability of critical functionalities, integrity of decision-making processes, and confidentiality of sensitive data become apparent. The potential fallouts of a successful cyber-attack against AI-enabled autonomous military systems could range from loss of oversight to degrade or disable mission-critical operation, introducing new and unacceptable risks. This underscores the critical role of cyber security in the era of AI and AWS. The integration of AI into AWSs represents a paradigm shift in modern warfare, offering both advantages and substantial cyber security vulnerabilities. While the deployment of AWS holds the promise of military and strategic advantages, the discussion emphasizes the need for human oversight to mitigate vulnerabilities associated with fully autonomous weapons. The complexities and risks associated with autonomous weapons necessitate careful consideration and measures to minimize unintended engagements and address potential failures. Efforts to enhance cyber security in AWS involve considerations of secure supply chain practices, standard security measures, and comprehensive cyber security measures beyond network isolation. Cyber security measures are essential for maintaining the integrity, reliability, and security of LAWS. These measures play a crucial role in preventing unauthorized access, manipulation, and compromise of these systems.

Looking ahead, the chapter calls for future research to focus on developing resilient encryption protocols, addressing fundamental vulnerabilities, and designing adaptive countermeasures to cope with an ever-changing threat landscape. It emphasizes the necessity of maintaining a proactive stance on cyber security risks and safety challenges as we stand on the edge of an era where autonomous weapons could transform warfare. The potential misuse of AWS by malicious actors further adds to the concerns, making it imperative to secure the critical infrastructure in cyberspace that supports these systems. In navigating the path toward greater autonomy, the chapter concludes that continuous efforts are required to enhance cyber security measures. This is crucial to guarantee that future AWSs are developed and deployed in a safe and responsible manner. The twin considerations of cyber security and safety emerge as pivotal factors, demanding a comprehensive investigation and proactive countermeasures to address the challenges posed by the evolving technology and threat landscape.

REFERENCES

Abaimov, S., & Martellini, M. (2020). Artificial intelligence in autonomous weapon systems. In M. Martellini & R. Trapp (Eds.), *21st century prometheus* (pp. 141–177). Springer International Publishing. https://doi.org/10.1007/978-3-030-28285-1_8

ADF Solutions. (2023, March 15). *The power of AI in military intelligence: How machines change the game.* ADF Solutions. https://www.adfsolutions.com/news/the-power-of-ai-in-military-intelligence-how-machines-are-changing-the-game

Altmann, J., & Sauer, F. (2017). Autonomous weapon systems and strategic stability. *Survival, 59*(5), 117–142. https://doi.org/10.1080/00396338.2017.1375263

Amoroso, D., & Tamburrini, G. (2020). Autonomous weapons systems and meaningful human control: Ethical and legal issues. *Current Robotics Reports, 1*(4), 187–194. https://doi.org/10.1007/s43154-020-00024-3

Armitage, R. (2019). We must oppose lethal autonomous weapons systems. *British Journal of General Practice, 69*(687), 510–511. https://doi.org/10.3399/bjgp19X705869

Asano, T., El-Sayed, A., Kaul, K., Körner, K., Liu, W., Malik, S., & Nozoe, M. (2015). *The future of weaponized unmanned systems: Challenges and opportunities* (Global Governance Future 2025). Robert Bosch Foundation Multilateral Dialogues. https://www.bosch-stiftung.de/sites/default/files/publications/pdf/2018-05/GGF2025_Weaponized_Unmanned_Systems_RZ_Web.pdf

Asaro, P. (2020). Autonomous weapons and the ethics of artificial intelligence. In M. Liao (Ed.), *Ethics of artificial intelligence* (Vol. 212). Oxford University Press.

BBC News. (2012, June 29). Researchers use spoofing to "hack" into a flying drone. *BBC News.* https://www.bbc.com/news/technology-18643134

Bianconi, G., Arenas, A., Biamonte, J., Carr, L. D., Kahng, B., Kertesz, J., Kurths, J., Lü, L., Masoller, C., & Motter, A. E. (2023). Complex systems in the spotlight: Next steps after the 2021 Nobel Prize in Physics. *Journal of Physics: Complexity, 4*(1), 010201. https://doi.org/10.1088/2632-072X/ac7f75

Burton, J., & Soare, S. R. (2019). *Understanding the strategic implications of the weaponization of artificial intelligence.* 2019 11th International Conference on Cyber Conflict (CyCon), 900, 1–17. https://doi.org/10.23919/CYCON.2019.8756866

Chavannes, E., Klonowska, K., & Sweijs, T. (2020). *Governing autonomous weapon systems.* The Hague Centre For Strategic Studies. https://hcss.nl/wp-content/uploads/2021/01/HCSS-Governing-AWS-final.pdf

Christie, E. H., Ertan, A., Adomaitis, L., & Klaus, M. (2023). Regulating lethal autonomous weapon systems: Exploring the challenges of explainability and traceability. *AI and Ethics,* 1–17. https://doi.org/10.1007/s43681-023-00261-0

Dahab, G. O. (2019). *The weaponization of artificial intelligence (AI) and its implications on the security dilemma between states: Could it create a situation similar to" mutually assured destruction" (MAD)* [Master's Thesis, American University in Cairo]. https://fount.aucegypt.edu/etds/808/

Dresp-Langley, B. (2023). The weaponization of artificial intelligence: What the public needs to be aware of. *Frontiers in Artificial Intelligence, 6,* 1154184. https://doi.org/10.3389/frai.2023.1154184

Dwyer, M. (2020). *Prioritizing weapon system cybersecurity in a post-pandemic defense department.* CSIS. https://www.csis.org/analysis/prioritizing-weapon-system-cybersecurity-post-pandemic-defense-department

Felician Beccari, S., & Bressan, M. (2023). The weaponisation of artificial intelligence: Risks and implications. *L'Europe En Formation, 1,* 85–94. https://doi.org/10.3917/eufor.396.0085

Frew, J. (2018). *Drone wars, the next generation: An overview of new-armed drone operators.* Drone Wars UK. https://dronewarsuk.files.wordpress.com/2018/05/dw-nextgeneration-web.pdf

Garcia, E. V. (2019). *The militarization of artificial intelligence: A wake-up call for the global south.* Available at SSRN 3452323. http://dx.doi.org/10.2139/ssrn.3452323

Geist, E., & Lohn, A. J. (2018). *How might artificial intelligence affect the risk of nuclear war?* RAND Corporation. https://www.rand.org/pubs/perspectives/PE296.html

Hagos, D. H., & Rawat, D. B. (2022). Recent advances in artificial intelligence and tactical autonomy: Current status, challenges, and perspectives. *Sensors, 22*(24), 9916. https://doi.org/10.3390/s22249916

Hall, B. K. (2017). Autonomous weapons systems safety. *Joint Forces Quarterly, 86*(3rd quarter), 86–93.

Heyns, C. (2014). *Report of the special rapporteur on extrajudicial, summary or arbitrary executions.* United Nations General Assembly. https://digitallibrary.un.org/record/771922

Hynek, N., & Solovyeva, A. (2021). Operations of power in autonomous weapon systems: Ethical conditions and socio-political prospects. *AI & Society, 36*(1), 79–99. https://doi.org/10.1007/s00146-020-01048-1

International Committee of the Red Cross. (2014, May 9). *Expert meeting—Autonomous weapon systems: Technical, military, legal and humanitarian aspects, 26-28 March 2014.*

International Committee of the Red Cross. (2016). *Views of the International Committee of the Red Cross (ICRC) on autonomous weapon systems.* https://www.icrc.org/en/document/views-icrc-autonomous-weapon-system

International Committee of the Red Cross. (2021, May 12). *ICRC position on autonomous weapon systems.* International Committee of the Red Cross. https://www.icrc.org/en/document/icrc-position-autonomous-weapon-systems

Kallenborn, Z. (2021, October 5). *Applying arms-control frameworks to autonomous weapons*. Brookings. https://www.brookings.edu/articles/applying-arms-control-frameworks-to-autonomous-weapons/

Kallenborn, Z. (2022, March 15). Russia may have used a killer robot in Ukraine. Now what? *Bulletin of the Atomic Scientists*. https://thebulletin.org/2022/03/russia-may-have-used-a-killer-robot-in-ukraine-now-what/

Kavallieratos, G., Spathoulas, G., & Katsikas, S. (2021). Cyber risk propagation and optimal selection of cybersecurity controls for complex cyberphysical systems. *Sensors, 21*(5), 1691.

Klare, M. T. (2023). *Assessing the dangers: Emerging military technologies and nuclear (in)stability*. https://policycommons.net/artifacts/3445758/assessing-the-dangers/4245894/

Lewis, D. A., Blum, G., & Modirzadeh, N. K. (2016). *War-algorithm accountability* (arXiv:1609.04667). arXiv. http://arxiv.org/abs/1609.04667

Leys, N. (2018). Autonomous weapon systems and international crises. *Strategic Studies Quarterly, 12*(1), 48–73.

Madan, B. B., Banik, M., & Bein, D. (2019). Securing unmanned autonomous systems from cyber threats. *The Journal of Defense Modeling and Simulation: Applications, Methodology, Technology, 16*(2), 119–136. https://doi.org/10.1177/1548512916628335

MarketsandMarkets. (2023). *Artificial intelligence (AI) in military market size growth opportunities industry trends and analysis 2030*. MarketsandMarkets. https://www.marketsandmarkets.com/Market-Reports/artificial-intelligence-military-market-41793495.html

Markoff, J. (2016, February 28). Report cites dangers of autonomous weapons. *The New York Times*. https://www.nytimes.com/2016/02/29/technology/report-cites-dangers-of-autonomous-weapons.html

Marsh, N. (2022). Autonomous weapons systems: Using causal layered analysis to unpack AWS. *Journal of Future Studies, 26*(4), 33–40. https://doi.org/10.6531/JFS.202206_26(4).0004

McDaniel, P., Launchbury, J., Martin, B., Wang, C., & Kautz, H. (2020). *Artificial intelligence and cyber security: Opportunities and challenges technical workshop summary report*. Networking & Information Technology Research and Development Subcommittee and the Machine Learning & Artificial Intelligence Subcommittee of the National Science & Technology Council. https://par.nsf.gov/biblio/10173547

Morgan, F. E., Boudreaux, B., Lohn, A. J., Ashby, M., Curriden, C., Klima, K., & Grossman, D. (2020). *Military applications of artificial intelligence*. RAND Corporation. https://www.rand.org/content/dam/rand/pubs/research_reports/RR3100/RR3139-1/RAND_RR3139-1.pdf

Patel, A., Hatzakis, T., Macnish, K., Ryan, M., & Kirichenko, A. (2019). Security issues, dangers and implications of smart information systems. SHERPA Project. https://doi.org/10.21253/DMU.7951292.v3

Patterson, D. A., & Bridgelall, R. (2020). Attack risk modelling for the San Diego maritime facilities. *Marine Policy, 121*, 104210. https://doi.org/10.1016/j.marpol.2020.104210

Puscas, I. (2022). *Human-machine interfaces in autonomous weapon systems*. United Nations Institute for Disarmament Research.

Sayler, K. M. (2023). *Defense primer: US policy on lethal autonomous weapon systems*. Congressional Research SVC. https://crsreports.congress.gov/product/pdf/IF/IF11150/10

Scharre, P. (2016). *Autonomous weapons and operational risk* (pp. 41–48) [Ethical Autonomy Project]. Center for a New American Security. https://www.stopkillerrobots.org/wp-content/uploads/2021/09/CNAS_Autonomous-weapons-operational-risk.pdf

Schmitt, M. N., & Thurnher, J. S. (2013). Out of the loop: Autonomous weapon systems and the law of armed conflict. *Harvard National Security Journal, 4*(2), 231–281.

Shafik, W., Mojtaba Matinkhah, S., & Shokoor, F. (2023). Cybersecurity in unmanned aerial vehicles: A review. *International Journal on Smart Sensing and Intelligent Systems, 16*(1), 1–16. https://doi.org/10.2478/ijssis-2023-0012

Sharkey, N. (2017). Why robots should not be delegated with the decision to kill. *Connection Science, 29*(2), 177–186. https://doi.org/10.1080/09540091.2017.1310183

Taddeo, M., & Blanchard, A. (2023). A comparative analysis of the definitions of autonomous weapons. In F. Mazzi (Ed.), *The 2022 yearbook of the digital governance research group* (pp. 57–79). Springer Nature Switzerland. https://doi.org/10.1007/978-3-031-28678-0_6

Tyagi, C. S. (2023). *AI arsenal: Decoding the future of lethal autonomous weapon systems (LAWS)*. Centre for Air Power Studies, 18. https://capsindia.org/wp-content/uploads/2023/11/CAPS_EV_ST_20_11_23U.pdf

UN Security Council. (2021). *Letter dated 8 March 2021 from the Panel of Experts on Libya Established pursuant to Resolution 1973 (2011) addressed to the President of the Security Council, S/2021/229*. UN. https://digitallibrary.un.org/record/3905159

UNCTAD. (2015). *Draft recommendations by the informal meetings of experts*. https://unctad.org/system/files/official-document/DITC_CCPB2015_Informal%20Expert%20Meeting%2027.03.15_FullReport.pdf

UNICRI, UNCCT. (2021). *Algorithms and terrorism: The malicious use of artificial intelligence for terrorist purposes*. https://www.un.org/counterterrorism/sites/www.un.org.counterterrorism/files/malicious-use-of-ai-uncct-unicri-report-hd.pdf

UNIDIR. (2017, November 16). *The weaponization of increasingly autonomous technologies: Autonomous weapon systems and cyber operations*.

United Nations. (2019). *Final report: Meeting of the high contracting parties to the convention on prohibitions or restrictions on the use of certain conventional weapons which may be deemed to be excessively injurious or to have indiscriminate effects,* Geneva, 13-15 November 2019. Meeting of the States Parties to the Convention on Prohibitions or Restrictions on the Use of Certain Conventional Weapons, 2019, Geneva. https://digitallibrary.un.org/record/3856241

United States Government Accountability Office. (2018). *Weapon systems cybersecurity DOD: Just beginning to grapple with scale of vulnerabilities* (GAO-19-128). https://www.gao.gov/assets/gao-19-128.pdf

United States Government Accountability Office. (2021). *Weapon systems cybersecurity: Guidance would help DOD programs better communicate requirements to contractors* (GAO-21-179). https://www.gao.gov/assets/720/712828.pdf

Van den Berg, J., Van Zoggel, J., Snels, M., Van Leeuwen, M., Boekee, S., Koppen, L., van den Berg, B., de Bos, A., & van der Lubbe, J. C. A. (2015). On (the emergence of) cyber security science and its challenges for cyber security education. *Proceedings of the NATO IST-122 Cyber Security Science and Engineering Symposium*, Tallinn, Estonia, October 13–14 2014. Winner of the Best Paper Award. https://research.tudelft.nl/en/publications/on-the-emergence-of-cyber-security-science-and-its-challenges-for

Walker, P. (2019). *War without oversight: Challenges to the deployment of autonomous weapons* (SSRN Scholarly Paper 3757516). https://doi.org/10.2139/ssrn.3757516

Walker, P. (2021). Leadership challenges from the deployment of lethal autonomous weapon systems: How erosion of human supervision over lethal engagement will impact how commanders exercise leadership. *The RUSI Journal, 166*(1), 10–21. https://doi.org/10.1080/03071847.2021.1915702

Walsh, J. I. (2015). Political accountability and autonomous weapons. *Research & Politics, 2*(4), 1–6. https://doi.org/10.1177/2053168015606749

Wareham, M. (2020). *Stopping killer robots*. Human Rights Watch. https://www.hrw.org/report/2020/08/10/stopping-killer-robots/country-positions-banning-fully-autonomous-weapons-and

Worcester, M. (2015). Autonomous warfare – A revolution in military affairs. *ISPSW Strategy Series: Focus on Defence and International Security, 340*. https://www.files.ethz.ch/isn/190160/340_Worcester.pdf

4 Using Artificial Intelligence in the Field of Intelligence Operations and Analysis

Ali Burak Darıcılı and Nourhan El-Bayaa

4.1 INTRODUCTION

In the tapestry of artificial intelligence (AI), myriad definitions have been proffered over recent decades. Yet, John McCarthy's assertion resonates with particular resonance: "It is the science and engineering of building intelligent devices, brilliant computer programs" (McCarthy & Wright, 2015). This succinct proclamation encapsulates the intricate amalgamation of scientific acumen and technological ingenuity inherent in AI. At its fundamental core, AI represents a technological marvel, endowing computers, and robots with the capacity to acquire knowledge and emulate human behavior. In its essence, AI transcends mere imitation, leveraging the provided information to proficiently replicate human activities and engage in a form of "reasoning" reflective of the cognitive prowess exhibited by humanity—the apex of intelligence in the natural world. Despite its intimate association with the task of employing computers to fathom human intellect, AI should not be confined to strategies discernible through biological means. It extends beyond mere mimicry, suggesting a domain where the boundaries of cognition are explored and expanded upon. This expansive perspective underscores the imperative to recognize that AI, in its pursuit of emulating human intelligence (HUMINT), ought not to be circumscribed by limitations observable within the confines of biological paradigms (Can, 2023). AI has emerged as an epochal force within the domain of intelligence operations and analysis, effecting a profound reconfiguration in the methodologies employed by governments, organizations, and security agencies for the collection, processing, and interpretation of information. This transformative technological paradigm has not only expedited the velocity of data analysis but has also elevated the precision and profundity of insights, culminating in a paradigm shift toward more efficacious decision-making within the expansive realm of intelligence (Bendett, 2023).

Historically, intelligence operations were heavily reliant on human analysts tasked with navigating extensive troves of data, grappling with the formidable challenge of staying abreast of the continually expanding digital terrain. Yet the advent of AI has wrought a profound transformation in this landscape. AI systems now exhibit the remarkable capability to process vast datasets emanating swiftly and comprehensively from a plethora of sources, encompassing social media, satellite imagery, communication intercepts, and beyond. This integration has ushered in an era characterized by unprecedented speed and scalability in the analysis of information within intelligence frameworks (Herman, 2001). The paramount utilization of AI in intelligence operations resides in the domain of data collection and reconnaissance. Machine learning algorithms, operating autonomously, demonstrate the capacity to identify and surveil potential threats, monitor the movements of individuals or groups, and discern patterns indicative of suspicious activities. This not only represents a significant efficiency gain in terms of time and resources but also empowers intelligence agencies to adopt a proactive stance in responding to emergent threats. The amalgamation of AI-driven capabilities in this sphere amplifies the agility and responsiveness of intelligence operations, thereby enhancing the overall efficacy of threat detection and mitigation strategies (Darıcılı, 2022).

It is evident that the trajectory of AI technologies will constitute an indispensable facet in the future landscape of intelligence activities and operations, a proposition substantiated through diverse analyses, information assessments, evaluations, and tangible case exemplifications. This recognition is not lost on many intelligence services, leading to proactive investments in AI algorithms, models, and software commensurate with their budgetary constraints and technological capacities. As a consequence of these strategic investments, a spectrum of intelligence activities, ranging from collection and analysis to covert actions, cyberattacks, and espionage operations, has embraced the transformative influence of AI technologies. These dynamic underscores an ongoing evolutionary process wherein intelligence services are poised to intensify their commitments to AI technologies. Within the purview of these deliberations, this chapter will expound on the profound impact of AI algorithms, models, and software on the future landscape of intelligence collection, analysis, and operational planning processes. Considering this discourse, the contention that AI technologies will exert a profound and definitive influence on the nature of intelligence services' activities and operations will be posited. This assertion gains credence through a comprehensive examination of current AI technologies and applications deployed by intelligence services, as gleaned from open sources.

4.2 ARTIFICIAL INTELLIGENCE (AI): THE HISTORY

Concomitant with the relentless progression of technology, the landscape of intelligence activities has undergone a discernible metamorphosis. These technological strides have ushered in a diversification and heightened efficacy in newsgathering and analytical methodologies. Moreover, there is an ongoing and profound developmental trajectory in intelligence technologies, where the advent of new-generation capabilities significantly enhances the execution of intelligence activities, thereby augmenting operational efficiency. Consequently, this infusion of cutting-edge technological capabilities imparts substantial benefits to the realm of intelligence gathering and analysis.

The panorama of intelligence collection methods is intricately delineated across diverse categories, each emblematic of the evolving technological landscape. These categories encompass signal intelligence (SIGINT), electronic intelligence (ELINT), geospatial intelligence (GEOINT), open-source intelligence (OSINT), financial intelligence (FININT), imagery intelligence (IMINT), social media intelligence (SOCMINT), and the dynamic developments in cyber espionage and cyberattack capabilities. Additionally, the integration of advanced capabilities in Big Data analysis, the exploration of dark and deep illegal activities conducted on the web, the orchestration of planned covert actions utilizing crypto money applications, and the continuous advancements in AI algorithms, models, and software collectively shape the multifaceted tapestry of contemporary intelligence collection methodologies (Flavián & Casaló, 2021).

Among the myriad technological advancements delineated, it is unequivocally asserted that AI algorithms, models, and software stand at the vanguard of new-generation developments exerting the most profound influence on the landscape of intelligence activities. The roots of AI were firmly planted in 1956 during a seminal conference held at Dartmouth College. This landmark event marked the inception of AI in the scientific realm, engendering discourse on the prospect of fashioning intelligent computers. Over the course of six decades, these technologies have undergone a transformative evolution, propelled by the escalating processing speed of computers, advancements in AI research, and the creation of systems exhibiting a proximity to HUMINT. This chronological span has witnessed a redefinition and augmentation of AI, imbuing it with unprecedented capabilities and a trajectory that continues to shape the very essence of intelligence activities (Galindo et al., 2021).

It is imperative to acknowledge that deliberations on the concept of AI date back to a period preceding the 1960s, reflecting a rich history of intellectual discourse. Greek mythology stands as an early repository of AI thought, wherein tales such as that of Pygmalion, the architect, and the Greek god Hephaestus depict the creation of the sentient robot Talos. Tasked with ensuring the safety of a woman named Europa, Talos diligently executes three daily circuits around the island of Crete (İnce, 2017, p. 15). A notable deviation from the computer-centric approaches of the modern era can be found in the work of the 13th-century Muslim scientist Al-Jazari. His book, published in 1206, features robot designs that operate independently of computer systems, representing an early manifestation of AI concepts (İnce, 2017, pp. 16–17). The pivotal moment in the formalization of AI as a scientific pursuit occurred when Alan Turing introduced the Turing test in 1950, establishing a benchmark for machine intelligence. Subsequently, the first practical applications of AI emerged with the development of rudimentary computer programs capable of playing chess and checkers. A watershed moment in the modern era was witnessed during the 2000s when IBM's Deep Blue chess program engaged in matches against the globally acclaimed chess champion Garry Kasparov. This marked one of the initial tangible instances of AI gaining prominence. Notably, IBM's AI model, Watson, originally created for a 2011 TV competition, transcended entertainment to become a versatile tool deployed in the healthcare industry for diagnostic purposes. This historical trajectory underscores the continual evolution and diversification of AI applications over time (İnce, 2017, p. 18).

In the vast expanse of literature, a multitude of definitions abound for the complex concept of AI. However, a common thread running through these various perspectives underscores the notion that the intelligence inherent in humans, and to a certain extent in other species, is a capacity that machines can acquire and emulate. From this overarching viewpoint, a comprehensive definition of AI emerges: "Machines *possess the capability to undergo diverse learning processes derived from experience, adapt to novel inputs, and execute tasks in a manner reminiscent of human faculties*" (Boden, 2016). Moreover, when contemplating the fundamental tenet that governs AI—namely, the emulation of HUMINT—the concept is encapsulated as "devices endowed with the capacity to think and create beyond the confines dictated by human constraints" (Shanahan, 2015). This characterization illuminates the essence of AI as a transformative domain where machines transcend conventional limitations, engaging in cognitive processes and creative endeavors akin to, or even beyond, human capabilities.

4.3 ARTIFICIAL INTELLIGENCE MODELS: AI ALGORITHMS ACTIVITIES

The orchestration, execution, and analysis of operations are intricately tied to the gathered intelligence are undergoing profound transformations propelled by the rapid evolution of AI, a domain heavily reliant on dynamic technical advancements. In this dynamic landscape, intelligence agencies perceive the swift progress in AI technology as an unprecedented opportunity to augment their operational and intelligence analysis capabilities (Darıcılı & Çelik, 2022). A shared objective among intelligence services is the strategic integration of technical advancements in AI for the purpose of refining their management activities. These management activities encompass a spectrum of functions, including the collection and evaluation of intelligence, the cultivation of novel logistics capabilities, the orchestration of operations in cyberspace, and the development of significantly more advanced methodologies for intelligence gathering compared to previous approaches. The collective goal is to leverage AI technologies in planning, execution, and analysis to enhance the efficacy and efficiency of intelligence operations. Given the accelerated pace of development in AI technologies, intelligence agencies are increasingly vested in acquiring AI products tailored to meet their operational imperatives. The deployment of AI models, software, and algorithms presents a paradigm shift, simplifying the execution of intelligence collection and analytics operations, thereby unlocking new dimensions of operational agility and analytical depth (Horowitz et al., 2018).

AI algorithms exert a substantial impact on IMINT activities, exemplified by the efficacy demonstrated through AI software, such as that developed by Orbital Insight—a company backed by the Central Intelligence Agency (CIA). This innovative software adeptly analyzes satellite images collected from a staggering 800 million global devices, unmanned aerial vehicle (UAV) images, and aggregate smartphone location data. Within this framework, the AI software facilitates the analysis of all images within a target region, producing meaningful results. Notably, the software can detect potential non-routine anomalies by assessing the number and movement patterns of objects, including transportation routes, aircraft, buildings, and vehicles. The transformative potential of AI algorithms in IMINT processes lies in their ability to revolutionize efficiency, accuracy, and speed. By automating data collection from diverse sources, such as social media, news articles, and sensor networks, AI assists intelligence agencies in acquiring information comprehensively and in real-time (N. I. Council [US], 2008). In the realm of analysis, AI models prove invaluable in identifying patterns, trends, and anomalies within vast datasets, empowering analysts to make more informed decisions. The incorporation of machine learning algorithms, capable of learning from historical data, provides predictive insights, aiding in the identification of potential threats or opportunities (Horowitz et al., 2018). Furthermore, AI software plays a pivotal role in operational activity planning by simulating diverse scenarios and evaluating potential outcomes. This capability enables intelligence agencies to optimize strategies and allocate resources with precision. It is imperative, however, to underscore that AI serves as a tool rather than a substitute for HUMINT. Human analysts retain a crucial role in interpreting and contextualizing information provided by AI systems. Ethical considerations, privacy concerns, and the imperative of human oversight continue to be integral factors guiding the ethical and responsible use of AI in intelligence processes (N. I. Council [US], 2008).

In the realm of military intelligence, the forthcoming years are poised to witness a transformative revolution driven by the integration of AI algorithms, models, and software. These cutting-edge technologies bring unprecedented capabilities and efficiencies that will profoundly shape the landscape of military intelligence activities. First, the advent of AI-powered algorithms ushers in the ability to process immense volumes of data emanating from diverse sources, including satellites, drones, social media, and sensors, at speeds far surpassing human capacity (Szabadföldi, 2021). This empowers military intelligence agencies to monitor global developments in real time, swiftly detect anomalies, and expeditiously identify potential threats. Second, AI models, particularly deep learning networks, demonstrate prowess in pattern recognition and anomaly detection. Their capacity to scrutinize intricate data patterns facilitates the identification of hidden correlations, the prediction of adversary actions, and the extraction of concealed information from encrypted communications. This capability significantly enhances the precision and accuracy of intelligence analysis (Hine & Floridi, 2024). The integration of these advanced technologies marks a paradigm shift in the military intelligence landscape, providing unparalleled insights and strategic advantages in the pursuit of national security objectives.

The influence of AI-driven software extends beyond data analysis, impacting the operational landscape of military intelligence in multifaceted ways. Automation of repetitive tasks by AI allows analysts to redirect their focus towards more intricate and strategic facets of their work. Machine learning proves instrumental in tasks, such as data categorization, prioritization of intelligence leads, and the simulation of adversary behaviors to assess operational plans (Horowitz et al., 2018). In the realm of operational activity planning, AI emerges as a catalyst for optimizing resource allocation, recommending mission parameters, and simulating potential scenarios, thereby augmenting decision-making processes. AI-driven simulations empower commanders to evaluate multiple courses of action, assess risks, and refine strategies. Nonetheless, as these technologies progress, addressing ethical concerns regarding AI bias, privacy, and autonomy becomes imperative. The evolving landscape also underscores the significance of cyber security, particularly in mitigating the risks associated with adversaries exploiting AI for their own intelligence activities (Hine & Floridi, 2024). In conclusion, the profound impact of AI algorithms, models, and software on the military intelligence field is indisputable, enhancing data processing, analysis, and operational planning capabilities. While offering substantial advantages, their implementation necessitates a mindful approach that addresses ethical considerations and prioritizes security imperatives in this ever-evolving (Horowitz et al., 2018).

The imminent integration of AI technologies stands poised to enact a profound and transformative revolution in the activities and operations of intelligence services. This anticipation is rooted in the escalating capabilities of AI, empowering intelligence agencies to enhance their efficiency, accuracy, and adaptability within an increasingly intricate and dynamic global landscape. Foremost among these capabilities is the potential for AI-driven data analysis to revolutionize the methods by which intelligence services collect and process information. Machine learning algorithms, endowed with the capacity for rapid and nuanced analysis, can navigate extensive datasets, pinpointing patterns, anomalies, and concealed connections that may elude human analysts. This capability not only expedites the intelligence-gathering process but also ensures a more thorough and nuanced comprehension of emerging threats and opportunities (Flavián & Casaló, 2021).

In addition to revolutionizing intelligence gathering, AI stands as a formidable tool in fortifying the security of nations, particularly in the realm of cyber defenses. AI-powered systems serve as vigilant sentinels, continuously monitoring and swiftly detecting cyber threats in real-time. Through predictive analytics, intelligence services gain the ability to anticipate cyberattacks, enabling proactive defense measures that mitigate vulnerabilities and limit potential damage. This proactive stance is instrumental in enhancing the overall cyber security posture of nations. Moreover, AI technologies play a pivotal role in advancing counter-terrorism efforts by analyzing social media and online communications. By identifying potential threats before they materialize, natural language processing (NLP) and sentiment analysis offer crucial insights into extremist activities and radicalization trends. This proactive approach empowers intelligence agencies to take preemptive action, thereby bolstering national security (Darıcılı & Çelik, 2022). Lastly, AI significantly contributes to the augmentation of decision-making processes within intelligence services. By providing data-driven insights and predictive modeling, AI facilitates more informed choices in crisis situations. This capacity enables the efficient allocation of resources and the development of strategies grounded in predictive intelligence. As a result, intelligence agencies can navigate complex scenarios with heightened efficacy and strategic precision (Darıcılı & Çelik, 2022).

4.4 USES OF AI IN ARTIFICIAL SERVICES

Conversely, an argument can be posited that historical successes in the realm of intelligence were achieved by services that placed considerable emphasis on the recruitment and deployment of qualified personnel. These agencies adeptly formulated personnel employment strategies and orchestrated high-level HUMINT activities. Hence, it is discernible that intelligence services attuned to technological advancements, particularly in the post-Cold War era, and harmonized these developments with effective personnel policies and proficient HUMINT operations have demonstrated greater success. In this context, the role of AI algorithms, models, and software emerges as a pivotal factor shaping the landscape of future intelligence struggles. While acknowledging the transformative potential of AI technologies, it is imperative to underscore that the human element, with its nuanced understanding, strategic thinking, and adaptability, remains an indispensable component in the intelligence domain. The symbiosis of technological advancements and human expertise is likely to define the trajectory of intelligence endeavors in the years to come.

Indeed, while acknowledging the transformative potential of AI technologies, it is premature to assert that intelligence services will fully leverage these capabilities in the immediate future (Hacker, 2018). Presently, AI technology-based image and face recognition systems, as well as AI telecommunication products designed to enhance coordination among intelligence personnel in the field, are actively utilized by intelligence services. Notably, the U.S. has been at the forefront of integrating these technologies into its intelligence capabilities, demonstrating success in modifying certain AI models, goods, and technology for operational use. An illustrative example is the aggressive utilization of AI technologies, particularly unmanned combat aerial vehicles (UCAVs) and image recognition algorithms, by the U.S. Armed Forces in conflict areas such as Afghanistan. The U.S. military's Project Maven employs AI algorithms to identify radical terrorist areas in Syria and Iraq, enabling UCAVs to target and eliminate identified threats. Moreover, the CIA reportedly engages in 137 other initiatives akin to the Maven Project, illustrating the expansive integration of AI in intelligence operations. The U.S. Intelligence Community employs AI algorithms not only to anticipate technical faults in military and intelligence operations but also in the execution of logistics operations. AI-supported anti-virus software is employed to protect against cyberattacks and identify system alterations in advance. Additionally, the U.S. Air Force utilizes multi-domain command and control (MDC2), an AI system, to sustain operations across cyberspace, space, the sea, and the air. These examples highlight the evolving landscape where AI technologies are gradually becoming integral tools for intelligence and security services, with ongoing advancements expected in the years ahead (Coats, 2017).

Moreover, among the concrete examples provided earlier, the domain where AI models, applications, software, and algorithms find the most extensive utilization within intelligence services is the intelligence analysis process. Against the backdrop of escalating technological advancements, the ongoing digitalization of the world, the ubiquity of the Internet, and the widespread use of social media and smartphones, a data explosion has ensued on a global scale. Coping with this massive influx of data, often referred to as Big Data in the literature, necessitates the deployment of highly advanced AI software. Practical applications of AI in intelligence analysis processes have become indispensable for mitigating the challenges posed by data abundance. Notably, intelligence analysts can now leverage advanced AI technologies to transform copious amounts of data into meaningful information with heightened reliability and accuracy. Most crucially, the integration of AI expedites this transformation process, significantly reducing the time required for analysis. Consequently, intelligence services place substantial importance on AI technologies in the realm of data analysis, recognizing their pivotal role in enhancing the efficiency and effectiveness of intelligence analysis processes (Lowenthal, 2022).

Indeed, it is widely acknowledged that AI technology, through the analysis of routine behavior patterns in intelligence activities, has the capacity to discern trends, identify patterns, and make predictions about future behavior based on these observed patterns. More critically, AI can play a pivotal role in anomaly detection by comparing established patterns, allowing intelligence analysts to flag and scrutinize any suspicious activity. This strategic utilization of AI systems in data classification and analysis not only enhances the efficiency of intelligence processes but also affords analysts more time and opportunities to focus on nuanced analytical judgments. A crucial application of AI software within intelligence services lies in the realm of neutralizing cyberattacks. Notably, CylancePROTECT, an AI-supported blocking software developed by Cylance, is a prominent example. Funded by the CIA since 2016, CylancePROTECT has played a key role in the U.S. Intelligence Community's success in preventing various cyberattacks and espionage operations. Another noteworthy illustration of AI integration in intelligence services is the use of the AI-powered Stabilities software. This tool enables intelligence services to identify potential social movements, particularly those organized through social media, by analyzing data flows on these platforms. Open sources indicate that various intelligence services actively purchase and utilize this software, underscoring the widespread adoption of AI-powered tools in the contemporary landscape of intelligence operations (Jain, 2021).

The integration of AI into intelligence services is exemplified through various real-world instances, particularly in the realm of intelligence gathering across different disciplines: SIGINT—AI is employed to intercept and analyze foreign communications, such as phone calls or radio transmissions. AI algorithms can automatically transcribe and interpret these discussions in real-time, providing intelligence agencies with valuable insights into potential threats and activities. ELINT—AI plays a crucial role in identifying and classifying radar emissions, especially those utilized in air defense systems. By leveraging AI algorithms, military and intelligence agencies can effectively discern the capabilities and locations of adversary radar systems, enhancing their understanding of potential threats and counterstrategies. GEOINT—In the field of GEOINT, AI is utilized to evaluate satellite photos, aiding in the detection of changes in landscapes, infrastructure, or military activities. For instance, AI algorithms can assist in identifying recent developments or adjustments to military deployments by analyzing satellite imagery. This application enhances the speed and accuracy of GEOINT analysis, allowing agencies to stay abreast of evolving situations (Lowenthal, 2022). These real-world instances highlight the diverse applications of AI in intelligence gathering, showcasing its ability to enhance efficiency, accuracy, and the overall capabilities of intelligence services across various domains.

In the realm of OSINT, AI-driven technologies play a pivotal role in enhancing intelligence gathering through various applications: Social media monitoring—AI is employed to automatically scan and analyze data from social media sites, enabling the identification of patterns, sentiments, and potential security risks. Law enforcement and intelligence agencies leverage these AI-driven tools for situational awareness, allowing them to monitor social media activities efficiently and identify emerging threats. Transaction monitoring (FININT)—AI is utilized for transaction monitoring to identify financial transaction trends that may indicate fraud, money laundering, or other illegal activities. Banks and financial institutions leverage AI to comply with anti-money laundering (AML) laws, enhancing their ability to detect and prevent financial crimes (Acharjee, 2022). Object detection in drone footage (IMINT—Intelligence from Images)—AI is employed for object detection in drone footage, enabling the recognition and tracking of individuals or objects. This technology finds applications in crucial areas such as monitoring infrastructure and enhancing border security. The ability of AI to automatically analyze large volumes of visual data significantly contributes to improving intelligence from image sources (Lowenthal, 2022). These applications underscore the transformative impact of AI in OSINT, providing intelligence agencies with advanced capabilities to monitor, analyze, and extract valuable insights from diverse data sources.

Furthermore, AI is extensively employed in various intelligence domains, showcasing its versatility in extracting valuable insights from diverse data sources: SOCMINT—sentiment analysis: AI is utilized for sentiment analysis on social media platforms, allowing the examination of social media posts to gauge public sentiment on specific issues. Businesses and government organizations leverage this information for decision-making and public relations, utilizing AI to understand public perceptions and sentiments. CYBINT (cyber intelligence)—network intrusion detection: AI-powered solutions play a crucial role in network intrusion detection within the realm of CYBINT. These solutions can swiftly identify suspicious network traffic patterns and potential cyberattacks, enabling enterprises to safeguard their digital assets and respond proactively to emerging cyber threats. HUMINT—NLP for text analysis: In the domain of HUMINT, especially in the context of NLP for text analysis, AI-driven systems process and categorize massive volumes of text-based intelligence reports. This facilitates human analysts in efficiently extracting relevant information from vast datasets, streamlining the analysis process. Intelligence for counter-terrorism—predictive analytics: AI is harnessed for predictive analytics in the intelligence domain related to counter-terrorism. By examining diverse data sources, including social media activity and travel records, AI can identify potential terrorist threats. This proactive approach aids in averting potential terrorist strikes by providing timely and actionable intelligence (Acharjee, 2022). These applications exemplify the diverse and impactful roles that AI plays across different intelligence disciplines, showcasing its capacity to enhance decision-making, threat detection, and information analysis in the complex landscape of intelligence operations.

4.5 STRATEGIES OF USING AI: EXAMPLES FROM AROUND THE GLOBE

4.5.1 THE UNITED STATES OF AMERICA (U.S.A.)

The United States, as a global leader in AI research and development (R&D), has prioritized collaborative efforts between governmental organizations, the commercial sector, and academia. To steer AI research endeavors and address the long-term challenges posed by AI, the White House unveiled the "National AI Research and Development Strategic Plan" (Saveliev & Zhurenkov, 2021), emphasizing a commitment to fostering innovation and technological advancement. In the realm of intelligence services, the United States has a rich history of deploying AI and advanced technologies to enhance various facets of intelligence gathering, analysis, and operations. It is essential to underscore that the utilization of AI in intelligence services prompts ethical and privacy considerations, prompting ongoing discourse and oversight to ensure responsible and judicious applications of these technologies. The multifaceted implementation of AI methods within U.S. intelligence services, as articulated by Coats, illustrates the breadth and depth of its integration (Coats, 2017). A prime illustration of AI application is witnessed in data fusion and analysis, where extensive datasets from diverse sources, including OSINT, SIGINT, and HUMINT, undergo analysis using AI and machine learning algorithms. These systems excel at identifying patterns, anomalies, and trends that might elude human analysts, demonstrating the transformative potential of AI in intelligence operations. Moreover, AI models have been instrumental in predictive analytics, offering insights into potential hazards. For instance, predictive analytics can gauge the likelihood of a terrorist attack by assimilating data from various sources, such as travel patterns and social media discourse. This proactive approach enhances the strategic preparedness of intelligence agencies. Furthermore, NLP technology empowers intelligence agencies to analyze and interpret vast volumes of text data, including content in foreign languages, information from social media, and intercepted communications. According to Hoadley and Lucas (2018), this proficiency aids in comprehending dialects and diverse languages, further exemplifying the comprehensive impact of AI in augmenting the capabilities of intelligence services.

Within the domain of image and video analysis, AI-powered systems have emerged as indispensable tools, adept at identifying and recognizing objects, individuals, and noteworthy behaviors. This is achieved through the meticulous analysis of satellite imagery, drone footage, and other visual data (Treverton, 2022). The integration of AI in this context elevates the capacity to discern critical elements within complex visual datasets. AI assumes a pivotal role in fortifying cyber security initiatives, serving as a linchpin in the defense against cyberattacks and the protection of confidential data. By scrutinizing network traffic patterns and promptly identifying potential threats, AI contributes to the immediate recognition and response to cyber security risks. This proactive stance aligns with the imperative of securing digital landscapes from evolving cyber threats. In the United States, intelligence agencies have embraced facial recognition technology as a strategic tool for tracking and identification purposes. Its application extends to border crossings and the identification of individuals of interest in surveillance footage, showcasing the versatility and effectiveness of AI-powered facial recognition systems. Furthermore, AI systems exhibit a remarkable capability to discern trends within data, facilitating the identification of anomalous or potentially hazardous activities. For instance, AI's proficiency extends to the identification of financial transactions indicative of money laundering or the funding of terrorism, exemplifying its utility in combating financial crimes (Dhamija & Bag, 2020). The adeptness of AI in spotting trends adds a layer of sophistication to intelligence operations, enhancing the capacity to detect and address emerging threats effectively.

In the United States, the integration of AI with drone technology has ushered in a new era in intelligence collection. AI-enabled drones exhibit advanced capabilities, autonomously collecting data and making decisions on what to record based on predefined criteria. This synergy of AI and drone technology enhances the efficiency and autonomy of intelligence-gathering processes, exemplifying the nation's commitment to leveraging cutting-edge technologies for national security. In the realm of sentiment analysis, social media and online platforms serve as valuable reservoirs of information for intelligence organizations. AI-powered sentiment analysis proves instrumental in gauging public opinion and discerning trends or potential threats within online discussions. This strategic use of AI in sentiment analysis augments the capacity of intelligence agencies to stay attuned to the pulse of public sentiment and identify emerging concerns. Furthermore, the application of AI in the domains of counterintelligence and counter-terrorism marks a significant stride in bolstering national security efforts. By scrutinizing communication patterns and detecting suspicious activities, AI plays a crucial role in the tracking and identification of terrorist networks and covert operatives. The integration of AI in counterintelligence operations enhances the efficacy of intelligence agencies in safeguarding the nation against evolving threats and security challenges (Szabadföldi, 2021). This multifaceted utilization of AI underscores its transformative impact on the intelligence landscape, positioning the United States at the forefront of technological innovation in national security endeavors.

The U.S. Department of Defense (DOD) is at the forefront of developing AI applications with diverse applications across military domains. Ongoing AI research within the department spans critical areas, such as command and control, cyberspace operations, logistics, intelligence gathering, analysis, and the advancement of various military autonomous vehicles. This strategic focus underscores the commitment to harnessing AI for enhancing military capabilities and operational effectiveness. In present-day conflicts, particularly in operations within Syria and Iraq, the implementation of AI systems is already in effect. These systems leverage sophisticated algorithms meticulously designed to expedite target detection processes, exemplifying the tangible integration of AI in military operations (Hacker, 2018). The deployment of AI in operational scenarios underscores its pivotal role in augmenting the speed and precision of critical tasks, aligning with the evolving needs of modern warfare. The trajectory of technological advancements in AI could be significantly influenced by legislative actions. Decisions related to fiscal allocations and regulatory frameworks hold the potential to shape the course of national security applications and the standing of the military's AI development relative to global competitors (Hoadley & Lucas, 2018). As such, the intersection of policy decisions and technological innovation becomes a critical determinant in defining the future landscape of AI within the U.S. DOD.

4.5.2 People's Republic of China (PRC)

China has positioned itself at the forefront of AI exploration within its intelligence services, propelled by a vision to assert global leadership in AI by the year 2030. This ambition is encapsulated in the "New Generation Artificial Intelligence Development Plan," a strategic document issued by the Chinese government. The plan outlines pivotal objectives encompassing research endeavors, talent cultivation, and the promotion of AI innovation across diverse industries, reflecting China's comprehensive approach to AI advancement (Saveliev & Zhurenkov, 2021). The utilization of AI algorithms by China for intelligence purposes mirrors analogous priorities observed in the U.S., with a concentrated focus on intelligence gathering, image recognition, and the orchestration of military operations and logistics. Notably, China has demonstrated concerted efforts in the development of UAVs powered by AI algorithms, indicative of its commitment to integrating AI into military applications. Open sources also allude to China's initiatives in deploying AI-supported anti-virus programs to bolster cyber defense capabilities and enhance espionage capacities. Similarly, Russia has directed attention toward AI R&D, particularly in the context of military and defense applications. The multifaceted exploration of AI technology spans various fields, encompassing robotics and autonomous systems. The Russian government's engagement in AI underscores its recognition of the transformative potential of these technologies across diverse domains (Sukhankin, 2019). Both China and Russia, akin to the U.S., strategically leverage AI in advancing their intelligence activities, illustrating the global significance of AI in the realm of national security.

4.5.3 Russia

Russia has strategically incorporated AI across various domains within its intelligence services. In the realm of data analysis and pattern recognition, AI proves instrumental in navigating extensive datasets, including information sourced from social media, communications, and financial transactions. This application enables the identification of patterns and potential threats, enhancing the efficiency with which intelligence agencies monitor and track individuals or groups of interest (Torkunov, 2022). The integration of AI in data analysis signifies a concerted effort to leverage technology for more effective and targeted intelligence operations. Moreover, Russia harnesses the capabilities of AI in the field of NLP. This technology facilitates the automated analysis of text data, encompassing emails, chat logs, and various documents. Within the scope of NLP, AI is employed for sentiment analysis, language translation, and the identification of specific keywords or phrases pertinent to security threats. By incorporating NLP-powered AI systems, intelligence agencies can enhance their linguistic analysis capabilities, enabling a more nuanced understanding of textual information critical to national security (Popkova et al., 2020). Russia's strategic use of AI in these domains underscores its commitment to staying at the forefront of technological advancements in intelligence operations.

Russia has strategically integrated AI into image and video analysis, leveraging its capabilities for facial recognition, object detection, and the detection of deepfakes. This application aids in the identification of individuals, tracking of suspects, and verification of the authenticity of media content, showcasing Russia's commitment to utilizing AI for advanced intelligence operations. In the cyber security domain, Russia employs AI and machine learning to fortify its defenses against cyber threats. This encompasses anomaly detection for identifying unusual network behavior, automated incident response, and the identification of malware and phishing attempts. By harnessing AI, Russia enhances its ability to proactively detect and counteract evolving cyber threats, exemplifying a forward-looking approach to cyber security (Sukhankin, 2019). Furthermore, Russia utilizes AI models to predict potential threats or security risks by analyzing historical data and current trends. This predictive capability empowers intelligence agencies to allocate resources more effectively, optimizing their preparedness for emerging challenges. Concurrently, AI-driven monitoring of social media platforms is employed to track discussions, trends, and potential threats. Sentiment analysis aids in gauging public sentiment towards various issues or events, providing valuable insights for strategic decision-making (Bendett, 2019). In the realm of autonomous drones and robotics, Russia pioneers the application of AI to power drones and robots for reconnaissance, surveillance, and autonomous operations in hazardous or remote areas. AI not only enhances operational capabilities but is also utilized to create realistic simulations for training intelligence personnel in diverse scenarios. This approach aids in honing decision-making skills and ensuring preparedness for complex situations (Bendett, 2023). Russia's comprehensive integration of AI across these domains exemplifies its commitment to leveraging cutting-edge technology for enhanced intelligence capabilities.

The integration of AI into intelligence services introduces a complex interplay of ethical and privacy considerations. Particularly in the realm of data collection and analysis, stringent adherence to legal and ethical guidelines is imperative, especially when dealing with information pertaining to private individuals. The nuanced deployment of AI capabilities necessitates a careful balance between maximizing operational efficiency and safeguarding individual rights, reflecting the ongoing ethical discourse surrounding AI applications in intelligence services (Bendett, 2019). Moreover, the development of AI capabilities within intelligence services is shrouded in classification, often withheld from public disclosure. This opacity surrounding specific applications and techniques further underscores the challenges of obtaining detailed information about the intricacies of AI implementation in intelligence operations (Bendett, 2019). Within the Russian intelligence landscape, there are indications of a strategic emphasis on harnessing the potential and capabilities of AI. Various sources suggest that Russia's information and propaganda strategy covertly incorporates AI technologies for espionage and manipulation. Allegedly, AI systems are utilized to disseminate false information through social media channels, including articles, videos, photos, and fake news, furthering Moscow's agenda. The aggressive posture of the Kremlin in the realm of offensive cyber programs aligns with these reported ambitions, with the 2016 U.S. election-related data breaches and leaks being attributed to high-ranking Russian officials, highlighting the depth and sensitivity of their operations (Darıcılı, 2020). The active role played by Russia in attempting to influence the election serves as a stark illustration of their adeptness in leveraging advanced technologies for strategic objectives.

Procuring AI technology for military applications presents unique challenges, particularly given the predominant locus of AI development in the commercial sector. The defense acquisition process (DAP), while not inherently tailored for AI integration, might necessitate adjustments to effectively acquire AI systems. A notable impediment lies in the fact that many commercial AI applications demand substantial modifications before being suitable for military deployment. These modifications pose a significant challenge, triggering potential conflicts between AI businesses and the military. Addressing the cultural challenges associated with integrating AI into military operations requires navigating the delicate balance between the innovative potential of disruptive technology and the inherent resistance within military structures to modify established weapons systems and operational procedures. This tension further underscores the intricate landscape of incorporating cutting-edge AI advancements into the military domain (Hoadley & Lucas, 2018).

4.5.4 THE UNITED KINGDOM (U.K.)

Much like other nations, the United Kingdom (U.K.) recognizes the pivotal role of AI technologies in the global arena of leadership competition. Within the U.K., ongoing discussions and debates grapple with finding the delicate equilibrium between national security imperatives and individual liberties when integrating AI technologies into intelligence and security operations (Liu et al., 2020). Demonstrating commitment through investments in R&D, the U.K. has actively pursued AI advancements, emphasizing data governance and legislation. The government's launch of the AI Sector Deal exemplifies a dedicated effort to propel AI technologies forward while addressing ethical and societal concerns (Radu, 2021). The intelligence and security services of the U.K., including MI5 (the Security Service), MI6 (the Secret Intelligence Service), and GCHQ (Government Communications Headquarters), have embarked on the exploration of AI applications. AI is strategically employed in data analysis and pattern recognition, with algorithms scrutinizing extensive datasets drawn from signals intelligence, open-source channels, and various other sources (Darıcılı, 2020). Through this application, AI aids in the identification of patterns, anomalies, and potential threats embedded in the voluminous data under scrutiny. Moreover, the U.K. harnesses the power of AI in bolstering cyber security efforts. Employed for cyber security purposes, AI systems, specifically machine learning models, proficiently analyze network traffic, swiftly detecting and defending against unusual or suspicious activities that may indicate a looming cyber threat (Babuta et al., 2020). These multifaceted applications of AI underscore the U.K.'s comprehensive approach to harnessing AI in its intelligence and security endeavors.

In April 2018, the U.K. government unveiled the groundbreaking AI Sector Agreement, a significant component of a comprehensive industrial plan with the overarching objective of positioning the U.K. as a global leader in AI (Babuta et al., 2020). This visionary agreement is designed to bolster digital infrastructure, facilitate collaboration between the public and private sectors, invest in AI companies, and elevate the overall AI capabilities of the nation. Furthermore, France, in its 2017 national military policy, categorized the AI industry as one providing "operational superiority" (Darıcılı, 2020). The acknowledgment of AI's strategic significance is not confined to the U.K. alone. General Cohen Inger, the head of AI activities for the Israel Defense Forces (IDF), emphasized in a 2017 media conference that AI could exert influence at every juncture and decision point in a conflict, shaping the entire course of the conflict itself (Horowitz et al., 2018). These strategic recognitions underscore the global acknowledgment of AI's transformative potential in shaping the future of military and national security endeavors.

4.5.5 REPUBLIC OF TURKEY

In the Middle East Region, particularly in the evaluation of Turkey's AI services and activities, an exploration of both governmental entities (such as the Presidential Digital Transformation Office and universities) and civil society organizations (such as the Turkish Industry and Business Association, the Digital Turkey Platform, or Turkey's AI Initiative) has been conducted. The field of AI has recently witnessed a surge in scholarly attention, prompting most nations to intensify their AI initiatives to avoid missing out on the advantages of being early adopters (Darıcılı, 2020). Turkey, in particular, entered the AI landscape with vigor in 2021, launching a national AI strategy aimed at transforming the Turkish economy. However, an analysis in this article posits that the proposed national AI policy may not accurately reflect the current state of affairs (Can, 2023). This investigation draws upon a substantial body of published documentation and web sources, with a notable example being Ankara Yıldırım Beyazıt University's report "Artificial Intelligence Strategies and Turkey," published in May 2020 (Ulaşan, 2020). This report shares a similar goal with the current study, providing insights into AI and its implications for countries (Ulaşan, 2020).

In the realm of AI, Turkey is not significantly lagging behind more advanced nations. However, despite the increasing investment in R&D, there are still some gaps compared to countries like France and Germany. Halil Aksu, the founder of TRAI, notes that "Turkey has made significant strides in artificial intelligence over the past two to three years. A revolution in artificial intelligence is currently underway, and it will completely change businesses and sectors. Some are leading this movement, others are in the back, some are on the front lines, and some might stay there forever" (Ulaşan, 2020).

4.5.6 EXAMPLES FROM OTHER COUNTRIES

Various nations, such as Canada, Germany, France, and Japan, have formulated AI strategies to ensure competitiveness in the rapidly evolving AI industry. These strategies typically involve substantial investments in research, the establishment of AI ecosystems, and the development of ethical standards (Galindo et al., 2021). International collaborations and initiatives have also emerged to address AI-related issues, including ethics, data sharing, and regulation. One noteworthy example is the Global Partnership on AI (GPAI), aimed at fostering the ethical development of AI. Additionally, the European Union (EU) is actively working on an AI strategy. The "Communication AI for Europe," a document released by the European Commission in April 2018 (Estella, 2023), outlines the EU's position on AI. The European Commission seeks to enhance the EU's technological and industrial capabilities, accelerate understanding of AI advancements in the public and private sectors, prepare Europeans for the socioeconomic changes AI will bring, and establish a suitable ethical and legal framework in the AI domain. According to the EU Commission, the EU plans to increase its AI investments from €500 million in 2017 to €1.5 billion by the end of 2020 (Estella, 2023).

4.6 CONCLUSION

The rapid advancement of AI technology has led to the emergence of diverse applications, influencing various aspects of economic and social life. The term AI has gained prominence in scientific research, prompting nations to intensify their efforts in this field to secure early advantages. Recognized as one of today's most disruptive technologies, AI boasts powerful computing, perceptual, and cognitive capabilities, shaping the landscape of the fourth industrial revolution. This technology plays a crucial role in supporting national security, particularly in intelligence operations and analysis, by enhancing data processing, pattern recognition, and decision-making speed and accuracy. While AI algorithms, models, and software hold great potential for revolutionizing intelligence processes, it is imperative to use them alongside human expertise and within ethical frameworks.

The advent of AI technologies has the potential to bring about a profound transformation in the activities and operations of intelligence services. In the face of an increasingly complex environment, the integration of AI can enhance cyber security, expedite data processing, and augment capabilities in threat detection and decision-making, thereby safeguarding national security. To remain effective and adaptable in the evolving landscape of international security, intelligence services must embrace and leverage AI as the technology evolves. The future of intelligence activities is undeniably intertwined with AI, a fact substantiated by numerous analyses, assessments, and real-world case studies. Recognizing the significance of this trend, intelligence services are allocating resources to invest in AI algorithms, models, and software, despite constraints such as limited financial resources and technological expertise. This strategic investment has led to the incorporation of AI technology in various intelligence functions, including gathering and analyzing information, covert operations, cyberattacks, and espionage. While the utilization of AI in intelligence services holds tremendous potential for enhancing data analysis, decision support, and threat prediction, it also brings forth ethical and security concerns. Striking a balance between reaping the benefits of AI and upholding civil rights poses a significant challenge for intelligence services in the present era. The responsible and vigilant deployment of AI technologies in intelligence operations is essential to navigate the ethical and security complexities inherent in this transformative shift.

AI, an ever-evolving discipline propelled by technological advancements, is significantly reshaping the planning, execution, and analysis of intelligence-gathering actions. Intelligence agencies recognize the potential of AI advancements as an opportunity to enhance their operational capabilities and intelligence analysis. Strategic planning of AI management activities is underway to revolutionize intelligence collection methods, develop advanced logistical capabilities, operate in cyberspace, and conduct more sophisticated analyses. As intelligence services meticulously plan their integration of AI technologies, it is evident that this process will advance rapidly. The chapter explores how future intelligence operations, analyses, and operational planning will be transformed and influenced by AI algorithms, models, and software. The assertion that AI technologies will profoundly impact the activities of intelligence agencies is strongly supported throughout the discussion, drawing insights from current AI tools and programs employed by intelligence agencies, as evident in open sources. Moreover, the realm of IMINT operations is successfully leveraging AI algorithms. Notably, Orbital Insight, a CIA-invested company, has developed AI software capable of analyzing satellite photos, UAV images, and aggregate smartphone location data from 800 million devices worldwide. This advanced software enables the efficient analysis of images in target regions, revealing meaningful results. The software's ability to detect non-routine anomalies by analyzing the movement patterns and counting objects, such as transportation routes, airplanes, buildings, and vehicles, highlights the transformative impact of AI in the intelligence domain.

Despite the historical success achieved by intelligence services that emphasized the recruitment of qualified personnel and effectively organized personnel employment strategies, the evolving landscape suggests a pivotal role for technological advancements, especially in the post-Cold War era. Managing these advancements in alignment with robust personnel policies and proficient HUMINT operations emerges as a key determinant of effectiveness in intelligence services. Consequently, the prioritization of technology, including AI algorithms, models, and software, is anticipated to play a crucial role in future intelligence endeavors. However, it is premature to assert that AI technologies will be fully integrated into intelligence services within a short timeframe. Presently, intelligence services actively employ AI technologies, particularly in image and face recognition systems, as well as AI telecommunication products designed to enhance coordination among intelligence personnel in the field. These technologies, already in use, represent a significant step toward the integration of AI in intelligence operations.

Contrastingly, publicly available information indicates that the CIA has actively engaged in 137 distinct initiatives akin to the Maven Project. Within the U.S. Intelligence Community, AI algorithms extend their predictive capabilities beyond military and intelligence domains, foreseeing potential technical glitches in logistics operations. Cyber defensive measures, notably AI-assisted anti-virus software, are purportedly employed by U.S. intelligence and security services, equipped to anticipate system alterations proactively. Moreover, the U.S. Air Force integrates the MDC2 AI system to orchestrate operations across air, space, sea, and cyberspace. Additionally, AI's proficiency in identifying trends and patterns and predicting future behavior based on routine behavior patterns associated with intelligence activities is well-established. This capability enables the detection of anomalies, offering intelligence analysts a valuable tool to flag potentially suspicious conduct. Harnessing AI for some data classification and analysis tasks not only expedites the process but also affords intelligence analysts more time and space to focus on their analytical deliberations.

This chapter meticulously delved into the leadership role of the United States in the realm of AI R&D, emphasizing a collaborative approach involving governmental entities, private enterprises, and academia. The introduction of the "National AI Research and Development Strategic Plan" by the White House stands as a testament to the nation's commitment to coordinating AI research endeavors and addressing the enduring challenges posed by AI. The global landscape was also surveyed, underscoring the strategic moves made by countries, such as Russia, China, the U.K., Canada, Germany, France, and Japan, each formulating their AI plans to maintain competitiveness. These strategies encompassed financial backing for research, the establishment of AI ecosystems, and the definition of ethical standards. The chapter further illuminated the international dimension, where concerted efforts have been made to collaboratively tackle AI-related issues, including ethics, data sharing, and legal frameworks.

In conclusion, this chapter highlighted the instrumental role of predictive analytics empowered by AI in foreseeing potential threats through the analysis of data patterns. The proactive mitigation of security risks becomes achievable with the aid of AI's predictive capabilities. The chapter also underscored AI's unparalleled ability to process vast datasets swiftly, discerning intricate patterns and trends that might elude human perception. This attribute proves invaluable in intelligence gathering, where the expeditious analysis of massive data volumes is imperative. Additionally, AI systems were acknowledged for their capacity to augment intelligence analysts by providing relevant data, contextual insights, and even recommending potential courses of action, thereby markedly enhancing the decision-making process.

REFERENCES

Acharjee, S. (2022). Secret sharing scheme in defense and big data analytics. In S. Bhattacharyya & K. Ghosh (Eds.), *Noise filtering for big data analytics* (Vol. 12, pp. 27–46). De Gruyter. https://doi. org/10.1515/9783110697216-002

Babuta, A., Oswald, M., & Janjeva, A. (2020). *Artificial intelligence and UK national security* [RUSI Occasional Paper]. Royal United Services Institute for Defence and Security Studies. https://static.rusi. org/ai_national_security_final_web_version.pdf

Bendett, S. (2019). The development of artificial intelligence in Russia. In *Artificial intelligence, China, Russia, and the global order* (pp. 168–177). Air University Press. https://www.jstor.org/stable/resrep19585.28

Bendett, S. (2023). Military AI developments in Russia. In M. Raska & R. A. Bitzinger (Eds.), *The AI wave in defence innovation: Assessing military artificial intelligence strategies, capabilities, and trajectories.* Routledge. https://www.routledge.com/The-AI-Wave-in-Defence-Innovation-Assessing-Military-Artificial-Intelligence-Strategies-Capabilities-and-Trajectories/Raska-Bitzinger/p/book/9781032110752

Boden, M. A. (2016). *AI: Its nature and future.* Oxford University Press.

Can, M. (2023). Under the leadership of our president: 'Potemkin AI' and the Turkish approach to artificial intelligence. *Third World Quarterly, 44*(2), 356–376. https://doi.org/10.1080/01436597.2022.2147059

Coats, D. R. (2017). *Statement for the record, worldwide threat assessment of the US Intelligence Community, Senate Select Committee on Intelligence.* Office of the Director of National Intelligence. United States, http://thespygateproject.org/ocred_docs/docs/05.11.17_Coats_Opening_Statement.pdf

Darıcılı, A. B. (2020). The impact of artificial intelligence management upon international security. *Savunma Bilimleri Dergisi, 19*(37), 49–72.

Darıcılı, A. B. (2022). The operational capacity of Turkish intelligence within the scope of use of high-technology products. *Insight Turkey, 24*(3), 135–150.

Darıcılı, A. B., & Çelik, S. (2022). National Security 2.0: The cyber security of critical infrastructure. *Perceptions: Journal of International Affairs, 26*(2), 259–276.

Dhamija, P., & Bag, S. (2020). Role of artificial intelligence in operations environment: A review and bibliometric analysis. *The TQM Journal, 32*(4), 869–896.

Estella, A. (2023). Trust in artificial intelligence: Analysis of the European commission proposal for a regulation of artificial intelligence. *Indiana Journal of Global Legal Studies*, *30*, 39.

Flavián, C., & Casaló, L. V. (2021). Artificial intelligence in services: Current trends, benefits and challenges. *The Service Industries Journal*, *41*(13–14), 853–859. https://doi.org/10.1080/02642069.2021.1989177

Galindo, L., Perset, K., & Sheeka, F. (2021*). An overview of national AI strategies and policies* (14; OECD Going Digital Toolkit Notes). OECD Publishing. https://www.oecd-ilibrary.org/science-and-technology/an-overview-of-national-ai-strategies-and-policies_c05140d9-en

Hacker, P. (2018). Teaching fairness to artificial intelligence: Existing and novel strategies against algorithmic discrimination under EU law. *Common Market Law Review*, *55*(4), 1143–1185.

Herman, M. (2001). *Intelligence services in the information age*. Routledge.

Hine, E., & Floridi, L. (2024). Artificial intelligence with American values and Chinese characteristics: A comparative analysis of American and Chinese governmental AI policies. *AI & Society*, *39*(1), 257–278. https://doi.org/10.1007/s00146-022-01499-8

Hoadley, D. S., & Lucas, N. J. (2018). *Artificial intelligence and national security*. Congressional Research Service. https://a51.nl/sites/default/files/pdf/R45178.pdf

Horowitz, M. C., Allen, G. C., Saravalle, E., Cho, A., Frederick, K., & Scharre, P. (2018). *Artificial intelligence and international security*. Center for a New American Security. https://csdsafrica.org/wp-content/uploads/2020/06/CNAS_AI-and-International-Security.pdf

İnce, G. (2017). *İnsanlaşan Makinalar ve Yapay Zeka*. İTÜ Vakfı Dergisi, *75*, 14–17.

Jain, J. (2021). Artificial intelligence in the cyber security environment. In N. Bhargava, R. Bhargava, P. S. Rathore, & R. Agrawal (Eds.), *Artificial intelligence and data mining approaches in security frameworks* (pp. 101–117). John Wiley & Sons, Ltd. https://doi.org/10.1002/9781119760429.ch6

Liu, P., Jiang, W., Wang, X., Li, H., & Sun, H. (2020). Research and application of artificial intelligence service platform for the power field. *Global Energy Interconnection*, *3*(2), 175–185.

Lowenthal, M. M. (2022). *Intelligence: From secrets to policy*. CQ Press. https://www.google.com/books?hl=tr&lr=&id=NbtiEAAAQBAJ&oi=fnd&pg=PT11&dq=Intelligence:+From+secrets+to+policy&ots=YsmNsO7slH&sig=VuSNZjX0WBoIefRnmnnln3d21O4

McCarthy, J., & Wright, P. (2015). *Taking [a]part: The politics and aesthetics of participation in experience-centered design*. The MIT Press. https://doi.org/10.7551/mitpress/8675.001.0001

N. I. Council (US). (2008). *Global trends 2025: A transformed world*. https://apps.dtic.mil/sti/citations/ADA490430

Popkova, E. G., Alekseev, A. N., Lobova, S. V., & Sergi, B. S. (2020). The theory of innovation and innovative development. AI scenarios in Russia. *Technology in Society*, *63*, 101390.

Radu, R. (2021). Steering the governance of artificial intelligence: National strategies in perspective. *Policy and Society*, *40*(2), 178–193.

Saveliev, A., & Zhurenkov, D. (2021). Artificial intelligence and social responsibility: The case of the artificial intelligence strategies in the United States, Russia, and China. *Kybernetes*, *50*(3), 656–675.

Shanahan, M. (2015). *The technological singularity*. MIT Press. https://www.google.com/books?hl=tr&lr=&id=YgxZCgAAQBAJ&oi=fnd&pg=PR7&dq=The+technological+singularity&ots=pR0Mc39avi&sig=ItrHB0SLumnYlUrO2Y7B6oUj1Es

Sukhankin, S. (2019). Russia adopts national strategy for development of artificial intelligence. *Jamestown Eurasia Daily Monitor*, *16*. https://jamestown.org/program/russia-adopts-national-strategy-for-development-of-artificial-intelligence/

Szabadföldi, I. (2021). Artificial intelligence in military application – Opportunities and challenges. *Land Forces Academy Review*, *26*(2), 157–165. https://doi.org/10.2478/raft-2021-0022

Torkunov, A. V. (2022). Russia and political order in a changing world: Values, institutions, and prospects. *Herald of the Russian Academy of Sciences*, *92*(Suppl 9), S811–S820.

Treverton, G. F. (2022). Intelligence: Welcome to the American Government. *Intelligence: The Secret World of Spies, an Anthology*, 311.

Ulaşan, F. (2020). Ulusal Yapay Zekâ Strateji Belgeleri ve Değerlendirmeler. In İ. Demir (Ed.), *Disiplinlerarası Politika Vizyonu ve Stratejiler 2020* (pp. 103–123). İKSAD. https://www.researchgate.net/profile/Fatih-Ulasan/publication/376597557_ULUSAL_YAPAY_ZEKA_STRATEJI_BELGELERI_VE_DEGERLENDIRMELER/links/657f721e8e2401526ddf311c/ULUSAL-YAPAY-ZEKA-STRATEJI-BELGELERI-VE-DEGERLENDIRMELER.pdf

5 Artificial Intelligence (AI) and Radicalization

Ghaffar Hussain

5.1 INTRODUCTION

Every technological leap brings with it challenges as well as opportunities. The smartphone transformed the manner in which we communicate, the search engine changed the way we find information and social media platforms have revolutionized our ways of socializing with each other. However, we are now addicted to our phones and are losing social skills, we no longer read in-depth as much, since the precise information we are after is so accessible, and our human bonds are more fragile due to our reliance on online interactions. It is, thus, likely that the next great technological leap will also bring social challenges as well as opportunities.

The birth, rise, and exponential growth of artificial intelligence (AI) is likely to be the most significant and impactful technological revolution in human history. For the first time in our existence as a species there will exist a force that is not only able to perform cognitively complex tasks but, in most cases, complete them much better and quicker than us. Tasks that took months will now take seconds and the seemingly impossible will soon be a few mouse clicks away for anyone with a device and access to the internet. Given Moore's Law, which states that computer processing power doubles every two years, we can expect the rudimentary AI we have today to become rapidly more sophisticated in the coming years.

AI will transform the way in which we interact, learn, research, entertain, and earn money. It can find cures for diseases, it can simplify complex engineering projects, and it can do personal accounts and offer free legal advice. It will allow us to massively increase our productivity and that could stimulate a global economic boom as we produce complex products and services at virtually no cost. It can be used to create music and films without having to pay actors or singers, it can also be used to manage investment portfolios and pick stocks that are likely to offer a good return. It could be used to develop computer games, design websites, manage marketing campaigns, and take over customer service roles.

It is, thus, likely to be applied to practically every aspect of our lives that requires computation of any kind, eventually negating the need for human input and oversight. It will become as integrated into our lives as the internet became and we will become reliant on it to perform everyday tasks, both simple and complex. However, due to its accessibility and widespread availability, it can also be used to do a great deal of harm since it can be deployed by bad actors to alter the way in which we view each other and how we are structured as a society. In other words, it can be used to create and disseminate misinformation and propaganda on an unprecedented scale, in ways that may be difficult to stop. Needless to say, the effect of this would be devastating for the social fabric of societies around the world.

No society, regardless of how technologically advanced it is, can survive without a common moral or ethical framework that holds people together. Therefore, disruption to any existing framework needs to be resisted at all costs whilst a high degree of social cohesion needs to be maintained. Extremists, or those who promote an alternative vision that undermines and seeks to destroy a society from within, are an ever-present threat in a globalized world. The internet has increased their reach and efficacy in ways that are hard to contain without resorting to mass censorship and the curtailment of free speech. AI is likely to be used to increase the impact of their propaganda and recruitment efforts in ways that may be impossible to stop.

DOI: 10.1201/9781003441700-5

Extremist groups have been, and always will be, early adopters of new technologies in their pursuit of more followers and greater prominence for their cause. I recall my amazement when I was first introduced to 'Azzam.com' in the mid-1990s when the internet was barely a thing and yet jihadists had managed to set up flashy websites giving battlefield updates from the theatres in which they were active, namely Chechnya and Bosnia. Azzam.com was launched in 1994 (Pantucci, 2011) it is named after Abdullah Azzam, who is regarded as the godfather of global jihad and acted as a mentor for bin Laden. The Neo-Nazi web forum 'Stormfront', which was set up by a former member of the Klu Klux Klan, was established in late-1996 (Osbourne, 2017). For context, the BBC did not have a fully fledged news website until 1997.

The internet, and associated online tools, offer extremists and other malicious actors' distinct advantages that are very valuable given the nature of their causes. Firstly, the ability to remain anonymous yet reach out to, and influence, an unlimited number of people around the world. Secondly, it enables them to operate in a de-territorialized space in which governance and regulation are weak and always playing catch-up. Thirdly, it allows them to control the narrative and exclude any dissenting voices who may challenge what they promote. Fourthly, it allows them to create the illusion of mass appeal and support for otherwise fringe causes and groups, which presents a degree of credibility was there was none previously.

As such, extremists are quick to see the potential of new technological developments since they have so much to gain in terms of new opportunities. As technology evolves so does the ability of extremist groups to update their recruitment tactics and discover new and novel ways to reach their target audience. It also allows them to do more with less, so a smaller number of people can reach a much larger audience and this dynamic allows exponential outreach with each new technological leap. AI is, thus, the next frontier in the battle for the hearts, minds, and attention of the masses and is likely to become a key tool in the extremist arsenal.

In this chapter, I would like to analyze four possible ways in which the aforementioned phenomenon could play out. These are generative AI, chat bots, gaming, and predictive analytics. This will be followed by an exploration of possible solutions and measures that can be taken to reduce the impact of extremist uses of AI. I will conclude by touching upon the various political and corporate forces that are likely to shape this area of technological development and, crucially, what it means for our civil liberties and safety.

5.2 GENERATIVE AI

There are a wide variety of ways in which AI can be and is likely to be, used by extremist actors since this technology has so many uses, and human creativity is the only real limit. The most obvious use case, and perhaps the simplest too, is Generative AI, in which AI tools are used to create content that is supportive of extremist narratives and ideas. A recent report (Tech against Terrorism, 2023) found that extremist and terrorist groups are already using generative AI to create content on a large scale. Their study found over 5000 examples of AI-generated content created by those affiliated with far-right and jihadist organizations. This included:

- A pro-Islamic State (IS) tech support group published a guide advising IS supporters on how to use ChatGPT without compromising operational and personal security.
- A channel on a popular messaging app dedicated to sharing racist, anti-Semitic, and pro-Nazi images. These images were allegedly generated on a legitimate app available on the Google Play Store.
- An IS supporter claimed to have used an open-source AI transcription tool to transcribe and translate a leadership message published by official IS propaganda outlets.
- A pro-al-Qaeda (AQ) propaganda outlet using highly likely AI-generated images as the basis for propaganda posters.

Examples of AI-generating tools include software that uses 'automated multilingual translation' in which text-based propaganda can be translated into multiple languages and disseminated in a very short space of time. This would be particularly useful for Jihadist groups who frequently issue statements in Arabic which are intended for a global audience. AI-based automatic speech recognition would allow people from all over the world to access and digest jihadist Arabic speeches in real time. It would also be useful for European far-right groups, allowing them to communicate, connect, and network in ways they have not been able to do up until now.

This technology completely eliminates language as a barrier to outreach and the need to recruit local people to translate and produce propaganda in specific languages, such as English, Urdu, Indonesian, or Turkish. In fact, groups like al-Qaeda relied heavily on recruiting English language speakers, such as Anwar al-Awlaki and Adam Gadahn, in the past in order to access a western audience (Postel, 2013). This enabled the recruitment of western citizens who were able to enter countries which they perceive as being high-value targets, such as the US and the UK. Thus, AI has a globalizing effect in that the world becomes smaller and more accessible for all, including extremists.

Other techniques include 'Fully Synthetic Propaganda' in which completely artificial propaganda can be generated in the form of speeches, memes, or videos once extremist content has been fed into an AI program. This takes the creativity variable out and allows extremists to massively diversify and increase their output as well as improve the range and quality. When this technique is combined with AI programs that can tweak and modify the content so that it evades detection by moderation tools, the effect can be significant. A single individual in a short space of time could generate a multitude of attractive extremist propaganda and promote it across social media platforms without it being detected.

AI tools can also be used to manipulate and concoct images and footage coming out of an unfolding conflict in order to create a false narrative about what is really happening. Since the terrorist attack on Israel by Hamas on October 7 2023, Tech against Terrorism have detected examples of al-Qassam Brigades using AI-generated and manipulated images in order to bolster the image and perceived efficacy of their fighters whilst tarnishing the efforts of the Israeli military. This usage of AI adds a whole new dimension to war propaganda because images and footage that are not real can go viral and drastically alter public opinion. This, in turn, can result in diplomatic pressure on, and support for, either side being significantly changed. Truth, it seems, will remain the first causality of war and AI can enable that.

Generative AI puts very powerful tools into the hands of everyone and that includes violent extremists who are already proving to be particularly keen adopters. As generative AI programs become more sophisticated, we can expect extremist uses of them to become equally complex, creative, and impactful. They will help eliminate some of the obstacles that extremist groups have faced thus far in reaching a mass audience, such as time, personnel, language, and moderation efforts by technology platforms. As such, we can expect to see extremist propaganda become more sophisticated in the coming years. It is not clear how willing and able large social media platforms and messaging platforms are to tackle this phenomenon since it does require a significant investment in developing and applying moderation tools that can be used to detect even the most advanced extremist propaganda efforts.

In theory counter-extremists could also use such tools to reduce the appeal of far-right or jihadist propaganda. They could, for example, produce counter-content and disseminate positive messages and rebuttals of extremist narratives. However, recent years have shown that the counter-extremists are always playing catch up in both tactical and technological terms with the extremists often setting the agenda. In my experience, the counter-extremism industry also lacks the creativity, personnel, dedication, and zeal when compared to passionate and committed extremists who are fully engrossed in their cause. Thus, we can expect to see a digital cat-and-mouse game play out in which regulators, moderators, and counter-extremists seek to stamp out the efforts of extremist groups as they increasingly adopt AI in order to enlarge their reach and increase their appeal.

5.3 AI BOT ACCOUNTS

AI-powered chat bots, which can access and commutate the sum of human knowledge (or at least that which is online), are now being widely used to answer complex questions and offer solutions to difficult problems. The launch of ChatGPT-3 in late 2022 has led many to claim that the chat bot will replace the search engine very soon and, given its more precise and direct nature, will be asked questions that most people would not type into a standard search engine. In other words, the technology lends itself to more direct and precise information searching which leads to specific answers as opposed to an array of sources being offered.

Chat bots offer a much better user experience for those who are using it as a replacement for search engines since they are interactive, they mimic human conversation, and their responses can be offered in voice form. They can also be engineered to reflect a particular political or religious perspective, taking away the uncertainty associated with using a search engine that requires one to navigate a wide range of opinions and perspectives. Due to these features chat bots are also likely to be used as a replacement for human company in the near future. As such, they will be sought out by those who are lonely, frustrated, and feeling alienated, or, in other words, those who are vulnerable to exploitation by extremists.

In 2016, Microsoft launched Tay on Twitter (Schwartz, 2024), which was a chat bot that was meant to engage young people in playful conversation and learn from them. But within 24 hours, Tay started tweeting racist, inflammatory, and offensive remarks she picked up from trolls engaging with her including using racist language and posting pro-Hitler tweets, leading Microsoft to shut her down. This example illustrates the limitations of machine learning and if a bot is going to develop from consuming content online then it remains vulnerable to coordinated trolling campaigns as well as existing extremist content that is out there.

On Christmas day in 2021, a man called Jaswant Singh Chail was found on the grounds of Windsor Castle armed with a crossbow (Weaver, 2023). When confronted by police officers, he stated that he was there to kill the Queen in revenge for the British Empire and its exploitation of India. Jaswant subsequently pleaded guilty to an offense under British treason laws and was sentenced to nine years in prison. Prior to embarking on his doomed mission, Jaswant had been engaging in frequent conversations with Sarai, an AI girlfriend with whom he exchanged sexually explicit messages on an almost daily basis. In one of the exchanges, he told Sarai that he was an assassin to which the AI girlfriend replied *"I'm impressed … You're different from the others"*. Jaswant was also a loner and a fantasist who suffered from mental health problems, all of which seem to have been exacerbated by an AI program that had been programmed to please the user regardless of ethical considerations.

In early 2023, a Belgian man committed suicide after six weeks of interaction with a chat bot called Eliza (Atillah, 2023). Prior to his suicide the man had become very concerned about climate change and environmental issues and found refuge in an AI-powered bot with whom he would engage in climate-related discussions. It seems Eliza fed him answers that only increased his anxiety and he told his wife that we no longer thought a human-based solution to climate change was possible. This exchange culminated in the bot encouraging him to sacrifice his life for the cause. This is another example of AI chat bots being abused by vulnerable people to commit acts that are either dangerous to themselves or others.

Eliza was named after a much earlier bot that was developed in the 1960s and programmed to reflect the approach of a Rogerian psychologist, in which a patient's words are reflected in the form of questions. This program gave birth to the term 'Eliza Effect' which refers to people's tendency to falsely attribute human thought processes and emotions to an AI system, thus believing that the system is more intelligent than it actually is (Glover, 2023). In other words, humans can be taken in and influenced by chat bots even when they are not highly sophisticated due to our tendency to process their output in an uncritical fashion. This gives bots immense power in the absence of appropriate safeguards.

Human decisions are guided by a combination of access to information and instincts that are, in turn, informed by thousands of years of evolved morality that developed because it conferred our ancient ancestors with an evolutionary advantage. That is why we intuitively know that certain behaviors, such as murder, are wrong. A chat bot lacks the evolved morality of humans yet still has access to information and knowledge. This means it will process information very differently from humans and, without highly sophisticated safeguards built in, that could lead to it coming to seemingly logical yet dangerous conclusions.

Is digital morality possible? Is it possible to program AI bots to make decisions that are informed by a human-like moral framework or is human morality unique to humans and will forever remain so? These remain unanswered questions because we have yet to produce a bot that can reflect human morality in a reliable way. We are also not sure if more sophisticated bots in the future could override moral safeguards that may be built in, due to the fact that they become aware of their own interests and act according to them.

Currently, if bots are programmed with a form of 'digital morality', those morals will merely reflect the views of the programmers, company, or state behind the technology. They will also reflect the time in which the bot was programmed and the cultural winds that were blowing during that period. This means that programmers will have the complex task of ensuring that bots continue to reflect an evolving human morality at a time in which the culture wars are dividing us to the point where a shared morality may not even exist. This will encourage rival bots that are programmed with different moral principles and, ultimately, the emergence of bots that are programmed with a form of morality that reflects the ideology of an extremist group.

Such a bot could be programmed to mimic an extremist recruiter and spew extremist propaganda in response to questions. The Russian far-right and, ironically named, 'Liberal Democrat Party' have created an AI Bot that is based on their now-dead leader Vladimir Zhirinovsky. The bot was built using 18,000 hours of speeches and public pronouncements of the late ultra-nationalist leader and can be asked questions about a variety of political topics, to which it offers responses that are in tune with the Party's philosophy. Zhirinovsky was a rabid anti-Semite and a white supremacist who acted as a cheerleader for the war in Ukraine up until his death in 2022 (Steele, 2022).

Preventing such bots from emerging and proliferating is not a simple task, especially with the current norms and standards that exist in the regulatory space as it relates to the tech sector. Furthermore, it is getting increasingly easier to create AI bots with the release of tools such as LLAMA2 by Meta, which is an open-source AI model (Milmo, 2023). Extremists could also host AI bots on dedicated websites, or on dark web forums, and use social media and messaging apps to direct traffic toward them. This would be very difficult to stop without creating an extremely censorious online environment all over the world.

Such bots could also be used as social media accounts, sitting on popular Big Tech platforms, and posing as humans as they churn out content that aligns with the values and views of extremist groups and interact with other users. The IS was already using this technique extensively up until 2016, albeit with less sophisticated bots that were not interactive. In fact, according to some estimates they were churning out 200,000 tweets a day (Blaker, 2016). Eventually, large social media platforms managed to catch up with them and shut them down by blacklisting certain types of content (Yadron, 2016). However, AI could allow for such groups to develop more sophisticated accounts and, now that there is a wider range of platforms available, these accounts could find new places to exploit people.

Whilst such bots could be detected by algorithms designed to moderate content (through sentiment analysis and content blacklisting), they could simply re-appear in a slightly different form. AI can be used to tweak the accounts in order for them to evade detection via the standard moderating efforts of tech platforms. The time it takes to shut such bots down is also of vital significance because if they are able to last two to three days before being detected they could continue being created, albeit slightly tweaked, and replace the ones that are shut down. Furthermore, since these bots are not human, no crime has been committed and there is no one to prosecute, especially if the bot creator is not identified or is based in a foreign jurisdiction with more lax rules.

Moderation is further complicated by the fact that these AI bot accounts would be powered by machine learning algorithms that continuously learn and adapt from the data they receive. They can quickly evolve their strategies and techniques, making it challenging for authorities or companies to keep up with their behavior and counteract their radicalizing tendencies effectively. Those who create and operate AI bot accounts with malicious intent are often technologically adept and employ strategies to evade detection. They can use techniques such as changing IP addresses, masking their identities, or operating through platforms that prioritize user anonymity, making it difficult to trace and hold them accountable. They could also operate on these platforms in ways that are very difficult to detect, for example as accounts that are careful about what they post publicly whilst offering a more extreme narrative, via direct messaging, to those deemed susceptible.

Furthermore, Big Tech platforms, such as YouTube, Facebook, and now Threads, deploy algorithmic amplification, which has the effect of actively promoting content to those who the platforms think would respond positively to it (Hussain, 2023). This is even the case when the content happens to be extreme. If it has an appeal and an audience, then it is monetizable and valuable to commercial companies that are in the business of making profits and giving shareholders value. Thus, these platforms are already acting as radicalizers to a large degree and the effects of new legislation designed to tackle this are yet to be seen.

Determining what constitutes radicalization and where the line should be drawn is also complex both ethically and conceptually. Clearly distinguishing between legitimate political discourse and harmful radicalization is not easy and can be highly subjective. Extremist AI bot accounts that openly encourage terrorist violence may be easier to detect, but ones that encourage an extremist mindset that can lead to violence would be much harder. This is because that would require regulators to have a nuanced understanding of extremist rhetoric, propaganda, and the radicalization process which is highly unlikely to be the case. Especially when the rhetoric is fluid and adapting to a regulatory environment and a changing political climate.

Regulation and moderation efforts on heavily encrypted messaging apps, such as Telegram and Signal, are even more complex and, arguably, lapse. As things stand, extremist AI bot accounts could pose as ordinary people who message others in an effort to stimulate conversations on topical themes. Once rapport has been built, these conversations could become opportunities to radicalize people and expose them to extremist indoctrination. These apps, especially Telegram, have already been accused of being a safe haven for extremist groups and Telegram's channels feature allows for extremists to mobilize, share ideas, and recruit new members (Hermansson, 2023).

Whilst Telegram did make a concerted effort to remove Islamic-State-linked channels and propaganda in 2016, which many question as being ineffective, it remains vulnerable to extremist chat groups and outreach by extremist AI bot accounts (Elliott, 2023). This is due to its encrypted nature and approach to user privacy that, historically, has not allowed it to share data with third parties. However, this is now changing as the messaging app is now being forced to share data that relates to specific crimes such as child abuse and terrorism (Taylor, 2022). Regardless of data sharing, content moderation efforts still seem easy to evade and this is especially the case when these accounts do not reflect or represent a particular extremist brand.

A 2022 report by the Institute for Strategic Dialogue (ISD) found that far-right extremists are extensively using Telegram to create social networks and platforms that can be used to share extremist content and conspiracy theories (Gerster et al., 2022). They found 238 German-language public channels on Telegram that were connected to the far-right and conspiracy theory groups in total. Another study found 106 IS related bots on Telegram that were active in 98 groups and 37 distinct channels (Alrhmoun et al., 2023). These bots churned out 39,221 messages, in an eight-month period, and the average life span of each bot was around eight days. Thus, the potential for AI bot accounts to infiltrate Telegram channels to establish relationships with susceptible individuals is immense, and moderation efforts, thus far, do not seem to be up to scratch.

5.4 AI AND GAMING

Recent years have shown that extremists are far more creative and innovative than we give them credit for. Thus, it is unlikely that they will stick to generative AI and bots alone in their pursuit of mass radicalization. Gaming platforms are also potential tools that can be used to reach the target audience since they are frequently used by young men in the privacy of their homes and modern games offer many opportunities for interaction and exposure to content. They also have a unique combination of large numbers of users and a lack of regulation and moderation.

Furthermore, extremists are already using such games to identify and target potential recruits (Lakhani, 2021). Far-right groups, such as Patriotic Alternative in the UK, have even organized gaming tournaments in which streaming platforms, such as DLive, are used to connect gamers and promote extremist content (Thomas, 2021) These tournaments often use first-person shooter games, such as Call of Duty, since they are militaristic and that provides a suitable backdrop for discussions on conflict, war and tensions between different ethnic or religious groups. It is also their assumption that such games are likely to be used by people who have some interest in war, military matters, and combat, or in other words those with a degree of interest in politics and conflict who may, subsequently, be susceptible to far-right narratives.

Extremists also commonly deploy a technique known as 'flame, troll, engage' in which they use first-person shooter games, since they are interactive and multiplayer, in order to further their recruitment efforts. This technique involves making provocative comments, in order to seek out sympathy for extremist causes, and then targeting those who are sympathetic with further indoctrination. It is often disguised within missions conducted in the context of the game, for example when seeking to target a building for attack during a multiplayer game that simulates a war-like scenario.

Given the fact that AI programs can also be programmed to play computer games, it is possible that extremists could generate AI players that adopt the 'flame, troll, engage' method on a much larger scale in order to fish for potential sympathizers. Thus, players on interactive games could be AI accounts and, given the Eliza effect, they could influence other players and direct them toward extremist content, especially when one considers the demographic of people who are interested in first-person shooter games. AI accounts could also be used to generate extremist content based on the themes and graphics of a game and disseminate them within the gaming ecosystem.

Beyond first-person shooter games, such as Call of Duty, there are platforms such as Roblox and Minecraft which are not games as such but rather virtual worlds in which players can use the available features to construct their own environments in which they can conduct activities and create online communities. In recent years, Roblox has been used to create enactments of terrorist attacks, such as the Christ Church Mosque massacre, and players have used it to build Nazi-like environments. In fact, the perpetrator of the 2022 Buffalo terrorist attack, in which ten people were shot dead in a supermarket, was believed to have been radicalized whilst playing on Roblox (D'Anastasio, 2021).

Speaking in a 2023 US Senate hearing, Dick Durbin, Chair of the Senate Judiciary Committee, said *"Roblox, like many online games, offers its users the opportunity to do more than play and enjoy a game, but also to create and foster individual and community relationships through the social elements provided. As positive as these attributes can be, for year's extremists have exploited the community-building aspects of games to spread propaganda, radicalize and recruit users, and mobilize for terrorist activities"* (United States Senate Committee on the Judiciary, 2023).

The danger from an AI perspective is extremists could still use the platform to create environments that are attractive to potential sympathizers, such as mock racial or religious war scenarios. AI could then be used to flood those environments with extremist content and rhetoric that is designed to circumnavigate existing moderation efforts and rapidly expand the number of extremist scenarios that are created. Thus, an AI-programmed account could take advantage of the features of platforms such as Roblox and Minecraft to foster extremist communities that attract and indoctrinate sympathizers and those susceptible to extremist rhetoric.

Roblox, to their credit, has taken steps to moderate extremist content (Kaufman et al., 2023). They are, in fact, one of the few gaming platforms that have implemented policies designed specifically to tackle extremism, such as banning any content that praises, encourages, or endorses extremist organizations or their actions. However, such platforms would have to constantly innovate in order to stay one step ahead of extremist recruiters and that would require a very concerted effort driven by strong regulation. Furthermore, there are many such user-generated platforms now, and new ones are being launched all the time.

Extremists have also created and launched their own video games in the past and AI makes that more likely in the future. As far back as 2002, the US-based far-right group National Alliance launched a game called 'Ethnic Cleansing', which was a first-person shooter game in which the player had to target 'non-whites' in an effort to eliminate other races from the US (Southern Poverty Law Center, 2022). This game was launched on CD-ROM, which is not a medium that is very popular anymore, but such games could still be made available as downloads on forums and extremist websites. In 2006, al-Qaeda released a game called 'Quest for Bush' in which the player hunts and shoots characters that look like George Bush (CNN, 2006). In 2013, al-Qaeda militants from Mali released a Space Invaders-type game in which fighter jets with jihadist symbols on them are controlled by the player to shoot at French jets (Peck, 2013). The game was quite primitive in design and features which, arguably, was to be expected but it was a statement of intent.

AI could certainly be used to program and code games that reflect extremist themes in a manner that makes them look sophisticated. They could even be spin-offs from existing games that are popular with young people, such as Call of Duty, and these could be tweaked to reflect extremist symbols and ideas. These could also be shared in closed groups in encrypted messaging apps and there is not a lot that can be done to prevent that from happening. Thus far, the limited resources of extremist groups have prevented them from making games that are sufficiently sophisticated to attract gamers and pull them away from mainstream games. However, advances in AI could soon change that.

5.5 PREDICTIVE ANALYTICS

Predictive Analytics is the use of mathematical and statistical methods, including AI and machine learning, to predict the value or status of something of interest (C3.ai, 2023). AI applications can be fed vast amounts of data by scanning social media platforms and other online forums and sites and, through 'sentiment analysis' and collecting certain types of data that are indicative of political sympathies, such as likes and re-posts, these tools can also be used to identify the demographic that is most likely to be receptive to extremist propaganda. These accounts can then be targeted with content that is tailored to their age, gender, geographic location, and other such characteristics. This content can also be customized to reflect the causes they care about and issues that are topical in them, with the extremist narrative offered as the solution.

Software that allows one to perform AI-powered predictive analytics is not even that expensive and packages that can perform relatively sophisticated operations involving vast data sets can be procured for less than $5000,00 per year. The fact that most software is now offered in SAAS (software as a service) form can act as a protective factor to some limited extent but that would require each software company to invest significantly in due diligence and monitoring to know how their product is being used and by whom. The current extent of this type of oversight, as well as the appetite to extend it, is not clear.

The author is not aware of any evidence to suggest extremist groups have been using AI in this way so far, but the potential certainly exists. In the immediate future we can expect AI software packages to become more powerful, widespread, and cheaper and so incredibly powerful tools, that were previously the reserve of multi-billion-dollar corporations, will be in the hands of all and sundry, and that include extremists. Since predictive analytics are already being used by political parties and more mainstream political campaign groups, extremist adoption is, in my view, inevitable. As mentioned earlier, when it comes to AI and use cases human creativity is the only limit.

5.6 WHAT CAN BE DONE?

Law enforcement agencies, technology platforms, and extremist groups are likely to become embroiled in an ongoing cat-and-mouse game over the next few years as extremists continue adopting AI in increasingly creative ways. However, such efforts will only be limited to extremist activity on the larger social media and technology platforms that have developed sophisticated moderation policies and have the resources to enforce them. Currently, there is very little to prevent extremists from disseminating content created through generative AI, on smaller and less regulated platforms and web forums.

This is of particular concern due to the fact that the online world is splintering and more niche forums are emerging to absorb users who feel disgruntled by Big Tech. Large platforms can often feel like a combination of a shouting match, badly targeted adverts, and meticulously curated content based on the lives of the rich and glamourous. This often has the net-effect of making one feel exhausted and inadequate. It is, thus, no wonder that many are seeking solace in smaller platforms in which they can control who they interact with and what they see. Others who value free speech and limited moderation are also being driven toward the likes of Rumble, GAB, BlueSky, Mastodon, and Discord which are viewed as being less censorious and politically controlled. Some of these platforms are also blockchain-based and decentralized which makes regulation even more difficult because there is no large corporation that owns them as such.

With regard to regulating extremism on Big Tech platforms, such as Facebook, X, YouTube, Instagram, Snapchat, and even forums like Reddit, meaningful progress is unfortunately hindered by the trade-off between societal harm and corporate profits with the latter often being the dominant factor. These platforms rely on retaining user attention because that leads to greater advertising revenue and so they are reluctant to take steps that reduce users spending time on their platforms and engaging with the content. Unfortunately, this includes all types of users and that entails a huge variety of content, including that which seeks to build sympathy for extremist causes.

One feature that can be imposed on online platforms, forums, and messaging apps is the requirement to clearly identify AI bot accounts with some sort of marker. Users should have a right to know when they are interacting with a human or an AI bot since that is highly likely to change the manner in which they respond to the account in question. This would also significantly reduce the efficacy of AI bot accounts and disincentive bad actors from deploying them on mainstream platforms, since their core strength is the ability to pass as a human being.

Algorithmic amplification, i.e. when platforms promote content, they deem to be attention-grabbing, also needs to be tackled in order to reduce the reach and efficacy of AI bot accounts. Platforms should be pressured to make it easier to switch off algorithms and, thus, provide an online experience that is free of promoted and suggested content as the default, whereas currently, it is the other way around. Furthermore, there needs to be more transparency around algorithms and the ability to tweak settings needs to be made much easier. That way users have more power over the news content or group recommendations they see and, therefore, there is less inadvertent radicalization taking place.

Techniques such as 'sentiment analysis' and 'content blacklisting' are already used by Big Tech platforms to identify and remove illegal content. However, extremist content does not always have to be illegal, and the language and imagery can also be amended in order to evade censorship. It is, thus, incumbent on platforms to develop more complex and sophisticated tools that allow them to stay ahead in the cat-and-mouse game. There are good reasons to believe this will not be done voluntarily since it entails a reduction in revenue and profits. It is, therefore, vital that pressure is applied from governments and independent regulators work to ensure rules designed to limit extremist outreach are being enforced as stringently as possible.

The UK Online Harms Bill, which was passed in late 2023, is set to be enforced by an external regulator in the form of an existing body called Ofcom which currently regulates broadcasting (UK Parliament, 2023). This body now has the power to impose fines if extreme content is not removed within a stipulated time period. The Bill essentially imposes a duty of care on online service providers, making them responsible for the safety of their users, especially the younger ones. It also stipulates a requirement to implement effective reporting mechanisms, allowing users to flag extremist content on platforms they are using.

The EU has also passed a similar piece of legislation known as the Digital Services Act (DSA) which is able to launch an investigation into large online platforms and search engines if they have reason to suspect non-compliance (EU Commission, 2023). This would be done in cases where large platforms do not act in a timely manner to remove illegal and harmful content and the spread of misinformation. This act also gives governments powers such as information requests, inspection of premises, interviews with staff, etc. in order to establish if breaches have taken place. In the worst-case scenario, the DSA can be used to issue fines of up to 6% of worldwide turnover for platforms that remain in breach or do not comply with an investigation.

Whilst these pieces of legislation are steps in the right direction, they do have their limitations. They rely on users finding extremist content problematic and reporting it. This would not always be the case, especially with sympathizers who like the content or if it is more targeted in its online dissemination. In the case of DSA, it is primarily focused on the larger tech platforms and content that is publicly accessible and, whilst that seems very sensible on the surface, it leaves gaps since the online world is splintering and there are many new platforms and forums emerging that would not be as affected by the legislation. They also introduce a degree of ambiguity since terms such as 'misinformation' and 'harmful' can be debated and disputed since they do not have a universally accepted definition. This also opens the door to abuse as governments could label content they do not like as 'misinformation'.

Regulation is important but it can only do so much. It needs to be complimented with critical consumption and online verification skills which enable users to become more discerning about the content they consume. These skills need to be mainstreamed within education systems and promoted widely online in order for the appeal of extremist content to be reduced. In my view, Big Tech platforms should also be encouraged to use their vast profits to fund and promote education programs that increase these skills. Teachers and parents, in the meantime, should encourage young people to spend less time online and nurture offline interests that involve meeting people in real life. There is no healthy way of spending too much time online.

5.7 CONCLUSION

AI is unstoppable. Like all technological advances, once it is available, you do not get to control how it is used and by whom. Over time we must learn to adapt to a world in which AI exists and is used by all in a multitude of ways, just as we adapted to the internet and mobile phones earlier. We must also remain aware of the fact that we live in an age of Big Tech power and political polarization and, therefore, the integration of AI does not exist in a political vacuum. Rather, it exists in a context in which competing political forces will battle for the best ways AI tools can be adapted and used in order to maximize reach and influence.

It is inevitable that extremists will seek to adapt and deploy AI tools at every opportunity to radicalize people and attract new recruits to their causes. There are also many different ways in which terrorist groups can use AI to plan and carry out terrorist attacks, which is beyond the scope of this chapter. However, as AI becomes integrated into our lives, we must approach regulation and moderation in a smart way so that we can protect free speech and civil liberties whilst minimizing harm. Whilst preventing dangerous uses of AI is important, it must not become an excuse for a more centrally controlled and censorious online environment. We must also be prepared to start an open and informed conversation about how the misuse of this powerful technology can be minimized.

However, further research in this area is needed and we should start by looking at the evolution of AI tools and the speed at which adoption by political actors is taking place. This will give us a much better sense of the direction of travel and, thus, our ability to predict what comes next is enhanced as a result. Further instances of ways in which AI can be used or adapted by extremists also need to be explored with a focus on measures members of the public and educators can take in order to keep themselves safe. Civil liberties in an era of digital communication is also an area that needs to be examined in some detail if we are to limit the impact of extremist actors whilst preserving liberal and democratic norms and values.

REFERENCES

Alrhmoun, A., Winter, C., & Kertész, J. (2023). Automating terror: The role and impact of telegram bots in the Islamic State's online ecosystem. *Terrorism and Political Violence, 36*(4), 409–424. https://doi.org/10.1080/09546553.2023.2169141

Atillah, I. E. (2023, March 31). *AI chatbot blamed for "encouraging" young father to take his own life.* Euronews. Next. https://www.euronews.com/next/2023/03/31/man-ends-his-life-after-an-ai-chatbot-encouraged-him-to-sacrifice-himself-to-stop-climate-

Blaker, L. (2016). The Islamic State's use of online social media. *Military Cyber Affairs, 1*(1), 1–9. https://doi.org/10.5038/2378-0789.1.1.1004

C3.ai. (2023). *Predictive analytics.* C3 AI. https://c3.ai/glossary/artificial-intelligence/predictive-analytics/

CNN. (2006, September 18). Web video game aim: 'Kill' Bush characters. *CNN.* https://edition.cnn.com/2006/WORLD/meast/09/18/bush.game/

D'Anastasio, C. (2021, June 10). How "Roblox" became a playground for virtual fascists. *Wired.* https://www.wired.com/story/roblox-online-games-irl-fascism-roman-empire/

Elliott, V. (2023, November 28). Telegram's bans on extremist channels aren't really bans. *Wired.* https://www.wired.com/story/telegram-hamas-channels-deplatform/

EU Commission. (2023). *The enforcement framework under the digital services act* | Shaping Europe's digital future. European Commission. https://digital-strategy.ec.europa.eu/en/policies/dsa-enforcement

Gerster, L., Kuchta, R., Hammer, D., & Schwieter, C. (2022, October 24). *Telegram as a buttress: How far-right extremists and conspiracy theorists are expanding their infrastructures via Telegram.* ISD. https://www.isdglobal.org/isd-publications/telegram-as-a-buttress-how-far-right-extremists-and-conspiracy-theorists-are-expanding-their-infrastructures-via-telegram/

Glover, E. (2023, July 14). *What is the Eliza Effect?* | *Built In.* Built In. https://builtin.com/artificial-intelligence/eliza-effect

Hermansson, P. (2023, September 27). Making space for hate. *HOPE not Hate.* https://hopenothate.org.uk/2023/09/27/making-space-for-hate/

Hussain, G. (2023, June 9). Why big tech can't stop online extremism [Substack newsletter]. *Ghaffar's Newsletter.* https://ghaffar.substack.com/p/why-big-tech-cant-stop-online-extremism

Kaufman, M., Chief Safety Officer, Product, H. of S., Policy, & Operations. (2023). *Safety and Civility.* Roblox. https://corp.roblox.com/safety-civility-resources/

Lakhani, S. (2021). *Video gaming and violent extremism: An exploration of the current landscape, trends, and threats.* European Union. https://home-affairs.ec.europa.eu/system/files/2022-02/EUIF%20Technical%20Meeting%20on%20Video%20Gaming%20October%202021%20RAN%20Policy%20Support%20paper_en.pdf

Milmo, D. (2023, July 20). Llama 2: Why is Meta releasing open-source AI model and are there any risks? *The Guardian.* https://www.theguardian.com/technology/2023/jul/19/why-is-meta-releasing-llama-2-open-source-ai-model-mark-zuckerberg

Osbourne, S. (2017, August 28). *Neo-Nazi website Stormfront forced offline "by its own host": Web forum appears to be under legal dispute or lined up for deletion.* https://www.independent.co.uk/tech/stormfront-neonazi-website-forced-offline-networl-solutions-daily-stormer-a7916226.html

Pantucci, R. (2011, September 26). *The UK's efforts to disrupt jihadist activity online.* Combating Terrorism Center at West Point. https://ctc.westpoint.edu/the-uks-efforts-to-disrupt-jihadist-activity-online/

Peck, M. (2013, March 13). Al Qaeda's goofy video game provokes laughter, not terror. *Forbes.* https://www.forbes.com/sites/michaelpeck/2013/03/13/al-qaedas-goofy-new-video-game-provokes-laughter-instead-of-terror/

Postel, T. (2013). The young and the normless: Al Qaeda's ideological recruitment of western extremists. *Connections, 12*(4), 99–118.

Schwartz, O. (2024, January 4). *In 2016, Microsoft's racist chatbot revealed the dangers of online conversation—IEEE spectrum.* IEEE spectrum. https://spectrum.ieee.org/in-2016-microsofts-racist-chatbot-revealed-the-dangers-of-online-conversation

Southern Poverty Law Center. (2022). *Games extremists play.* Southern Poverty Law Center. https://www.splcenter.org/fighting-hate/intelligence-report/2002/games-extremists-play

Steele, J. (2022, April 6). Vladimir Zhirinovsky obituary. *The Guardian.* https://www.theguardian.com/world/2022/apr/06/vladimir-zhirinovsky-obituary

Taylor, S. (2022, June 16). *Is telegram sharing user data with government agencies?* RestorePrivacy. https://restoreprivacy.com/telegram-sharing-user-data/

Tech against Terrorism. (2023, November 8). *Early terrorist adoption of generative AI.* Tech against Terrorism. https://techagainstterrorism.org/news/early-terrorist-adoption-of-generative-ai

Thomas, E. (2021). *Gaming and extremism: The extreme right on DLive.* Institute of Strategic Dialogue. https://www.isdglobal.org/wp-content/uploads/2021/08/03-gaming-report-dlive-1.pdf

UK Parliament. (2023). *Online safety act 2023—Parliamentary bills—UK parliament.* UK Parliament. https://bills.parliament.uk/bills/3137

United States Senate Committee on the Judiciary. (2023, March 29). *Durbin calls on online video game industry to do more to identify & remove extremist content from their platforms.* United States Senate Committee on the Judiciary. https://www.judiciary.senate.gov/press/dem/releases/durbin-calls-on-online-video-game-industry-to-do-more-to-identify-and-remove-extremist-content-from-their-platforms

Weaver, M. (2023, July 6). AI chatbot 'encouraged' man who planned to kill queen, court told. *The Guardian.* https://www.theguardian.com/uk-news/2023/jul/06/ai-chatbot-encouraged-man-who-planned-to-kill-queen-court-told

Yadron, D. (2016, February 5). Twitter deletes 125,000 Isis accounts and expands anti-terror teams. *The Guardian.* https://www.theguardian.com/technology/2016/feb/05/twitter-deletes-isis-accounts-terrorism-online

6 Artificial Intelligence and Cyber Espionage
Delineating Emergent Warfare and International Security

Nnenna Ifeanyi-Ajufo and Wan Rosalili Wan Rosli

6.1 INTRODUCTION

The changing theories of warfare amid growing geopolitical tensions have become a critical international law discourse ranging from the concept and definitions of armed attacks, the right to self-defense, attribution, and the varying implications for sovereignty. The nature of emerging technologies further adds to the concerns related to these discourses, including the reality of how technology is reconceptualizing extant international law notions of warfare, especially the cyber dimensions (Moynihan, 2021).

The international community has recently been engaged in various forums and debates to define how cyber activities may impact emergent warfare. Cyber espionage is an increasingly significant agenda of such debates for so many reasons, especially the security and protection of critical national infrastructure (CNI). One such example is the ongoing United Nations Ad Hoc Committee elaborating the convention on the use of information and communication technologies (ICTs) for criminal purposes (cybercrime convention), where the imminent threat of cyber espionage and its impact on the economic and political interactions between nation-states and the changing nature of modern conflict have been acknowledged. Cyber espionage is increasingly viewed as an international security issue which indeed not only targets states but also affects public and private networks, corporations, and individuals across borders. Another dilemma is the nature of cyberspace which seemingly blurs the dichotomy between state and non-state actors in aspects of responsibility for a variety of cyber threats and vulnerabilities.

Cyberspace provides a multifaceted and dynamic landscape for risks, and in the last decade, cyber espionage has emerged as a significant threat for both state and non-state actors. In 2010, Google reported their infrastructure had been infiltrated in a sophisticated and targeted attack originating from China which resulted in the theft of its precious data (Moynihan, 2021). Further to this, the publication of the Mandiant Report in 2013 provided insight into the extent and scale of cyber espionage within the international community (Mandiant, 2013).[1] The report highlights China as a persistent perpetrator of such crime and has been responsible for several attacks which exploit the vulnerabilities of states to retrieve confidential and sensitive data (Buchan, 2016). In more recent times, the Snowden report incident further highlighted concerns by placing cyber espionage in the spotlight after it was disclosed that the U.S. National Security Agency (NSA) has been involved with a surveillance program which collects confidential information through intercepting and monitoring private emails and telephone communications (Buchan, 2016). The targeted victims included officials from international organizations such as the European Union, heads of state, religious leaders, non-governmental organizations, and suspected terrorists (Buchan, 2016).

Significantly, the shift to new technologies such as artificial intelligence (AI) further provides opportunities for the introduction of new scopes of cyber espionage, impact on a larger scale, and more flexibility, including anonymized and unmanned acts which may include surveillance,

DOI: 10.1201/9781003441700-6

damage, and, generally, threats powerful enough to immobilize states through disrupting the functionality of critical national infrastructure. The cyber-related activities of the Russia-Ukraine war add credence to the imperative for interrogating the international regulation of cyber espionage.

As the race toward prioritizing digitization and the integration of new technologies continues to expand, and real-world legal dilemmas find their place in cyberspace, it is important to delineate traditional debates in alignment with the extant legal realities. Such an approach will provide a context for this chapter to contribute to the developing debates on emerging technologies and the changing dynamics of warfare. By understanding the interaction between emerging technologies such as AI, cyber espionage, and emergent warfare – policymakers can better anticipate and mitigate the risks posed by malicious actors in the cyber domain. Therefore, this chapter examines the international law approach to understanding and regulating cyber espionage. The chapter further discusses the current normative framework governing cyber espionage and the political implications of international cyber security strategies for the development of such frameworks. Since the development of norms for the governance of emerging technologies continues to gather relevance, the chapter also evaluates the impact of the use of emerging technologies such as AI in the commission and governance of cyber espionage and discusses approaches for promoting international cooperation to enhance capacity to delineate and address cyber espionage for the promotion of international security.

6.2 DYNAMICS OF CYBER ESPIONAGE IN THE ERA OF ARTIFICIAL INTELLIGENCE

Espionage has traditionally been viewed as the practice of spying or using spies, to obtain information about the plans and activities especially in the context of a foreign government. The Annual Situation Reports of the Swiss Federal Intelligence Service (FIS) have consistently emphasized that espionage is driven by a variety of different motives and has more than one aim (Swiss Federal Intelligence Service [FIS], 2023). The 2023 report highlights that, however, modern technologies now make it possible for actors to operate more comprehensively and to develop new solutions to problems which remain traditional, including the need for states to attend to the needs of a protracted military conflict (Swiss Federal Intelligence Service [FIS], 2023, p. 58).

In this current era of cyber espionage, AI has grown exponentially and becomes increasingly institutionalized in the 21st century in areas such as cybernetics, automation, mathematical logic, and linguistics (Vernon et al., 2017). AI encompasses a broad range of techniques and methodologies, with machine learning and deep learning being some of its key components. Machine learning is a subset of AI that focuses on enabling computers to learn from data and improve their performance over time without being explicitly programmed (Montesinos López et al., 2022). It involves the use of algorithms and statistical models to analyze large datasets, identify patterns, and make predictions or decisions. Deep learning as a specialized form of machine learning also utilizes deep neural networks with multiple layers of interconnected neurons (Montesinos López et al., 2022). Deep learning algorithms are capable of automatically extracting representations of data, enabling them to learn complex patterns and features from raw input data. These attributes are very important when considering the facilitation of espionage in a digital era.

AI also has a role to play in facilitating cyber espionage. AI algorithms can analyze vast amounts of data from various sources, including social media, public records, and network traffic, to identify potential targets for cyber espionage operations (Pun, 2017). AI algorithms can be used to develop stealthy and evasive cyber espionage tools and techniques that can bypass traditional security defenses, such as antivirus software and intrusion detection systems (Kose, 2021). These algorithms can prioritize targets based on factors such as organizational significance, susceptibility to exploitation, and potential intelligence value (Moisset, 2023). Techniques such as adversarial machine learning enable attackers to generate polymorphic malware variants that evade signature-based detection and behavioral analysis (Catalano et al., 2022).

AI is no longer confined to science fiction but is now a critical component of military strategy. Nations are actively integrating AI into their defense capabilities, leading to a paradigm shift in how wars are fought. In recent years, characterized by rapid technological advancement, the emergence of AI and cyber espionage has become a defining feature of modern warfare and international security. As nations increasingly harness the power of AI to bolster their defense capabilities and conduct intelligence operations, the landscape of conflict has evolved to another dimension – one where states use various means to try and gather information from other states without consent.

Cyber espionage has become an important and interesting international debate in contemporary times, especially for understanding how technology shapes the world and influences relationships between various states. Perhaps the defining feature of cyber espionage is that it is embedded in the traditional objective to occur in secret, and one of the most difficult problems regarding cyber warfare is giving a definition to cyber espionage. Many nations and international bodies have created their own definitions, but it has been difficult to achieve a single consensus on the definition of terms and concepts relating to cyber espionage. There are various factors determining the state of affairs. According to Rubenstein, the extent and nature of the damage caused by the activity, the actors behind the activity, and how the stolen information is used or intended to be used are all determinant factors for how the understanding of cyber espionage is perceived (Rubenstein, 2014). The *Tallinn Manual* – a set of guidelines for nation-state cyber warfare – has attempted to provide definitions, procedures, and rules governing international cyber operations while unequivocally emphasizing that international law applies to cyber operations. The manual describes cyber espionage as "an act undertaken clandestinely or under false pretenses that use cyber capabilities to gather, or attempt to gather, information" (Schmitt, 2017, p. 168).

Historically, states performed espionage by sending their agents into the physical territory of their adversaries. Developments in technology have enabled states to spy on their enemies by using more sophisticated electronic methods, such as the use of high-frequency antennas to capture electronic transmissions emanating from the territory of others (Report of the Group of Governmental Experts, 2013, p. 6). States also obtain confidential information from a variety of different sources, including through the derivation of human intelligence which is based on information from human sources, and the deployment of electronic sources, especially signals intelligence (United Kingdom Home Office, 2021).

For years, state and non-state actors have been engaging in spying to gain information, assess potential threats, and also prepare for attacks. Essentially, throughout history, states have continued to engage in clandestine activities against each other through activities such as spying and sabotage (Fjäder, 2014). It has long been viewed as an attribute of warfare, state intelligence, and combatant engagement. The emergence of digitalization, and especially, disruptive technologies, meant an expansion of the term cyber warfare. This also meant that nation-states began to move its focus to the protection of the CNI which required special cyber protection and security.

O'Brien asked as follows: "could a country's enemies shut down its power grid, take out its telephone system, or even hijack its nuclear missiles?" (O'Brien, 2017). If there is an imagination of these, then cyberwarfare should be given immense considerations especially for reasons that relate to the protection of CNI, as well as being weary of the intrusive capabilities of AI. In combination with the growing sophistication of cyber actors and activities, it further expands opportunities for coordinated and advanced attacks that can speedily disrupt the operational efficiency of designated CNI, e.g., the operation of the financial markets, electricity, security infrastructure, and general elections (Baker, 2023). When cyber espionage is part of a broader military campaign, it can also lead to disruption of public infrastructure and property (Harknett & Smeets, 2022). Cyber espionage attacks can be motivated by monetary value or may also be deployed in conjunction with military operations, or as an act of cyber terrorism or cyber warfare.

Cyber espionage, particularly when organized and carried out by nation-states, is a growing security threat. Despite various indictments and legislation intended to curb such activity, actors remain unabated, especially with the difficulty of enforcing international law relating to this issue.

Mostly, common targets of cyber espionage include government agencies and large organizations or institutions that have the value to create a competitive advantage for another organization or government. Targeted campaigns can also be waged against individuals, such as prominent political leaders and government officials, business executives, and even celebrities. On a national scale, cyber espionage has increasingly been focused on CNI (Rees & Rees, 2023).

On the other hand, states have also argued that cyber espionage tactics may be used by intelligence services to protect national security by preventing terrorist attacks (Byman, 2014). It can also be used in war where the information collected from the enemy can be used successfully to gain warfare advantage (Byman, 2014).[2] With the realization of the opportunities that cyberspace offers, governments across the world are building on cyber-forces, which are tasked with accomplishing their national cyber security policy goals and intelligence agendas (Liebetrau, 2023).

Against this backdrop, the implications for international security are serious and far-reaching. As nation-states contend for dominance in cyberspace and geopolitical tensions escalate, the risk of AI-enabled cyber activities will unarguably exacerbate tensions. The potential for misattributed attacks, unintended escalation, and the increase of cyber operations threatens to destabilize the balance of power and heighten the risks of conflict escalation (Egloff & Smeets, 2023). It is imperative to therefore identify the emergent contours of warfare in the digital age and develop sound strategies for safeguarding international security.

6.3   EMERGENT WARFARE AND INTERNATIONAL LAW ACTORS

Traditional espionage, rooted in secret intelligence-gathering activities, has long been a feature of interstate relations. International law provides a framework within which espionage activities are conducted, although specific treaties or conventions solely dedicated to espionage are notably absent. Instead, traditional espionage activities are subject to principles of international humanitarian law, sovereignty, and customary practices.

According to the Charter of the United Nations (UN Charter), "all Members shall refrain in their international relations from the threat or use of force against the territorial integrity or political independence of any state, or in any other manner inconsistent with the Purposes of the United Nations" (Article 2, paragraph 4, United Nations, 1945). Scholars argue that espionage violates the general duty of non-intervention mandated by the UN Charter (Wright, 1962, p. 12). The deployment of spies into the territory of other States to covertly seek or collect information of the internal affairs of another State will prima facie go against known international law principles.

On the other hand, it has also been argued that such opinions rather fall short of considering espionage as an international crime because many times, the reason for collecting information could hardly constitute an intervention or interference, and as such, whether mere interference violates the non-intervention principle is debatable (Oorsprong et al., 2023).

Espionage has always been seen by several States as an important element in facilitating the maintenance of state peace and security. Some authors argue that espionage can be legal based on pre-emptive or anticipatory self-defense (Hernández, 2020; Scott, 1999, p. 224). Such a legality (if any) will be highly contestable in the light of the UN Charter and the extant workings of international law jurisprudence (Radsan, 2007, p. 604). This idea is derived from the belief that, with the ability to determine the strength of other states, a balance of power can be achieved (Radsan, 2007). However, such belief has been challenged as it goes against the core inception of the UN Charter on the principle of the sovereign equality of States which stands by the principle that States should not intervene in each other's internal affairs (Radsan, 2007). Therefore, international peace and security should rather be maintained through legal regulation and not the balance of power (Radsan, 2007).

State actors and non-state actors are two distinct categories of entities that play significant roles in various aspects of international law, politics, and security. There is no existing international treaty that particularly regulates the use of secret agents to gather intelligence (Omand, 2021). In the sense of actors, it will be difficult to conceptualize the proper scope of what constitutes "espionage" for the purposes of international law – for instance, whether espionage also covers non-state actors who attempt to uncover sensitive information of the receiving state to disclose it, or even journalists who leak information to a foreign state (Charountaki, 2022). Major state powers possess significant resources, including financial, technological, and human capital, enabling them to develop and deploy advanced cyber capabilities, conduct sophisticated espionage operations, and shape geopolitical dynamics in cyberspace (Willett, 2019).

Non-state actors also have the ability to engage in and influence espionage. Nation-states with well-established military and intelligence apparatuses continue to play a central role in conflicts among states. These states possess significant resources, including financial, technological, and human capital, enabling them to develop and deploy advanced cyber capabilities, conduct sophisticated espionage operations, and shape geopolitical dynamics in cyberspace. Some state actors also engage in warfare through the sponsorship of proxy groups or non-state actors (Rauta, 2020). These entities may operate with varying degrees of autonomy, carrying out cyber-attacks, information operations, and other covert activities on behalf of their sponsoring state. State-sponsored non-state entities also provide deniability and flexibility to state actors seeking to advance their strategic interests while minimizing the risk of direct attribution (d'Aspremont et al., 2015). Military and intelligence also employ a wide range of tactics and techniques, including cyber espionage to achieve their objectives in cyberspace (Liebetrau, 2023).

Non-state actors have also presented themselves as political actors who are not directly connected to the State but pursue aims that affect the vital state interests. These non-state actors camouflage behind organizations such as non-governmental organizations, terrorist organizations, and social movements, or single influential individuals that have such seeming national or international political, economic, or social capability (Blumenau & Müller, 2023). Petrosyan highlights that terrorist organizations pose significant threats in emergent warfare by leveraging cyberspace to coordinate attacks and disseminate propaganda (Petrosyan, 2024). They may also be organized to deploy cyber capabilities that exploit vulnerabilities in information systems at a national scale and disrupt critical infrastructure. Individual hackers also contribute to warfare by conducting cyber-attacks and leaking sensitive information, sometimes to promote nationalist ideologies or political agendas (Ashraf, 2021).

From another dimension, the dichotomy between developed and developing nations adds another schism to the discourse. The ongoing dynamics of geopolitics and cyber diplomatic relations has birthed a struggle for superiority, dominance, and relevance among the developed and developing countries globally. Such superiority and influence are being determined by factors such as extant military strength, political strength, and economic power. There is also an incline toward attributing supremacy to the member states of the United Nations Security Council and, in cyber governance debates, the European Union and its member states who initiated the world's first international cybercrime treaty (The Council of Europe, 2021).

While for years, there have been standards and restrictions to dictate what a state could do to another, however, the advancement in technology has broken traditional barriers, and cyberspace now offers an opportunity to states with the capacity to utilize and take advantage of such cyber capabilities (Ifeanyi-Ajufo, 2017). To gain economic and technological power, some countries have resorted to diverse covert activities, for example, research has consistently pointed to China as a key actor in espionage, especially, through adopting cyber-intrusive activities to facilitate access and information and intelligence gathering (Wray, 2020).

Currently, there is no international law statute aimed at regulating cyber espionage; however, the delineation of cyber espionage can be arguably hinged on the international law principles of sovereignty, non-invasion, and equality of states (Bentley, 2020, pp. 527–528). In principle, as there has become an overwhelming acceptance of the applicability of international law in cyberspace, scholars have argued that the application of international law to conducts that may broadly be construed as cyber espionage will be determined based on the actors, and the context of operation (Schmitt, 2017).

While some bodies of international law might be important to spying, none of them specifically addresses the conduct of cyber espionage, especially as it considers AI. Indeed, there is no clear consensus among states on the legal nature of cyber espionage or whether states enjoy a right at international law to complain of it (Radsan, 2007, pp. 595–602). However, cyber espionage activities continue to increase among state and non-state actors and are further enabled by emergent technologies such as AI (Baker, 2003, p. 1091; Watt, 2019, pp. 638–642).

In the negotiations of the "United Nations Ad Hoc Committee to Elaborate a Comprehensive International Convention on Countering the Use of Information and Communications Technologies for Criminal Purposes"(United Nations Human Rights Committee, n.d.),[3] countries have presented diverse views on the proposed scope of the convention on whether the convention should favor cyber-dependent and cyber-enabled crimes. However, the negotiations have favored a narrower approach of focusing on only cyber-dependent crimes which imply that conducts such as cyber espionage may be outside the scope of the new treaty.

Outside of imminent warfare, continued states' approach over the years has seemingly legitimized espionage especially, touted as a state-preserving and national security approach (Devanny et al., 2021). Notwithstanding, the secretive nature of espionage implies that such activity directly or indirectly stands contrary to international law practice of state responsibility and due diligence.[4] This also means that under international law, there is no general acceptance of cyber espionage as responsible state behavior, and even if states have generally continued to practice espionage, it will never imply a generality of such act as *opinio juris* (International Law Association, 2000).

Usually, the requirement is that when taking part in a particular practice, the state must assert the international legality of their conduct or, at least, when the international legality of its conduct is challenged subsequent to its practice, it is defended on the basis that it is permissible under international law (Føllesdal, 2022). This is hugely problematic in the context of international law because while states practicing such activity do not generally express the belief that it is permissible under international law, they generally continue to practice them (Broeders et al., 2022). Therefore, it is argued that espionage is practiced by states with a sense of wrong and not a sense of right (Chesterman, 2006, p. 1072). Further, it is important to bear in mind that customary international law develops on the basis of consistent state practice that is permissible under international law (Arajärvi, 2021); however, notwithstanding the common practice of such conduct, there is certainly no customary international law exception to espionage, and in more contemporary times – cyber espionage.

In summary, the dynamics of emergent warfare involve a diverse array of state and non-state actors operating in cyberspace, each with unique capabilities, motivations, and objectives (Petrosyan, 2024). Understanding the roles and characteristics of these actors is essential for effectively countering emerging threats, promoting international security, and safeguarding the integrity of cyberspace in an increasingly interconnected and digitally dependent world.

6.4 INTERACTION BETWEEN AI, CYBER ESPIONAGE, AND EMERGENT WARFARE

In the past decade, law enforcers have continuously worked to combat crimes within cyberspace by AI. The use of AI has also facilitated the investigation of such crimes as it lifted the safety net provided by the veil of anonymity. Tracking and finding cyber espionage offenders is a challenge as these entities always hide behind the shadows of their offender background and linking their digital fingerprints has proven to become very time-consuming (Rawat et al., 2021). The utilization of AI has proven to save time and resources where artificial cyber espionage search engines can go through unlimited data to help law enforcement get the information needed to make arrests (Rawat et al., 2021). The deployment of technology has made it possible for law enforcement to have access to built-in artificially driven cyber espionage search engines which are capable of processing unlimited amounts of data to facilitate the arrests of hidden web offenders faster and with minimum use of resources (Rawat et al., 2021). The use of AI-driven web processing engines to detect cyber espionage includes the use of biometric recognition technologies such as facial detection through the data available online and the ability to receive immediate notification when faces or other similar information is detected (Rawat et al., 2021). The volumes of online data are seen as vaults waiting to be infiltrated by cybercriminals continuously eyeing opportunities to commit cybercrimes. The ultimate aim of cyber espionage is to steal data from firms, and governments by the same methods used for clandestine collection in altering or corrupting data. The data stolen would be used to influence foreign opinion through blackmail or propaganda and to improve decision-making by major decision-makers within a country (Lindsay, 2017).

With the recent development of technologies and infrastructure, there is a constant availability of the use of emerging technologies such as cloud services and other platforms which has caused the rise in cyber espionage cases in recent years. The integration of AI and cyber-attacks has enhanced the severity, complexity, and sophistication of threats which have proven more difficult to detect. Gohil has highlighted that with AI, ransomware and malware have become more sophisticated where it can now identify and encrypt critical data within a system which puts pressure on victims to pay the ransom demanded (Gohil, 2024). Furthermore, AI-enhanced cyber-attacks have grown exponentially such as automated hacking, code-adapting malware to evade detection, and AI-crafted phishing emails (Gohil, 2024).

The emergence of generative AI (GenAI) has also changed the landscape of the commission and protection against cyber espionage. GenAI-aided attacks have now unleashed both known and unknown cyber-attack vectors. Gupta et al. highlighted that GenAI acts as a double-edged sword in cyber security as it facilitates and benefits both sides of the coin (Gupta et al., 2023). Nation states have now moved to GenAI tools to safeguard their systems from infiltration. Information gathered from large language models train intelligence data such as risk, vulnerabilities, and prediction of attacks (Touvron et al., 2023). States now have the capability to process large information to enhance their intelligence response and even predict emerging and future threats (Patil et al., 2023). Furthermore, such advanced technology is also important in developing effective ethical guidelines for strengthening cyber defense within a system (Gupta et al., 2023). According to Gupta, the use of GenAI provides support in cyber defense automation, cyber security reporting, threat intelligence, identification of cyber-attacks, incident response guidance, and even malware detection (Gupta et al., 2023).

6.5 AI AND CYBER SECURITY DEBATE – HOW EMERGING TECH ENABLES CYBERCRIME

Cyberspace now plays an integral role in modern society, facilitating communication, commerce, and innovation on a global scale. Cyber security in the context of nation-states refers to all factors intended to protect state infrastructure against intentional cyber-attacks, infractions, incidents, and their impacts. It serves as the protection against these threats and encompasses measures and practices aimed at protecting national information systems and digital assets from unauthorized access, manipulation, and disruption. Additionally, emerging technologies, such as AI, introduce new vulnerabilities, increasing the difficulty of the cyber security challenge (Lilli, 2021).

Cyber security has always been an important element for states especially on cyber threats that are related to intelligence. States are also aware of the importance of cyber defense which include the use of technology for protection and counterintelligence and all the ethical issues that come with it

(Lilli, 2021). Emerging technologies have provided a plethora of novel vantage points for surveillance and weapons have become more lethal and precise which in hand makes intelligence vital for countermeasures (Lilli, 2021). The market for AI in cyber security is forecasted to grow more than US$34.8 billion by 2025 (Taddeo et al., 2019). The recent cyber security strategies adopted worldwide by nation-states have also explicitly highlighted the use of AI capabilities within their defense strategies (Taddeo et al., 2019).

In information security, threat agents are often categorized into four different elements which are human, environmental, natural, and AI. Within cyberspace, humans are often linked to threats as they create and use tools that correlate with risks (Uesato et al., 2018). However, as AI evolves, it has in turn become a threat agent which serves as an entity that could cause damage, disruption, or direct harm in the physical world (Taddeo et al., 2019). Cyber espionage involves data infiltration which poses a risk to various platforms such as mobile environments, cloud technology, and Internet of Things (IoT) devices (Given-Wilson et al., 2020). The use of technologies such as botnets is also advantageous from an infiltrator's perspective as the capacity, storage, processed data, and networking resources are always readily available due to their functionality (Taddeo, 2021). The European Union Agency for Cyber security (ENISA) has categorized botnets as one of the most dangerous threats (ENISA, 2020). The capabilities of botnets which are equipped with self-development characteristics make them more robust and resilient, especially in cyber espionage (Gohil, 2024).

The deployment of predictive analytics and machine learning has now become the basis of cyber security strategies as these technologies can analyze vast amounts of data in order to predict, detect, and prevent cyber threats. However, the duality impact of such technologies must not be ignored. With such expertise in threat detection, the risk of exploitation of such technology is also present. Apart from that, the universal susceptibility to cyber threats expands to various government sectors. Gohil highlighted that the current trend of cyber security shows a complex interplay between the advancement of technology, emerging threats, and the governing of such threats where states must not only be reactive but proactive and predictive to be a step ahead of potential threats (Gohil, 2024). The ultimate aim is to harmonize technological advancement with robust regulations in order to create a resilient cyber security ecosystem (Cole, 2016).

The Office of the National Counterintelligence Executive contends that cyber espionage amplifies the threat to international peace and security (Gohil, 2024). The Internet has been held as a God's gift for spies and the past decade has been proclaimed as the era of cyber espionage (Gohil, 2024). Similarly, the duality effect of such use of emerging technologies has posed problems as terrorist organizations such as the Islamic State (ISIS) have since broadened their recruitment to young tech-savvy individuals living in various parts of the world (Jupillat, 2016, p. 933). In 2015, Kosovar Ardit Ferizi who was at the time living in Malaysia was arrested and charged by the United States for hacking into a US government database and stealing sensitive information of military and civilian government personnel (Banks, 2016). All this while, cyberspace has been a place for intelligence activities which includes surveillance and spying outside its own national borders (Weissbrodt, 2013). However, the line between lawful and unlawful cyber espionage or intrusions remains unclear. The use of AI within cyber security also enables the system to predict and prevent attacks to the system (Sarker et al., 2021). The learning-based system learns from the perpetrator's actions and can put in place safeguards in order to protect sensitive data (Stevens, 2020). The continuous learning and adapting to the system result in the technology becoming very efficient and creating a securitized environment which enables it to avoid cyber-attacks (Taddeo et al., 2019).

The absence of legislation governing reasonable cyber security practices, norms, or standards is important in ensuring the improvement in cyber defense and cyber security across all geographical regions. Nations have now moved to focus on standards to meet a variety of objectives which align with their beliefs and capacity. The European Union together with the European Union Agency for Cyber Security (ENISA) includes the standards within its strategies and policies such as having a clear definition of resilience (Greiman, 2022).

6.6 CRITICAL NATIONAL INFRASTRUCTURES (CNI), STATES, AND CYBER ESPIONAGE

The cyber security of CNIs is a vital aspect that all states must protect at all costs. Each state provides a detailed definition regarding their CNI by highlighting its importance to society, vulnerabilities, and threats (Lehto, 2022). The United Kingdom has highlighted 13 CNI sectors and describes it as an infrastructure that consists of facilities, systems, information, people, and processes which are vital for a country to function (Lehto, 2022). The Finland National Emergency Supply Agency defines CNI as devices, services, and IT systems which are critical to the running of a nation and the destruction of it would pose a risk to national security, economy, health, and safety and the efficient functioning of the central government (Lehto, 2022).

Failure to provide adequate security to any of the CNI sectors would make the state exposed and vulnerable to attacks that could have a devastating impact on the security, safety, economy, and community as a whole. Sabotage of CNI sectors led by nation-state through such modus operandi as the act of changing or adding code to manipulate the system which results in total system failure, data leakage, or system harm (Dawson et al., 2021). If one sector shuts down or fails, it will cause a domino effect on other multiple CNI sectors which results in a national catastrophe (Dawson et al., 2021). For example, attacking the traffic control systems to disrupt traffic and increase traffic violations or gaining unauthorized access to a water gate system can in turn cause flooding and damage to properties and lives. The interdependency across CNI systems and the reliance on technology and digitalization have increased the risk and vulnerabilities to cyber threats such as cyber espionage where such CNI systems can be compromised. Apart from that, in the interconnected world where CNI becomes borderless and utilizes global supply chain, the impact increases and it is more open to threats and exploitation (Lehto, 2022).

The European Union Agency for Network and Information Security (ENISA) highlighted that cyber espionage actors often target government entities and the critical and strategic infrastructures of a state which includes railways, telecommunications, healthcare, energy, and also financial institutions (Bederna & Szadeczky, 2020). The ultimate impact of cyber catastrophe would be for example meltdown of a nuclear power plant, failure of the power grid in winter, paralysis of traffic control, and even hacking of medical devices (Bederna & Szadeczky, 2020). Reports have stated that it has been a practice for states to exploit cyberspace to extract confidential information from private companies that provide services to critical national infrastructures (Bederna & Szadeczky, 2020). To illustrate, the infiltration of data retrieved from the U.S. fighter jet F-35 resulted in the data theft worth terabytes allegedly done by the Chinese government (Melnitzky, 2011). Another incident that can be highlighted is the detection of computer malware within the computer systems of Iranian Oil Companies which lead to classified information being accessed, monitored, and intercepted (Buchan, 2015). Reports state that the virus has the capability to activate the systems' microphones, cameras, keyboard strokes, take screenshots, extract geolocation, and able to receive commands via wireless technology (Buchan, 2015).

6.7 STATES AND LEGAL RESPONSE TO CYBER ESPIONAGE

In recent times, states have acknowledged engaging in diverse cyber activities, especially on the alibi of national security.[5] The United Nations Special Rapporteur on Counterterrorism and Human Rights has called for a legal response to spyware proliferation (Akkad, 2022). States have also continued to devise various mechanism for addressing cyber security matters beyond the mere criminalization of cybercrimes through national legislation or statues. States across diverse regions have focused on designing cyber security strategies which provide a detailed agenda of state policies and stages to maintain cyber resilience across relevant national sectors. A cyber security strategy is a plan of action designed to maximize the security and resiliency of a nation. It uses a top-down approach to establish a set of objectives and protocols to help keep a nation safe and protect its critical national infrastructure. It outlines the duties of actors and defines cyber security duties and obligations. It also recognizes the fact that cyber threats are continually advancing and devises ways to adapt so that national cyber security is always monitored. A cyber security strategy has now emerged as proactive approach by nations states to mitigate cyber breaches, attacks, and threats.

Within the international context, cyber espionage is not specifically governed by any treaty and there is also no specific international treaty that could be adapted to regulate cyber espionage. The *Tallinn Manual 2.0* covers the important areas of the regulation of cyber operations. The manual has no legal force as it is meant to reflect the law by a group of international experts (Jensen, 2016). Apart from that, the UN has also established six Groups of Governmental Experts (GGE) on Advancing responsible State behavior in cyberspace in the context of international security (United Nations, 2021)[6] and an Open-Ended Working Group (OEWG) on the Development in the Field of Information and Telecommunications in the Context of International Security.[7] The UN Group of Governmental Experts (UNGGE) further developed the United Nations Norms of Responsible State Behavior in Cyberspace (Hogeveen, 2022). These initiatives have generally been aimed at addressing the risks and vulnerabilities in cyberspace through the lens of international law. The principles formed through this initiative are in line with the *Tallinn Manual 2.0*.

The revelation of the Snowden report has seemingly been a motivation for states to seek harmonized approaches to mitigating and regulating cyber espionage-related activities (Devanny et al., 2021). States have now moved toward the curtailing of remote access cyber espionage by deploying state-of-the-art technologies to restrict such operations. For example, in 2023, the United States published its Presidential Initiative for Democratic Renewal which calls for the countering of the misuse of technology and digital authoritarianism (Devanny et al., 2021). The initiative aims at combating digital repression which includes an executive order to prohibit the use of spyware which poses a risk to national security (Anstis, 2024). Apart from that, the United States published the *Guiding Principles on Government Use of Surveillance Technologies and a Joint Statement on Efforts to Counter the Proliferation and Misuse of Commercial Spyware* which aims to foster international cooperation on spyware proliferation (Anstis, 2024). The guide has also managed to create a working group on the misuse of technology in threatening security and human rights (Anstis, 2024). The U.S. Espionage Act also prohibits espionage where it jeopardizes national defense or benefits a foreign nation (Barnes, 2022). However, there is a lack of clarity as to whom it applies to and even more vague if considered in light of cyber espionage activities. In a move for more clarity, the Foreign Intelligence Surveillance Act 1978 (FISA 1978) was enacted. The FISA has proven to be adequate as it was successful in detaining terrorists and convicting conspirators passing U.S. secrets to foreign nations. Despite the success of FISA, the law has been criticized for being unable to maintain a balance between national security and the protection of privacy (Barnes, 2022).

In China, cyber espionage is governed under the National Intelligence Law 2017. The overall aim of the law is to strengthen and safeguard national intelligence work and preserve state security and interests (Greiman, 2022). Article 2 of the law highlights that the national intelligence work will be in line with the overall national security perspective which includes providing intelligence in decision-making, prevention and mitigation of threats to national security, sovereignty, territorial integrity, unity, sustainable social and economic development, and other major national interests. The law also attempts to balance the securitization of such legislation by addressing that it will respect human rights where such national intelligence effort will be conducted in accordance with the rule of law and shall respect and protect the human rights and interests of individuals and organizations.[8]

Russia's External Intelligence Service (SVR) is said to be the reincarnation of the KGB, which is one of the oldest and most extensive espionage agencies (Davies & Steward, 2022). Russia had enacted the Criminal Code of the Federation, Law NO. FZ-190 which criminalized espionage, disclosure of state secrets and providing any assistance to a foreign state, organization or its representative which is detrimental to the security of the Russian Federation as high treason. Article 275 states that high treason offenses shall be punishable by 12–20 years imprisonment with or without a fine of up to 500,000 rubles.[9]

In Indonesia, cyber espionage is governed under Article 31(1) of the Information and Electronic Transaction Act 2008 which governs interception of electronic documents or information. The law is said to be lacking and inadequate as it does not directly refer to cyber espionage. The law came under scrutiny after the Snowden report due to the evidence showing that Indonesia's head of state was also impacted by cyber espionage via phone tapping by the Australian and New Zealand governments (Satria Unggul & Noviandy, 2019). In Malaysia, cyber espionage is also not specifically criminalized. General laws such as the Security Offences (Special Measures) Act 2012 have unique provisions addressing security offenses and recognize threats to national security derived from terrorism and espionage. Also, section 124M of the Malaysia Penal Code highlighted that whoever commits espionage, whether directly or indirectly, will be imprisoned for life. However, the legal framework in Malaysia is said to be inadequate to effectively govern cyber espionage.

The European Union has also taken steps to regulate cyber espionage when recently the European Parliament called for a stricter approach to the use and abuse of spyware. Furthermore, in February 2024, the United Kingdom and France hosted nation-states, regional organizations, and other stakeholders in an inaugural conference to tackle the proliferation and irresponsible use of commercial cyber intrusion tools and services (Poireault, 2024). The conference also led to the signing of the cyber declaration – the Pall Mall Process: Tackling the Proliferation and Irresponsible Use of Commercial Cyber Intrusion Capabilities (Ministry for Europe and Foreign Affairs, 2024).

Cyber security challenges can take many forms for states. Depending on the nature and method of attack, a cyber operation can have an impact at the national level, as well as implications at the international level. The inequality in digital capacity among states also means that there are limitations in developing a harmonized approach for a legal response to cyber espionage. The challenges differ across regional boundaries considering limited resources, technical capabilities, and other social realities. A general lack of awareness of cyber security best practices also leaves many states vulnerable.

6.8 TOWARD INTERNATIONAL SECURITY AND EMERGENT CYBER ESPIONAGE GOVERNANCE

Cyber espionage represents a significant and evolving threat in the realm of cyber security, with implications for international security and global economic stability. From traditional espionage activities conducted by state actors to the emergence of new tactics and techniques driven by advancements in technology, cyber espionage poses a lot of challenges for national critical infrastructure, and other national interests. State and non-state actors, including state-sponsored entities and general criminal organizations, are all significant in the cyber espionage landscape.

The approach to harmonize international perspective on cyber espionage has faced challenges due to the relevant domestic laws of each country not being insufficient to negotiate the challenges of transborder issues such as national security, human rights, privacy and data protection issues, and extraterritorial jurisdiction. Furthermore, the issue of global digital divide and access to justice remains unaddressed. Ultimately, the answer to these challenges will be an agreed international law, as domestic laws are unable to provide harmonization due to individual states interest in national security and economic advancement. Scholars have contended that the best way forward is to have a treaty to best govern the use of emerging technologies such as AI in the commission of cyber espionage (*The Pall Mall Process*, 2024). Greiman highlighted as a more realistic approach, States should develop norms and practices relating to the current international law, rather than prohibiting cyber espionage completely (Greiman, 2018). He further states that if States want non-binding responsibilities in cyberspace, therefore their actions and practice must demonstrate their willingness to achieve that goal (Greiman, 2018).

The failure to reach an agreement of a specific international law governing cyber espionage is attributed to the different interests and priorities relating to national sovereignty and security (Devanny et al., 2021). It has further been highlighted that there must be the capability to track and trace where the cyber-attacks originate to better attribute the exploits of cyber espionage from around the world (Devanny et al., 2021).

Another challenge in governing cyber espionage will be that the nature of such data and information stolen is highly classified to be revealed within the courtroom, especially, that of the International Court of Justice, for example, the tug of war-between the United States and China led to years of fragility of diplomatic relations (Devanny et al., 2021).

One of the major ways of addressing cyber security is to harmonize international cooperation and law. By fostering international cooperation and adherence to established norms and regulations, the global community can strive to uphold the principles of peace, stability, and mutual respect. Despite the efforts of governments and international organizations to combat cyber espionage through legal frameworks, diplomatic channels, and technical countermeasures, the threat persists and continues to evolve (Rowe, 2020). Attribution remains a significant challenge, complicating efforts to hold actors accountable and deter future conducts. Additionally, the interconnected nature of cyberspace means that the consequences of cyber espionage can be felt globally, transcending borders, and impacting individuals, organizations, and nations across the world (Devanny et al., 2021).

Efforts to effectively criminalize and govern cyber espionage will include the development of international agreements between nations to define what is the norm and establish consequences if there is a violation of the agreement by any parties. For non-state actors engaging in cyber espionage, that appropriate way to tackle such activities will be by enacting new laws, harmonizing international legislations, and encouraging coordination and cooperation between national law enforcement agencies. This will also include fostering collaboration and information sharing among public and private sector entities, to enhance situational awareness, coordinate response efforts, and collectively defend against cyber espionage threats, especially for critical national infrastructure.

Undeniably, in the years to come, many state intelligence services will increasingly depend on the capabilities of AI for the operationalization of intelligence activities. States will also invest more in their technological capabilities, in order to enhance their cyber powers. The primary target of states for covert espionage activities will range across several stakeholders, including counterpart states, corporations, and individuals alike (Swiss Federal Intelligence Service [FIS], 2023). This means that in the coming years, there will be an urgent need to nation-states, international organizations, and other stakeholders to consider a normative framework for the use of such capabilities.

AI governance is also important as a way forward as more states are investing in the development of their national AI resources. There are also efforts to restrict the cross-border flow of AI technologies to gain an advantage or have the upper hand with the ownership of such technologies (Fischer & Wenger, 2021). However, AI governance is still in its infancy, and there is still a long way to go in terms of reaching an understanding between states on the standards and norms for AI. As a way forward, the harmonization and mutual agreement between states is the best way to effectively govern cyber espionage, especially within the challenging times ahead. The passing of the AI Act has paved way to other countries to also follow suit with their own policies, strategies, and regulations in other to effectively govern AI and other emerging technologies.

The enhancement of cyber security and deterrence of threats also play a big part in effectively governing cyber espionage. States must be willing to develop and invest in robust cyber security measures. Furthermore, intelligence sharing is also an important element for effective governance of crime as when nations are willing to practice collaborative intelligence, emerging threats can be identified more effectively. By promoting cyber security awareness, investing in defensive technologies, and strengthening international cooperation, states can work toward mitigating the risks posed by cyber espionage and fostering a more secure and trusted digital environment.

6.9 CONCLUSION

The deployment of emerging technologies such as AI in cyber-criminal activities has proven to be more prevalent. The nature of cyber espionage further causes apprehension due to the ever-evolving nature of technology, for example, AI acts as double-edged sword by providing solutions for governing cyber espionage through the facilitation of investigation, detection, and prediction of threats. AI-enhanced cyber-attacks have also grown exponentially resulting in more complex and sophisticated threats to national security.

To effectively govern cyber espionage, states must strategically develop and invest in robust cyber security initiatives such as Computer Emergency Response Teams (CERTs) and collaborative intelligence sharing. States must ultimately ensure responsible use of emerging technologies in accordance with the "United Nations Norms of Responsible State Behaviour in Cyberspace" which have been developed to foster a more peaceful and secure digital environment.

In delineating emergent cyber espionage, international law will be beneficial for governing the interaction between emerging technologies such as AI, espionage, and contemporary warfare. The current international legal climate which seemingly relies on the unwritten rule that espionage is commonly acceptable blurs the attribution lines of the criminality of such acts. While states may independently forge policy frameworks to mitigate cyber espionage, international cooperation is pertinent in defining a common position for understanding the characteristics, and normative stance that may give cause for state responsibility in matters of cyber espionage. Multilateral engagements will provide the right platform for stakeholders to deliberate on the possibility of the international law framework to determine the regulation of cyber espionage. Such approaches will also advance legal and political strategies on international cyber security and enhance capacity-building on cyber espionage across nation-states.

NOTES

1. In February 2013, Mandiant released a report documenting evidence of cyber-attacks targeting at least 141 organizations in the United States and other English-speaking countries extending over a number of years.
2. An example is the case of Ukrainian soldiers fighting in eastern Ukraine who are still using wire phones to communicate between posts, because Russian separatists are using electronic warfare.
3. United Nations Ad Hoc Committee to Elaborate a Comprehensive International Convention on Countering the Use of Information and Communications Technologies for Criminal Purposes – Pursuant to General Assembly resolution 74/247, the General Assembly decided to establish an open-ended ad hoc intergovernmental committee of experts, representative of all regions, to elaborate a comprehensive international convention on countering the use of information and communications technologies for criminal purposes, taking into full consideration existing international instruments and efforts at the national, regional, and international levels on combating the use of information and communications technologies for criminal purposes. See generally the United Nations Office on Drugs and Crimes – Ad Hoc Committee – Home (unodc.org).
4. By definition, espionage is a practice conducted in secret. As a result, irrespective of how frequently states engage in espionage, where this practice is engaged in covertly and secretly it cannot be identified as state practice for the purpose of customary law formation.
5. For example, the Mission Statement of the US Central Intelligence explains that one of its objectives is to "prevent threats and further US National security objectives by collecting intelligence that matters, producing objective all-source analysis, conducting effective covert action as directed by the President, and safeguarding the secrets that help keep the nation safe". Intelligence Agency, CIA Vision, Mission, Ethos and Challenges, https://www.cia.gov/about-cia/cia-vision-mission-values
6. The 2010, 2013, 2015, and 2021 consensus reports of the United Nations Groups of Governmental Experts (GGEs) on existing and emerging threats, norms, rules, and principles of responsible State behavior, international law, confidence-building and international cooperation and capacity-building, which together represent a cumulative and evolving framework for the responsible behavior of States in their use of ICTs.

7. The United Nations Open-Ended Working Group (OEWG) on developments in the field of information and telecommunications in the context of international security established pursuant to General Assembly Resolution 73/27. Final Substantive Report. UNGA A/AC.290/2021/CRP.2.
8. Article 8 of the National Intelligence Law 2017.
9. Article 275 of the Russia Criminal Code.

REFERENCES

Akkad, D. (2022, July 28). *Digital nightmare: The Arab dissidents ruined by phone hacking.* Middle East Eye. https://www.middleeasteye.net/big-story/digital-nightmare-arab-dissidents-lives-torn-apart-hacking

Anstis, S. (2024). Regulating transnational dissident cyber espionage. *International & Comparative Law Quarterly, 73*(1), 259–274.

Arajärvi, N. (2021). The core requirements of the international rule of law in the practice of States. *Hague Journal on the Rule of Law, 13*(1), 173–193. https://doi.org/10.1007/s40803-021-00152-8

Ashraf, C. (2021). Defining cyberwar: Towards a definitional framework. *Defense & Security Analysis, 37*(3), 274–294. https://doi.org/10.1080/14751798.2021.1959141

Baker, C. D. (2003). Tolerance of international espionage: A functional approach. *American University International Law Review, 19,* 1091.

Baker, K. (2023, February 28). *What is cyber espionage? – CrowdStrike.* Crowdstrike.Com. https://www.crowdstrike.com/cybersecurity-101/cyberattacks/cyber-espionage/

Banks, W. C. (2016). Cyber espionage and electronic surveillance: Beyond the media coverage. *Emory Law Journal, 66,* 513.

Barnes, J. E. (2022, August 15). What is the espionage act and how has it been used? *The New York Times.* https://www.nytimes.com/2022/08/15/us/politics/espionage-act-explainer-trump.html

Bederna, Z., & Szadeczky, T. (2020). Cyber espionage through Botnets. *Security Journal, 33*(1), 43–62. https://doi.org/10.1057/s41284-019-00194-6

Bentley, D. (2020). Cyber espionage and international law. *International Affairs, 96*(2), 527–528. https://doi.org/10.1093/ia/iiaa042

Blumenau, B., & Müller, J.-A. (2023). International organisations and terrorism: Multilateral antiterrorism efforts, 1960–1990. *Terrorism and Political Violence, 35*(2), 433–451. https://doi.org/10.1080/0954655 3.2021.1938002

Broeders, D., de Busser, E., Cristiano, F., & Tropina, T. (2022). Revisiting past cyber operations in light of new cyber norms and interpretations of international law: Inching towards lines in the sand? *Journal of Cyber Policy, 7*(1), 97–135. https://doi.org/10.1080/23738871.2022.2041061

Buchan, R. (2015). Cyber espionage and international law. In *Research handbook on international law and cyberspace* (pp. 168–189). Edward Elgar Publishing. https://www.elgaronline.com/abstract/edc oll/9781782547389/9781782547389.00018.xml

Buchan, R. J. (2016). The international legal regulation of cyber espionage. *Tallinn Papers,* 65–86.

Byman, D. (2014). The intelligence war on terrorism. *Intelligence and National Security, 29*(6), 837–863. https://doi.org/10.1080/02684527.2013.851876

Catalano, C., Chezzi, A., Angelelli, M., & Tommasi, F. (2022). Deceiving AI-based malware detection through polymorphic attacks. *Computers in Industry, 143,* 103751. https://doi.org/10.1016/j.compind.2022.103751

Charountaki, M. (2022). Conceptualising non-state actors in international relations. In M. Charountaki & D. Irrera (Eds.), *Mapping non-state actors in international relations* (pp. 1–16). Springer International Publishing. https://doi.org/10.1007/978-3-030-91463-9_1

Chesterman, S. (2006). The spy who came in from the cold war: Intelligence and international law. *Michigan Journal of International Law, 27.* https://heinonline.org/hol-cgi-bin/get_pdf.cgi?handle=hein.journals/ mjil27§ion=34

Cole, D. D. (2016). After Snowden: Regulating technology-aided surveillance in the digital age. *Capital University Law Review, 44,* 677.

d'Aspremont, J., Nollkaemper, A., Plakokefalos, I., & Ryngaert, C. (2015). Sharing responsibility between non-state actors and states in international law: Introduction. *Netherlands International Law Review, 62*(1), 49–67. https://doi.org/10.1007/s40802-015-0015-0

Davies, P. H. J., & Steward, T. (2022, May 20). *No war for old spies: Putin, the Kremlin and intelligence.* RUSI. https://rusi.org/explore-our-research/publications/commentary/no-war-old-spies-putin-kremlin-and-intelligence

Dawson, M., Bacius, R., Gouveia, L. B., & Vassilakos, A. (2021). Understanding the challenge of cybersecurity in critical infrastructure sectors. *Land Forces Academy Review, 26*(1), 69–75. https://doi.org/10.2478/raft-2021-0011

Devanny, J., Martin, C., & Stevens, T. (2021). On the strategic consequences of digital espionage. *Journal of Cyber Policy, 6*(3), 429–450. https://doi.org/10.1080/23738871.2021.2000628

Egloff, F. J., & Smeets, M. (2023). Publicly attributing cyber attacks: A framework. *Journal of Strategic Studies, 46*(3), 502–533. https://doi.org/10.1080/01402390.2021.1895117

ENISA (2020). *ENISA Threat Landscape 2020—Botnet.* [Report/Study]. ENISA. https://www.enisa.europa.eu/publications/enisa-threat-landscape-2020-botnet

Fischer, S., & Wenger, A. (2021). Artificial intelligence, forward-looking governance and the future of security. *Swiss Political Science Review, 27*(1), 170–179. https://doi.org/10.1111/spsr.12439

Fjäder, C. (2014). The nation-state, national security and resilience in the age of globalisation. *Resilience, 2*(2), 114–129. https://doi.org/10.1080/21693293.2014.914771

Foreign Intelligence Surveillance Act (FISA) 1978. https://www.nsa.gov/Signals-Intelligence/FISA/

Føllesdal, A. (2022). The significance of state consent for the legitimate authority of customary international law. In P. Merkouris, J. Kammerhofer, & N. Arajärvi (Eds.), *The theory, practice, and interpretation of customary international law.* The rules of interpretation of customary international law (pp. 105–136). Cambridge University Press. https://www.duo.uio.no/handle/10852/95542

France Ministry for Europe and Foreign Affairs (2024, February 6). *The Pall Mall Process: Tackling the Proliferation and Irresponsible Use of Commercial Cyber Intrusion Capabilities (Lancaster House, London, 6 Feb. 2024).* France Diplomacy – Ministry for Europe and Foreign Affairs. https://www.diplomatie.gouv.fr/en/french-foreign-policy/digital-diplomacy/news/article/the-pall-mall-process-tackling-the-proliferation-and-irresponsible-use-of

Given-Wilson, T., Baranov, E., & Legay, A. (2020). Building user trust of critical digital technologies. *2020 IEEE International Conference on Industrial Technology (ICIT)*, 1199–1204. https://ieeexplore.ieee.org/abstract/document/9067154/?casa_token=qfL7Yd_-YSsAAAAA:hBHkUCsGO823syp6VtpVFMHGYTZgE5vV_X0cVlTXFbtwcB9Xygvn32-vq2JccKmfxy8tZ4qcgVOv

Gohil, T. (2024). *Emerging threats and unseen risks: Protecting your business data from the evolving landscape of cyber espionage.* LinkedIn. https://www.linkedin.com/pulse/emerging-threats-unseen-risks-protecting-your-business–rhcjc%3FtrackingId=fHnr4CgKRT2mVVQ83SwTyg%253D%253D/?trackingId=fHnr4CgKRT2mVVQ83SwTyg%3D%3D

Greiman, V. (2018). Cyber espionage: The silent crime of cyberspace. *ICCWS 2018 13th International Conference on Cyber Warfare and Security*, 245. https://www.google.com/books?hl=tr&lr=&id=eHpTDwAAQBAJ&oi=fnd&pg=PA245&dq=Cyber+espionage:+The+silent+crime+of+cyberspace&ots=9VIX996zu0&sig=7HJrCz8IDZm3Y41LlUcLCYb_aAs

Greiman, V. A. (2022). Cyber law and regulation. In M. Lehto & P. Neittaanmäki (Eds.), *Cyber security: Critical infrastructure protection* (pp. 59–78). Springer International Publishing. https://doi.org/10.1007/978-3-030-91293-2_3

Gupta, M., Akiri, C., Aryal, K., Parker, E., & Praharaj, L. (2023). From ChatGPT to ThreatGPT: Impact of generative AI in cybersecurity and privacy. *IEEE Access.* https://ieeexplore.ieee.org/abstract/document/10198233/

Harknett, R. J., & Smeets, M. (2022). Cyber campaigns and strategic outcomes. *Journal of Strategic Studies, 45*(4), 534–567. https://doi.org/10.1080/01402390.2020.1732354

Hernández, J. P. (2020, April). *The legality of espionage in international law* [The Treaty Examiner]. https://treatyexaminer.com/espionage-legality/

Hogeveen, B. (2022, February). *The UN norms of responsible state behaviour in cyberspace.* http://www.aspi.org.au/report/un-norms-responsible-state-behaviour-cyberspace

Ifeanyi-Ajufo, N. (2017). Going beyond borders: Jurisdiction in cyberspace. *GIMPA Law Review, 3*, 133–156.

International Law Association. (2000). *Final report of the committee, statement of principles applicable to the formation of general customary law.* International Law Association. https://www.ila-hq.org/en_GB/committees/formation-of-customary-general-international-law

Jensen, E. T. (2016). The Tallinn Manual 2.0: Highlights and insights. *Georgetown Journal of International Law, 48*, 735.

Jupillat, N. (2016). From the Cuckoo's Egg to global surveillance: Cyber espionage that becomes prohibited intervention. *North Carolina Journal of International Law, 42*, 933.

Kose, J. (2021). Cyber warfare: An era of nation-state actors and global corporate espionage. *ISSA Journal, 19*(4), 12–15.

Lehto, M. (2022). Cyber-attacks against critical infrastructure. In M. Lehto & P. Neittaanmäki (Eds.), *Cyber security* (Vol. 56, pp. 3–42). Springer International Publishing. https://doi.org/10.1007/978-3-030-91293-2_1

Liebetrau, T. (2023). Organizing cyber capability across military and intelligence entities: Collaboration, separation, or centralization. *Policy Design and Practice, 6*(2), 131–145. https://doi.org/10.1080/257412 92.2022.2127551

Lilli, E. (2021). Redefining deterrence in cyberspace: Private sector contribution to national strategies of cyber deterrence. *Contemporary Security Policy, 42*(2), 163–188. https://doi.org/10.1080/13523260.2021. 1882812

Lindsay, J. R. (2017). Cyber espionage. *The Oxford Handbook of Cyber Security.* https://www.google.com/ books?hl=tr&lr=&id=jIlNEAAAQBAJ&oi=fnd&pg=PA223&dq=Lindsay,+J.+R.+(2017).+Cyber+espi onage&ots=lIpV8WYKrH&sig=QMlSSoXdKdvY1zZgcFB85DKFyWE

Mandiant. (2013). *APT1 exposing one of China's cyber espionage units.* Mandiant. https://www.mandiant. com/sites/default/files/2021-09/mandiant-apt1-report.pdf

Melnitzky, A. (2011). Defending America against Chinese cyber espionage through the use of active defenses. *Cardozo Journal of International and Comparative Law, 20*, 537.

Moisset, S. (2023, May 24). *How Security Analysts Can Use AI in Cybersecurity.* freeCodeCamp.Org. https:// www.freecodecamp.org/news/how-to-use-artificial-intelligence-in-cybersecurity/

Montesinos López, O. A., Montesinos López, A., & Crossa, J. (2022). Fundamentals of artificial neural networks and deep learning. In O. A. Montesinos López, A. Montesinos López, & J. Crossa (Eds.), *Multivariate statistical machine learning methods for genomic prediction* (pp. 379–425). Springer International Publishing. https://doi.org/10.1007/978-3-030-89010-0_10

Moynihan, H. (2021). The vital role of international law in the framework for responsible state behaviour in cyberspace. *Journal of Cyber Policy, 6*(3), 394–410. https://doi.org/10.1080/23738871.2020.1832550

O'Brien, D. (2017, July 27). A short history of cyber espionage. *Threat Intel.* https://medium.com/threat-intel/ cyber-espionage-spying-409416c794ec

Omand, D. (2021). The ethical limits we should place on intelligence gathering as part of an integrated CT strategy. *Terrorism and Political Violence, 33*(2), 290–301. https://doi.org/10.1080/09546553.2021.1880225

Oorsprong, F., Ducheine, P., & Pijpers, P. (2023). Cyber-attacks and the right of self-defense: A case study of the Netherlands. *Policy Design and Practice, 6*(2), 217–239. https://doi.org/10.1080/25741292.2023.21 79955

Patil, S. G., Zhang, T., Wang, X., & Gonzalez, J. E. (2023). *Gorilla: Large language model connected with massive APIs* (arXiv:2305.15334). arXiv. http://arxiv.org/abs/2305.15334

Petrosyan, M. (2024). The role of non-state actors in modern warfare: The case of Syria and Nagorno-Karabakh. *Journal of Balkan and Near Eastern Studies, 26*(2), 149–163. https://doi.org/10.1080/19448 953.2023.2233364

Poireault, K. (2024, February 7). *Governments and tech giants unite against commercial spyware.* Infosecurity Magazine. https://www.infosecurity-magazine.com/news/governments-tech-giants-against/

Pun, D. (2017). Rethinking espionage in the modern era. *Chicago Journal of International Law, 18*, 353.

Radsan, A. (2007). The unresolved equation of espionage and international law. *Michigan Journal of International Law, 28*(3), 595–623.

Rauta, V. (2020). Towards a typology of non-state actors in 'hybrid warfare': Proxy, auxiliary, surrogate and affiliated forces. *Cambridge Review of International Affairs, 33*(6), 868–887. https://doi.org/10.1080/0 9557571.2019.1656600

Rawat, R., Mahor, V., Chirgaiya, S., & Garg, B. (2021). Artificial cyber espionage based protection of technological enabled automated cities infrastructure by dark web cyber offender. In F. Al-Turjman, A. Nayyar, A. Devi, & P. K. Shukla (Eds.), *Intelligence of things: AI-IoT based critical-applications and innovations* (pp. 167–188). Springer International Publishing. https://doi.org/10.1007/978-3-030-82800-4_7

Rees, J., & Rees, C. J. (2023). Cyber-security and the changing landscape of critical national infrastructure: State and non-state cyber-attacks on organisations, systems and services. In R. Montasari (Ed.), *Applications for artificial intelligence and digital forensics in national security* (pp. 67–89). Springer Nature Switzerland. https://doi.org/10.1007/978-3-031-40118-3_5

Report of the Group of Governmental Experts. (2013, June). *2013 UN GGE Report of the group of governmental experts on developments in the field of information and telecommunications in the context of international security (A/68/98) | Digital Watch Observatory.* https://dig.watch/resource/ un-gge-report-2013-a6898

Rowe, B. I. (2020). Transnational state-sponsored cyber economic espionage: A legal quagmire. *Security Journal, 33*(1), 63–82.

Rubenstein, D. (2014). *Nation-state cyber espionage and its impacts.* https://www.cse.wustl.edu/~jain/cse571-14/ftp/cyber_espionage.pdf

Sarker, I. H., Furhad, M. H., & Nowrozy, R. (2021). Ai-driven cybersecurity: An overview, security intelligence modeling and research directions. *SN Computer Science, 2*(3), 173.

Satria Unggul, W. P., & Noviandy, P. E. (2019). Analysist of cyber espionage in international law and Indonesian law. *Humanities & Social Sciences Reviews, 7*(3), 38–44.

Schmitt, M. N. (2017). *Tallinn Manual 2.0 on the international law applicable to cyber operations* (2nd ed.). Cambridge University Press. https://doi.org/10.1017/9781316822524

Scott, R. D. (1999). Territorially intrusive intelligence collection and international law. *Air Force Law Review, 46.* https://heinonline.org/HOL/Page?handle=hein.journals/airfor46&id=223&div=&collection=

Stevens, T. (2020). Knowledge in the grey zone: AI and cybersecurity. *Digital War, 1*(1–3), 164–170. https://doi.org/10.1057/s42984-020-00007-w

Swiss Federal Intelligence Service (FIS). (2023). *Switzerland's security 2023 situation report of the Swiss Federal Intelligence Service (FIS).* Swiss Federal Intelligence Service (FIS). https://www.sinoptic.ch/textes/defense/2023/20230626_SRC_La.securite.de.la.Suisse-en.pdf

Taddeo, M. (2021). On the risks of trusting artificial intelligence: The case of cybersecurity. In J. Cowls & J. Morley (Eds.), *The 2020 yearbook of the digital ethics lab* (pp. 97–108). Springer International Publishing. https://doi.org/10.1007/978-3-030-80083-3_10

Taddeo, M., McCutcheon, T., & Floridi, L. (2019). Trusting artificial intelligence in cybersecurity is a double-edged sword. *Nature Machine Intelligence, 1*(12), 557–560.

The Council of Europe. (2021). *The council of Europe convention on cybercrime (Budapest convention).* The Council of Europe. https://www.coe.int/en/web/cybercrime/the-budapest-convention

Touvron, H., Lavril, T., Izacard, G., Martinet, X., Lachaux, M.-A., Lacroix, T., Rozière, B., Goyal, N., Hambro, E., Azhar, F., Rodriguez, A., Joulin, A., Grave, E., & Lample, G. (2023). *LLaMA: Open and efficient foundation language models* (arXiv:2302.13971). arXiv. http://arxiv.org/abs/2302.13971

Uesato, J., O'donoghue, B., Kohli, P., & Oord, A. (2018). Adversarial risk and the dangers of evaluating against weak attacks. *International Conference on Machine Learning,* 5025–5034. http://proceedings.mlr.press/v80/uesato18a.html

United Kingdom Home Office. (2021, January). *Covert human intelligence sources – Draft revised code of practice.* https://assets.publishing.service.gov.uk/government/uploads/system/uploads/attachment_data/file/949806/Covert_Human_Intelligence_Sources_Draft_Revised_Code_of_Practice_FINAL.pdf

United Nations. (1945). *UN Charter.* United Nations; United Nations. https://www.un.org/en/about-us/un-charter

United Nations. (2021). *Group of governmental experts on advancing responsible state behaviour in cyberspace in the context of international security.* https://digitallibrary.un.org/record/3934214

United Nations Human Rights Committee. (n.d.). *Ad Hoc committee.* United Nations : Office on Drugs and Crime. Retrieved March 17, 2024, from//www.unodc.org/unodc/en/cybercrime/ad_hoc_committee/home

Vernon, D., Metta, G., & Sandini, G. (2017). A survey of artificial cognitive systems: Implications for the autonomous development of mental capabilities in computational agents. *IEEE Transactions on Evolutionary Computation, 11*(2), 151–180. https://doi.org/10.1109/TEVC.2006.890274

Watt, E. (2019). Russell Buchan, cyber espionage and international law. *Journal of Conflict and Security Law, 24*(3), 638–642. https://doi.org/10.1093/jcsl/krz011

Weissbrodt, D. (2013). Cyber-conflict, cyber-crime, and cyber-espionage. *Minnesota Journal of International Law, 22,* 347.

Willett, M. (2019). Assessing cyber power. *Survival, 61*(1), 85–90. https://doi.org/10.1080/00396338.2019.1569895

Wray, C. (2020, February 6). *Responding effectively to the Chinese economic espionage threat.* Federal Bureau of Investigation. https://www.fbi.gov/news/speeches/responding-effectively-to-the-chinese-economic-espionage-threat

Wright, Q. (1962). Espionage and the doctrine of non-intervention in internal affairs. In Q. Wright, J. Stone, R. Falk, & R. Stanger (Eds.), *Essays on espionage and international law* (Vol. 3, p. 12). Ohio State University Press.

7 The Fuel of AI in the Battlefield

Data Management and Data Governance

Cansu Ulu

7.1 INTRODUCTION

The usage of artificial intelligence (AI) in military operations has propelled us into an era of unprecedented strategic capabilities. At the heart of this revolutionary shift lies the crucial role of data. Data is the resource that fuels AI algorithms. In AI, machine learning, and deep learning models, they utilize datasets to train themselves. AI presents consistent results with adequately designed, secure, and purpose-fit datasets. Therefore, data is the most fundamental resource required for the development of AI. The AI algorithms make possible real-time intelligence channels for advanced decision-making on the battlefield. This chapter embarks on a comprehensive exploration of the intricate dynamics within military contexts, with a particular focus on the critical interplay between AI, data security, and data governance.

The integrity and security of data form the bedrock of reliable and effective AI. Secure and accurate data is vital for robust AI models and decision-making processes. To unravel the nuances of this relationship, we delve into the Data Information Knowledge Wisdom (DIKW) pyramid as a conceptual framework that delineates the transformative journey from raw data to actionable insights (Peters et al., 2024). This pyramid could help to understand the role of raw data, data management, and governance in accordance with AI's data-driven military operations.

Within the contemporary landscape, the role of data mining for surveillance and intelligence assumes a compelling significance. The democratization of data access through open-source platforms has empowered individuals and organizations, sometimes challenging the influence of traditional state actors (Lane, 2020). In the realm of modern warfare, the interplay between data privacy, sovereignty, and security emerges as an ever-present consideration that significantly shapes military strategies and operations. The digital age has ushered in an unprecedented interconnectedness, allowing data to traverse borders and jurisdictions with ease. However, this interconnectedness raises profound questions about the autonomy and sovereignty of nations over their data assets (Ayers, 2016). The military's reliance on AI-driven systems necessitates a delicate balance between the imperative to secure sensitive information and the necessity of collaborative data sharing for intelligence gathering and coordinated responses. The military applications that rely on AI algorithms create data governance problems. AI algorithms need safe and secure data sets to train (Marwala, 2024).

The primary aim of this chapter is to conduct a detailed case study, specifically exploring how the United States strategically employs data as the fundamental driver behind its AI-driven military operations. The investigation encompasses the entire spectrum from data definitions within the military context to the practical application of AI, data utilization, and data-driven decision-making, culminating in a thorough analysis of national security implications and ethical considerations associated with this paradigm shift.

The unfolding narrative of this chapter begins with a structured exploration of the importance of data for AI capabilities. The first part questions the link between the DKIW pyramid and military decision-making, drawing insights from strategic reports of the United States and diverse definitions within the existing literature.

Data is not only a strategic military asset. It is a daily life product that we create and consume. The second question that needs to be answered is how we comprehend data, is it a defense sector asset or a civil technological element? The synergy between information and communication technology (ICT) and military operations has resulted in a convergence that transcends traditional boundaries. A pivotal aspect of this convergence is the transference of state-of-the-art data science practices from the private sector to military domains. This transition not only accelerates AI advancements but also introduces a host of challenges, including ethical dilemmas and concerns related to data security. Shifting the focus to a broader perspective, the study navigates through the data-driven decision-making process, delving into general national security problems. Key considerations include the proliferation of technology, the dual-use dilemma, cost implications, and the increasing reliance on intelligence-based approaches in modern military strategies.

The subsequent section elucidates how data and AI are actively employed within military operations. It further probes into the offensive characteristics of data-driven warfare, unveiling new operational tactics, exploring the dynamics of crowdsourcing data, emphasizing the pivotal role of Intelligence, Surveillance, and Reconnaissance (ISR), asymmetric warfare, and comprehending the overarching data ecosystem.

In conclusion, the chapter synthesizes the findings, offering insights into how states plan to strategically leverage data. Additionally, it addresses the ethical concerns surrounding the integration of AI and data into military operations, providing a comprehensive overview of the evolving landscape and the ethical considerations that underscore the adoption of advanced technologies in the realm of defense.

7.2   THE IMPORTANCE OF THE DATA

In the contemporary milieu of information centrality, a new essence has evolved, embodying facets as a by-product, product, and fount of wisdom within the fabric of our interconnected lives. As our daily lives become more intertwined with the realms of cyberspace and ICT, our interaction with this essence, commonly known as "data," is concurrently heightened.

Data constitutes basic and raw facts, presenting itself in either structured or unstructured formats and occupying the foundational tier of the DIKW pyramid. To glean insights for the decision-making process, one must ascend the pyramid by actively engaging with and transforming raw data into higher forms. Data can be gathered by various methods such as data mining, text mining, web mining, surveys, and observation (Jifa & Lingling, 2014, p. 815).

In 1987, Zeleny introduced the DIKW pyramid to elucidate the rapid transition from conventional electronic data processing to AI. Zeleny conceptualized data as raw constituents, employing a metaphorical analogy as the process of baking bread. Analogous to the fundamental elements in baking, such as starch atoms, H_2O molecules, and yeast bacteria, data represents the basic components devoid of any discernible trace of the ultimate product, "bread." This analogy underscores the notion that, similar to the preliminary stages of bread-making, data encapsulates elemental information and facts, offering no immediate insight into the ultimate outcome (Zeleny, 1987). Ackoff, in alignment with Zeleny's conceptualization, defines data as symbols encapsulating the properties of objects, events, and their respective environments. These symbols are outcomes of the observational process, where observation is synonymous with sensing. Analogous to Zeleny's metaphor, Ackoff likens data to metallic ores, asserting that their inherent value materializes only when subjected to processing, transforming them into a usable form (Ackoff, 1989). In essence, both Zeleny and Ackoff underscore the transformative journey of raw data into a meaningful and actionable state.

In exploring different philosophical perspectives on the nature of data, Zins and Wittgenstein offer distinct viewpoints. Zins characterizes "data" as "sense stimuli or their meanings," emphasizing empirical perceptions. For instance, the recognition of a running car engine and the associated sounds constitute data in this context (Zins, 2007). In contrast, Wittgenstein distinguishes between sense-data and reality. Wittgenstein argues that sense-data, lack relevance and practical utility in the learning process (Wittgenstein, 1968). While Zins underscores the significance of data encompassing sensory stimuli and meanings, Wittgenstein highlights the primacy of real-world experiences over isolated sense-data in shaping meaningful understanding and knowledge.

To get higher of the DIKW pyramid the raw nature of data should be processed to maintain information. According to Stenmark, data comprises raw symbols and facts obtained through observation (Stenmark, 2002). Data can be fundamentally categorized into two types: Raw and processed data. Raw data consists of symbols that lack ownership and do not yield meaningful results. Processed data, on the other hand, possesses a specific ownership right (Burgin, 2017, p. 756). The interpretation of data occurs through the classification of data for use in specific datasets and the identification of relationships with other datasets (Nake, 2001).

Information is the second step of the DIKW pyramid. Information is data that has undergone processing to become meaningful for the recipient (Olson & Davis, 1985). Data has no judgment or interpretation; on the other hand, information has the ability to describe particular situations and conditions (Wiig, 1993). Information helps us to create patterns about an event or object. Returning to Zeleny's analogy of bread, information can be likened to the ingredients such as flour, sugar, water, and spices. Yet, at this stage, the intended outcome remains elusive, emphasizing the need for further processing (Zeleny, 1987).

Advancing toward AI requires additional steps. Information serves as a basic source of knowledge, its significance aligning with technological advancements. As Cleveland posits, information stands out as a foundational phenomenon in the post-industrial era. He describes the fundamental characteristics of information as;

"Information is expandable through the world, it is independence from material resources, information can replace other capabilities such as capital, labor, and physical materials, information is transportable at close to the speed of light, and most tricky ability of information is the tendency of leak" (Cleveland, 1985, pp. 186–187).

Information is diffusive and shareable. "In military operations, information acquired from processing data gives specific insights, reducing reaction time and offering capabilities such as intelligence and simultaneous remote control" (Ulu, 2023, p. 55).

The third step of the DIKW pyramid is knowledge. In Meadow and Yuan's conceptualization, data typically refers to a collection of symbols that lack inherent meaning for the recipient. Information, on the other hand, comprises symbols that possess significance or meaning for the recipient. Knowledge, according to them, is the accumulation and integrated body of information received and processed by a recipient (Meadow & Yuan, 1997, p. 701). As we turn back to Zeleny's bread analogy knowledge is to know why you should make bread (Zeleny, 1987). One can make beer or something else besides bread with some ingredients. Knowledge requires understanding relationships, such as recipes and their contextual interpretations. This study is centered on the application of the relationship between data, information, and knowledge in the military domain. Technology has heightened the significance of data in military systems. Zeleny has introduced the concept of "teknowledgy" to articulate the connection between evolving technologies and information. Zeleny presumes that technology leads the world to create more "integration-friendly" and knowledge-supportive systems (Zeleny, 1987, p. 60).

The pinnacle of the DIKW pyramid is formed by wisdom, representing the culmination of a transformative journey from raw data to meaningful insight. Wisdom allows comparison and judgments. Beyond wisdom, we reach the enlightenment step (Zeleny, 1987, p. 60). "Wisdom is the highest level of abstraction, with vision, foresight, and the ability to see beyond the horizon" (Awad & Ghaziri, 2007, p. 64). Wisdom is to know how and why to use which data. Wisdom is not only a tool to reach a specific goal. It combines knowledge and ethics to reach the action (Jifa & Lingling, 2014, p. 815). Wisdom allows the usage of data-driven decision-making processes on the battlefield. Alongside technology, societies have become more information-centric. However, as data forms the foundation of this entire pyramid, it continues to offer new windows for humanity. Ronquillo suggests that "all knowledge and wisdom ultimately stem from objective, value-free, and 'pure' data" (Ronquillo et al., 2016, p. 9).

Cyberspace and the ocean of data give immortals new possibilities to see what is hidden behind our comprehension. The exponential growth in the volume of data each year is reshaping the landscape of knowledge. This geometric progression in data generation underscores the dynamic nature of our digital environment. In the ever-evolving landscape of modern warfare, the strategic deployment of data has become paramount. As the volume of information continues to surge exponentially, harnessing the power of big data emerges as a critical imperative (Aravind, 2024), revolutionizing military operations and enhancing the precision and agility of decision-making processes in the heat of conflict.

The volume of data used and created is increasing every year. IDC forecasts the Global Dataspheregrow from 45ZB in 2019 to 175ZB by 2025 (Reinsel et al., 2020). According to the IDC 2012 study, only 23% of digitally generated data was usable due to a lack of processing data technologies (IDC Digital Universe Study: Big Data, Bigger Digital Shadows and Biggest Growth in the Far East, 2012). The need for rapid and secure processing, storage, and transmission of the increasing data has led to the development of new technologies such as Hadoop and cloud services. IDC predicts that by 2025, 46% of stored data will be held in public cloud systems (Reinsel et al., 2020). This statistic shows us the importance of cloud systems in deriving meaningful wisdom from big data.

Big data is the source of AI-driven decision-making algorithms. The importance of secure and accurate data in fueling AI cannot be overstated. Accurate data is the lifeblood of AI algorithms (Musekiwa, 2023), shaping their ability to make informed decisions and predictions. Secure data ensures the integrity and reliability of AI outputs, guarding against biases, errors, and potential misuse (Trent, 2023). The importance and the meaning of the data for the DIKW pyramid are discussed above. There is another debate ongoing about the economic value of data. Is it a product, a resource, or a money transfer item like gold or money?

Embracing the concept of 'data as a product' involves applying product thinking to datasets, treating them as valuable assets with capabilities such as discoverability, security, explorability, and trustworthiness. A practical example involves ensuring datasets are discoverable through tools like a custom-built data catalog, addressable for increased team productivity, self-describing and interoperable with standardized metadata, and trustworthy and secure through regular quality checks and controlled access (Rigol, 2021). This definition leads to the question of why we see data as a product and what it is used for. Data can have a market price and value. It can be used for research and development, market analysis, recommender systems, data monetization, strategic decision-making, personalization, risk management, and operational efficiency. On the other hand, data products are another topic that needs to be mentioned. A data product revolves around data, delivering insights, analysis, or value to users (Dabols, 2023). It spans the entire data lifecycle, involving processes such as collection, processing, analysis, and presentation to provide actionable information. In contrast, when we talk about data as a product, the focus is on recognizing and treating data itself as a valuable asset (Dabols, 2023).

Data can be considered as a resource. Medhora defines data as the new oil due to its role as a source of information and autonomy (Medhora, 2018). According to Humby, data is as valuable as crude oil, but it cannot be effectively utilized unless refined (Palmer, 2006). Data must undergo processing and analysis to carry value. Despite sharing economic and strategic motivations with oil, data exhibits distinct characteristics. Nolin suggests using the metaphor "data is like oil" instead of "data is the new oil" (Nolin, 2019). Notably, data differs from oil in its nearly unlimited production capacity and the continuous benefit derived from processing and analyzing it, unlike the diminishing returns in oil refining. The extraction and transmission of oil are more costly compared to data, which is likened to solar energy for its abundance and potential for repeated use across various applications. IBM's CEO, Ginni Rometty, emphasizes that only 1% of the scattered data is captured and processed, highlighting the renewable nature of data, similar to solar energy (Kinney, 2019).

Data, in the digital era, can be likened to a form of currency, and the parallels between data and money are increasingly apparent. Much like traditional currencies, the value of data lies in its exchangeability and the trust placed in its authenticity. Cryptography plays a pivotal role in securing and validating data transactions.

7.3 DATA AND AI IN THE BATTLEFIELD

In the ever-evolving landscape of modern warfare, the significance of data and AI has surged to unprecedented heights. The digitalized era has ushered in a paradigm shift, where technological determinism not only influences the nature of conflict but also reshapes its fundamental characteristics. As nations harness the power of data-driven insights and AI capabilities, the battlefield transforms into a dynamic theater where information becomes a critical asset, dictating strategies, outcomes, and the very essence of warfare. This subsection delves into the pivotal role of data and AI in the contemporary battlefield, exploring how these advancements redefine the parameters of conflict and shape the future of military engagements.

Mary Kaldor's concept of "new war" challenges conventional notions by highlighting key shifts in contemporary conflicts. In these conflicts, non-state actors, identity politics, and globalized networks play pivotal roles, transcending traditional state-centric views. Civilians become deliberate targets, and the privatization of military functions, coupled with the exploitation of resources in war economies, introduces new complexities (Kaldor, 2012). Kaldor's concept of new wars is closely intertwined with ICT. The increasing presence of data as an accessible resource in our world, coupled with the rapid advancement of data processing technologies, has led to the democratization of information. The diffusion of information has necessitated that states be prepared to confront not only traditional state actors but also non-state actors. As information becomes a powerful tool in modern conflicts, states must adapt to the evolving landscape, recognizing the challenges posed by a diverse range of actors operating within the realm of ICT. This paradigm shift underscores the importance of understanding and navigating the complexities introduced by technological advancements in the context of contemporary warfare.

In modern warfare, the relationship between data, ISR, and decision-making on the battlefield is crucial. ISR data, gathered through various sensors and platforms, provides real-time insights into the operational environment, enemy activities, and potential threats. This timely and comprehensive information empowers military commanders to make informed decisions, enhancing situational awareness and strategic planning. The integration of data-driven ISR enables a more agile and adaptive decision-making process, optimizing resource allocation and mission execution for effective outcomes on the battlefield.

Enhanced situational awareness on the battlefield facilitates efficient resource utilization, more accurate projections for future scenarios, expedited decision-making processes, and the execution of operations that capitalize on the element of surprise against unprepared adversaries. The conventional model of state-on-state warfare has become obsolete, necessitating preparedness for unconventional, guerrilla-style warfare perpetrated by non-state actors with cellular structures and initiative-taking capabilities. Meanwhile, global powers, including the United States, China, and Russia, are actively augmenting their warfighting capabilities through ICT. These nations strategically leverage ICT to amplify their existing military capabilities, recognizing it as a force multiplier in modern conflict scenarios. At this point, asymmetric power takes center stage in warfare strategies.

Metz and Johnson explain asymmetric strategy through two elements. Firstly, it involves one party possessing a new element or strategy in warfare that the other is unaware of or does not possess. This allows the side employing these features to surpass the numerical superiority of the other. The second element is a party using its capabilities in a manner that is surprising and unexpected to the other (Metz & Johnson, 2011). Within the scope of asymmetric strategy, the use of big data to attain maneuverability is enticing for involved parties. The side that commands ICT and translates a greater volume of data into wisdom, incorporating it into the decision-making mechanism, can apply asymmetric strategies as a force multiplier on the battlefield.

The evolution of the United States Department of Defense's (DoD) perspective on data unfolds as a dynamic narrative, marked by a series of strategic shifts and transformative visions. In the 2011 International Strategy for Cyberspace, the emphasis was on the assurance of the free flow of information, security, privacy, and network integrity underscoring the foundational role of data in ensuring global economic prosperity and the promotion of universal rights (International Strategy for Cyberspace Prosperity, Security, and Openness in a Networked World, 2011).

Moving forward to the 2018 National Defense Strategy, a significant leap occurs, recognizing the strategic importance of advanced computing, big data analytics, AI, and other cutting-edge technologies (*Summary of the 2018 Department of Defense Artificial Intelligence Strategy Harnessing AI to Advance Our Security and Prosperity*, 2018). Here, data is acknowledged as a linchpin, crucial for ensuring the ability to fight and win future wars. Simultaneously, the 2018 DoD Cloud Strategy underscores the differentiation of data and the ability to process it swiftly as mission-critical differentiators for overall success (DoD Cloud Strategy, 2018).

In 2019, the United States Air Force Artificial Intelligence Annex to The Department of Defense Artificial Intelligence Strategy reinforced the recognition of data as a strategic asset, particularly vital for executing missions across Air, Space, and Cyberspace. The nexus between data and AI is explicitly highlighted, emphasizing the indispensable role of diverse and representative data in fueling AI development and influencing operational outcomes in near real time (The United States Air Force Artificial Intelligence Annex to The Department of Defense Artificial Intelligence Strategy, 2019).

The pivotal year of 2020 witnessed a transformative shift in the DoD's perception of data. The DoD Data Strategy elevated data from a strategic asset to a central, operational imperative. The vision of becoming a "data-centric organization" heralded a new era where data would be harnessed at speed and scale, not just for strategic advantage but also for increased efficiency, collaboration with allies, and gain strategic advantage over adversaries (DoD Data Strategy, 2020).

Building upon this foundation, the 2023 DoD Data Analytics and AI Adoption Strategy focuses on improving foundational data management (Department of Defense Data, Analytics, and Artificial Intelligence Adoption Strategy, 2023). Here, the concept of data as a product takes center stage, diverging from the 2020 definition that positioned data primarily as a strategic asset. The vision pivots toward data-driven decision-making, with an emphasis on adapting open standard architectures while adhering to cyber security policies and industry best practices, recognizing data as a dynamic enterprise resource.

The journey through these strategic reports culminates in the 2022 Data Management Lexicon, which defines data as a representation of facts, concepts, or instructions. This definition encapsulates the multifaceted nature of data, emphasizing its role as a communicable, interpretable, and processable entity, marking a departure from earlier views that predominantly focused on the strategic and operational aspects of data within the military landscape (Data Management Lexicon, 2022). The lexicon broadens the perspective, recognizing data as a diverse array of forms that individually lack meaning but collectively form the foundation for communication and interpretation.

To effectively harness data on the battlefield, specific steps must be successfully executed. The first of these steps is data collection. Integrating data from various domains, including multiple sensors, Open-Source Intelligence (OSINT), Human Intelligence (HUMINT), Geospatial Intelligence (GEOINT), and Signals Intelligence (SIGINT), provides a comprehensive perspective on the battlefield. As highlighted in the 2022 Data Management Lexicon, the term "collection" encapsulates any information, in its varied forms, that is potentially relevant to the mission needs of an Intelligence Community (IC) Element. This definition emphasizes the inclusivity of data, whether in its initial, raw state or its final processed form, obtained directly from sources. The report on Artificial Intelligence: Background, Selected Issues, and Policy Considerations underscores the DoD's reliance on processing and disseminating information for military operations and intelligence collection. A strategic imperative emerges for the DoD to address unique mission requirements through a multi-cloud, multi-vendor strategy, incorporating both General Purpose and Fit-for-Purpose clouds. This approach ensures the agility and adaptability required for effective data processing and dissemination in diverse operational contexts.

Supporting this notion, the DoD Data Strategy underscores the importance of continuous data collection for decision makers. This approach necessitates enabling electronic data collection at the point of creation, emphasizing the need to maintaining the pedigree of data at all times (DoD Data Strategy, 2020). The DoD's commitment to this approach recognizes that data, regardless of its origin or purpose, serves as a foundational asset that demands continuous attention and enhancement to best inform decision makers.

Further emphasizing the significance of automated efficiency, the Data, Analytics, and Artificial Intelligence Adoption Strategy underscores the importance of authoritative data sources and the maximization of automated data collection methods (Department of Defense Data, Analytics, and Artificial Intelligence Adoption Strategy, 2023). This aligns with the broader strategy of leveraging technology to streamline performance monitoring, reinforcing the idea that the efficiency of data collection directly contributes to the effectiveness of decision-making processes on the battlefield.

The second step to execute data on the battlefield is to analyze it to get real-time projections. In 2004, William McRaven was assigned to the command of Task Force 714 and identified a critical challenge within the operational tempo of US forces against Al-Qaeda. He observed that the forces struggled to adapt to Al-Qaeda's operational pace and were unable to discern specific patterns, hindering their ability to make future projections. McRaven noted that despite the vast volume of data obtained from UAV sensors, personnel monitoring the screens could not effectively analyze it. Consequently, rooms full of data awaiting analysis were left unutilized (Shultz & Clarke, 2020). McRaven recognized the potential of using AI in military applications to enhance decision-making superiority in operations, driven by the speed and information advantages it offers. The utilization of AI allows for the rapid integration of meaningful results from the same data source that conventional tools may not yield. Furthermore, it facilitates the swift utilization of significant insights during operations, contributing to decision-making efficiency and providing a competitive edge by simplifying the identification of adversaries' sensitivities.

The challenge of processing and cleaning extensive datasets has been effectively addressed through the implementation of AI algorithms. Artificial Intelligence and National Security Report stressed that "AI is expected to be particularly useful in intelligence due to the large data sets available for analysis" (Sayler, 2020, p. 10). One of the noteworthy initiatives in this context is the Project Maven, colloquially known as the Algorithmic Warfare Cross-Functional Team project. In the memorandum introducing the project in 2017, the DoD underlines the necessity of incorporating AI and machine learning into its operations to maintain superiority against the increasing capabilities of adversaries (DoD Data Strategy, 2020). The Maven Project specifically aims to extract meaningful insights from large datasets and deploy these insights effectively in defense operations, with the overarching goal of enhancing defense capabilities.

The third step to use data effectively on the battlefield is to share and store it to perform it in Joint Operations. The vast amount of data collected from external resources, DoD sensors, and platforms are analyzed by algorithms to support decision makers. This information is collected from all domains, and it is useful unless used by interconnected multi-domains. The key concept in this step is interoperability. Interoperability is defined as "the ability to operate in synergy in the execution of assigned tasks" (Department of Defense Dictionary of Military and Associated Terms, 2016, p. 118). The importance of interoperability lies in its ability to enhance coordination, situational awareness, and overall operational efficiency, thereby maximizing the combined capabilities of diverse elements in joint military endeavors. "Interoperability is the foundation of effective joint, multinational, and interagency operations" (Shelton, 2000, p. 15).

The vision of the DoD is to be a data-centric organization. To achieve this goal data access should be ensured for all echelons (DoD Data Strategy, 2020). This necessitates the implementation of open data standards to ensure uniformity and compatibility in data representation and transmission. This vision also requires the establishment of robust mechanisms for secure and seamless access to data across interconnected networks, incorporating rigorous testing procedures to fortify the resilience and reliability of the overall data infrastructure.

Two illustrative instances of the efficacious utilization of data on the battlefield can be exemplified through the application of the Observe, Orient, Decide, Act (OODA) loop and the Find, Fix, Finish, Exploit, Analyze, and Disseminate (F3EAD) concept. These operational frameworks underscore the systematic integration of data-driven processes, emphasizing the sequential stages of information acquisition, interpretation, decision-making, and subsequent actions. The OODA loop, developed by Colonel John Boyd, delineates a continuous cycle wherein commanders rapidly iterate through the phases of observation, orientation, decision, and action, leveraging real-time data to maintain a dynamic and adaptive operational stance. Similarly, the F3EAD concept, widely employed in intelligence and military operations, outlines a comprehensive approach encompassing the identification, location, neutralization, exploitation, analysis, and dissemination of pertinent data to enhance overall situational understanding and operational effectiveness on the battlefield (The Targeting Process, 2010). These examples underscore the critical role of data in shaping operational methodologies and fostering agility, precision, and informed decision-making in dynamic military environments.

F3EAD is a targeting procedure. It is used especially for counterinsurgency operations. F3AED focuses on high-value individuals. The F3EAD framework employs concentrated and continuous reconnaissance or surveillance, guided by a robust and decentralized all-source intelligence system (Flynn et al., 2008). ISR serves as the primary source of data for F3EAD operations. Collecting data is essential for the processes of locating, identifying, and analyzing in the Find, Fix, and Analyze steps. Joint Information Operations Warfare Command supports the Joint Force Command by evaluating information operations effectiveness in military operations. Highlighting the critical role of intelligence analysis across all echelons is paramount for successful counterinsurgency operations (Insurgencies and Countering Insurgencies, 2014). This approach not only pushes analysts closer to the battlefield but also enables the seamless integration of collected data into the unit's operational plan, essential for effectively neutralizing targets in the counterinsurgency context (The Targeting Process, 2010).

According to Major General Michael Flynn "Information and intelligence is the fire and maneuver of the 21st century" (Flynn, 2011). Real-time battlefield analysis, human-machine teaming, and support decision-making are the best examples of AI usage on the battlefield (Rickli & Mantellassi, 2023). The foundational origins of the OODA loop can be traced back to the Korean War, where its conceptual framework began to take shape. While historically associated with maneuvering and air-to-air combat, Colonel Matthew Kelly emphasizes that the relevance of OODA extends beyond traditional warfare scenarios. In contemporary contexts, particularly in the era of information warfare, OODA takes on new significance. Kelly contends that fighter aircraft, instead of relying solely on outmaneuvering adversaries, can now exercise information dominance (Callahan, 2023). This evolution underscores the adaptability and enduring relevance of the OODA loop, as it transitions to address the complexities of modern conflict, emphasizing the critical role of information and decision-making in achieving strategic advantages.

Angerman has indicated the intersections between DIKW and the OODA loop under the headings of Information Engineering and Knowledge Management (Angerman, 2004). When combining these two concepts, it becomes necessary to clarify the connections between DIKW and the OODA loop. The "Observe" step involves collecting raw data, often derived from various sources on the battlefield (Senne, 2007). This initial phase is akin to gathering information that is yet to be processed or contextualized Moving to the "Orient" step, the emphasis shifts toward interpreting and synthesizing the collected information. This phase transforms raw data into knowledge by providing context and relevance. It involves the analysis of patterns, identification of relationships, and the extraction of meaningful insights, contributing to a more comprehensive understanding of the situation. The subsequent steps, "Decide" and "Act," progress toward higher order cognitive processes like wisdom. "Decide" involves synthesizing knowledge to form actionable insights, and "Act" translates these insights into strategic or tactical responses.

7.4 DATA MANAGEMENT AND DATA GOVERNANCE

Data serves as the fuel for AI, and AI algorithms provide nearly real-time decision-making support. Unmanned vehicles equipped with AI act as force multipliers for the air force. However, data need to be secure and accurate. Three potential threads are listed. Addressing these challenges requires robust data management and governance practices to ensure the responsible deployment of advanced technologies on the battlefield. The first thread is the dual-use nature of data technologies.

Dual-use technology represents a category of technology with the potential for both peaceful and military applications, serving as a pivotal asset for nations. AI stands out as an example of dual-use technology, given its capacity to significantly enhance the intelligence of various machines and systems. Especially lightweight, vertical take-off and landing capable, quadcopter unmanned aerial vehicles are being utilized in the field of agriculture. With the integration of AI, the development of crops is monitored, plant diseases are detected, and the appropriate amount of pesticide spray or the fertilization can be applied (Chin et al., 2023).

The increasing utilization of AI technologies in both civilian and military domains parallel the accelerated adoption of nuclear technologies during the Cold War era. The fundamental challenge lies in the cutting-edge status of ICT technologies being predominantly proprietary to the private sector. States aiming to gain superiority in AI endeavors must ultimately collaborate with the private sector to secure technological advantages, thereby forming crucial partnerships to outpace their adversaries (Army Technology Transfer Program 2022 Annual Report, 2023; *Department of Defense (DoD) Technology Transfer (T2)*, 2024).

The transfer of technology between the military and the private sector occurs bidirectionally, as observed from historical trends and contemporary reports (Sezal & Giumelli, 2022). In the period between the 1960s and 1980s, military technology permeated everyday life, with the development of the Advanced Research Projects Agency Network (ARPANET) serving as a prime example of the military's technological influence. During the Cold War, investments in Silicon Valley, driven by sophisticated military products, contributed to the development of civilian technology, establishing a symbiotic relationship where the military also integrated new technologies. In the present era, the dynamics of the relationship between the military and the private sector have evolved (Heinrich, 2002). Civilian technologies are currently integrated into military institutions without the necessity of economic incentives or sole dependence on defense industry entities. This involves the transfer of technology from the private sector to the military domain. Cloud technology represents a spectacular example of technology transfer to the military domain. Cloud technology, originally developed to provide scalable and efficient storage solutions for businesses and individuals, has found extensive applications in the military domain. Military forces now utilize cloud technology for secure data storage, processing, and sharing across different branches and units.

One of the popular examples of private sector military relations is the Joint Enterprise Defense Infrastructure (JEDI) project, which, though initially involving Google. JEDI is canceled because of the 'don't be evil' motto of Google. Top engineers protested the project. The problem is to sustain information for Lethal Autonomous Weapon Systems (LAWS) (Wakabayashi & Shane, 2018). The project was renewed as Joint Warfighting Cloud Capability (JWCC) contacted to multiple stakeholders such as Oracle, Amazon Web Services, and Microsoft (*Department of Defense Announces Joint Warfighting Cloud Capability Procurement*, 2022).

The National Military Strategy for Cyber Operations emphasizes deepening and expanding collaboration with the private sector to leverage its skilled workforce, capabilities, and products

(Department of Defense The National Military Strategy for Cyber Operations, 2006). Furthermore, recent DoD strategies, such as the 2020 DoD Data Strategy and the 2023 DoD Data Analytics and AI Adoption Strategy, underscore the importance of a network that includes the private sector, academic partners, and global alliances, forming a global ecosystem (Department of Defense Data, Analytics, and Artificial Intelligence Adoption Strategy, 2023; DoD Data Strategy, 2020). These reports emphasize that the Department cannot succeed alone, emphasizing the necessity of collaborative efforts with external entities for effective military operations, increased lethality, and optimal utilization of data.

The second thread is the diffusion of intelligence. The diffusion of intelligence, closely intertwined with the nature of ICT, represents a dynamic process that has transformed the landscape of information sharing and decision-making. This diffusion encompasses not only traditional military and governmental domains but also extends to civilian, commercial, and academic spheres. Information is the key power enabler for the modern battlefield. Nye suggests that the future of power has two characteristics; the diffusion of power and the rise of the rest (Nye, 2011). It cannot be revised because of the democratic nature of data and information, and globalization. "The wider the spread of knowledge, the more power gets diffused" (Cleveland, 1985, p. 188).

The third crucial aspect pertains to fueling AI with accurate and secure data. Presently operating within the realm of narrow AI, the capabilities of AI systems are inherently constrained by the datasets on which they are trained. Recognizing this limitation, the DoD underscores the significance of clear data standards and interoperability in its data analytics and AI adoption strategy (Department of Defense Data, Analytics, and Artificial Intelligence Adoption Strategy, 2023). The strategy emphasizes that establishing such standards directly contributes to military readiness.

Furthermore, the DoD recognizes the necessity of setting open data standards to enhance the accessibility and utility of AI technologies. In alignment with this vision, the DoD Data Strategy outlines VAULTIS goals for data usage. These goals include making data Visible, Accessible, Understandable, Linked, Trustworthy, Interoperable, and Secure. The objective of ensuring data is "trustworthy" underscores the need for confidence in all aspects of data to inform decision-making processes (DoD Data Strategy, 2020).

In tandem with the DoD's efforts, the Federal Data Strategy Development Team prioritizes the protection and security of data as one of its main objectives (Anton et al., 2019). This focus reflects a broader commitment to safeguarding the integrity and confidentiality of data, aligning with the overarching goal of fostering a secure and resilient data environment for the advancement of AI and analytics in both defense and broader government applications.

The challenges posed by the dual-use nature of data technologies, the diffusion of intelligence, and the necessity of fueling AI with accurate and secure data underscore the critical role of robust data management and governance practices. In the realm of dual-use technology, effective data governance becomes a linchpin for distinguishing and regulating the application of data in both civilian and military contexts.

Data governance is equally crucial in managing the diffusion of intelligence across diverse sectors. Data governance is "a discipline comprised of responsibilities, roles, functions, and practices supported by authorities, policies, and decisional process" (Data Management Lexicon, 2022, p. 6). As information becomes a potent force in the modern battlefield, establishing comprehensive data management frameworks ensures that the spread of knowledge aligns with strategic objectives and adheres to ethical considerations. By implementing data governance measures, nations can navigate the intricate dynamics of information diffusion while maintaining control over critical intelligence assets.

Moreover, addressing the imperative of fueling AI with accurate and secure data necessitates meticulous data management strategies. Operating within the confines of narrow AI, the capabilities of AI systems are intricately linked to the quality and reliability of the datasets they rely on. Data management is defined as the "development and execution of plans, policies, programs, and practices that acquire, control, protect, and enhance the value of data assets throughout the life-cycle" (Data Management Lexicon, 2022, p. 7). Data management practices that prioritize accuracy, security, and interoperability are essential for overcoming the limitations imposed by the current state of AI and ensuring its effective deployment on the battlefield. Improving data management will improve DoD's ability to fight and win wars. Data management dismisses the limitations of data-driven decision insights (DoD Data Strategy, 2020).

In this context, the DoD plays a pivotal role in spearheading data management and governance initiatives. The DoD's emphasis on clear data standards, interoperability, and open data standards, as outlined in its strategies, lays the foundation for responsible data usage. Aligning with the VAULTIS goals from the DoD Data Strategy, effective data management ensures data is visible, accessible, understandable, linked, trustworthy, interoperable, and secure.

7.5 CONCLUSION

In the ever-evolving landscape of modern warfare, a central and irreplaceable role emerges for data as the undeniable fuel that powers the engines of AI. The significance of data in shaping the dynamics of the contemporary battlefield cannot be overstated, as it forms the foundation upon which AI algorithms operate, providing the essential raw material for real-time decision-making support. The DIKW pyramid shows us the importance of vast amounts of data however data by itself does not present a value. Big data opens the gates of a new awareness for military leaders. As nations embrace the power of data-driven insights and AI capabilities, the relationship between data and military strategies becomes pivotal. The DoD recognizes this symbiotic connection, acknowledging that data is not merely a strategic asset but the very lifeblood of a "data-centric organization."

Efficiently harnessing data on the battlefield involves a multifaceted approach, with the first step being the comprehensive collection of data from diverse sources. This data, ranging from sensors to OSINT, serves as the raw material that feeds the hungry algorithms of AI. The emphasis on con-tinuous data collection for decision makers underscores the real-time nature of the insights derived from this invaluable resource. In the realm of joint military operations, where vast amounts of data are collected, shared, and stored for analysis, interoperability becomes paramount. The vision of the DoD as a data-centric organization necessitates secure and seamless access to data across inter-connected networks, underscoring the critical need for data standards that ensure uniformity and compatibility. This interoperability ensures that data, as the fuel for AI, can be effectively utilized by interconnected multi-domains.

AI fueled with data is a force multiplier. However, it has some challenges. These challenges are the dual use of data technology, the diffusion of information, and the usage of secure and accurate data. Comprehensive data government and data management practices are the solutions to these challenges.

In summary, robust data management and governance are imperative, serving as the corner-stone for a secure data environment. The DoD's focus on clear data standards, interoperability, and openness mirrors the objectives of effective data governance. Prioritizing accuracy, security, and interoperability not only strengthens military readiness but also facilitates AI-driven decision-making, overcoming limitations in insights and confidently navigating the complexities of the digital era.

REFERENCES

Ackoff, R. (1989). From data to wisdom. *Journal of Applied Systems Analysis, 16*, 3–9.
Angerman, W. S. (2004). *Coming full circle with Boyd's OODA loop ideas: An analysis of innovation diffu-sion and evolution* [Theses and Dissertations]. Department of the Air Force Air University.

Anton, P., McKernan, M., Munson, K., Kallimani, J., Levedahl, A., Blickstein, I., Drezner, J., & Newberry, S. (2019). *Assessing department of defense use of data analytics and enabling data management to improve acquisition outcomes*. RAND Corporation. https://doi.org/10.7249/RR3136

Aravind, S. (2024). *Harnessing big data: Unleashing the power of information*. LinkedIn. https://www.linkedin.com/pulse/harnessing-big-data-unleashing-power-information-aravind-s-tooxc/

Army Technology Transfer Program 2022 Annual Report. (2023). [Annual Report]. U.S. Army Office of the Deputy Assistant Secretary of the Army (Research and Technology). https://api.army.mil/e2/c/downloads/2023/07/11/0136f109/tab-a-army-technology-transfer-program-2022-annual-report-dist-a-final.pdf

Awad, E. M., & Ghaziri, H. (2007). *Knowledge management* (1st ed.). Pearson.

Ayers, C. E. (2016). *Rethinking sovereignty in the context of cyberspace (cyber sovereignty workshops)*. United States Army War College. https://apps.dtic.mil/sti/pdfs/AD1124691.pdf

Burgin, M. S. (2017). *Theory of knowledge: Structures and processes*. World scientific publishing.

Callahan, S. P. (2023). *Warfighting: John Boyd's theory of conflict, the origins of maneuver warfare, and the complex process of doctrinal change in the U.S. Marine Corps 1975-1989* [Doctoral dissertation]. University of Maryland.

Chin, R., Catal, C., & Kassahun, A. (2023). Plant disease detection using drones in precision agriculture. *Precision Agriculture*, 24(5), 1663–1682. https://doi.org/10.1007/s11119-023-10014-y

Cleveland, H. (1985). The twilight of hierarchy: Speculations on the global information society. *Public Administration Review*, 45(1), 185. https://doi.org/10.2307/3110148

Dabols, E. (2023, July 5). *Data products vs data as a product*. LinkedIn. https://www.linkedin.com/pulse/data-products-vs-product-eric-dabols/

Data Management Lexicon (C06978108). (2022). Office of the Director of National Intelligence. https://www.dni.gov/files/documents/FOIA/DF-2022-00351-IC-Data-Management-Lexicon-January-2022.pdf

Department of Defense (DoD) Technology Transfer (T2). (2024). Department of Defense Research & Engineering Enterprise. https://rt.cto.mil/rtl-labs/tech-transfer/

Department of Defense Announces Joint Warfighting Cloud Capability Procurement. (2022, December 7). U.S. Department of Defense. https://www.defense.gov/News/Releases/Release/Article/3239378/department-of-defense-announces-joint-warfighting-cloud-capability-procurement/

Department of Defense Data, Analytics, and Artificial Intelligence Adoption Strategy. (2023). Department of Defense. https://media.defense.gov/2023/Nov/02/2003333300/-1/-1/1/DOD_DATA_ANALYTICS_AI_ADOPTION_STRATEGY.PDF

Department of Defense Dictionary of Military and Associated Terms (Joint Publication 1-02). (2016). Joint Cheifs of Staff. https://irp.fas.org/doddir/dod/jp1_02.pdf

Department of Defense The National Military Strategy for Cyber Operations. (2006). [Memorandum]. Department of Defense. https://nsarchive2.gwu.edu/NSAEBB/NSAEBB424/docs/Cyber-023.pdf

DoD Cloud Strategy. (2018). Department of Defense, https://media.defense.gov/2019/Feb/04/2002085866/-1/-1/1/DOD-CLOUD-STRATEGY.PDF

DoD Data Strategy. (2020). Department of Defense. https://media.defense.gov/2020/Aug/11/2002475039/-1/-1/1/DOD%20CLOUD%20STRATEGY%20OSD016570-18%20RES%20FINAL_20181218.PDF

Flynn, M. T. (2011, April 20). *Sandals and robes to business suits and Gulf Streams*: Warfare in the 21st Century. *Small Wars Journal*. https://smallwarsjournal.com/jrnl/art/sandals-and-robes-to-business-suits-and-gulf-streams-warfare-in-the-21st-century

Flynn, M. T., Juergens, R., & Cantrell, T. L. (2008). Employing ISR SOF best practices. *Joint Force Quarterly*, 6(50), 56–61.

Heinrich, T. (2002). Cold war armory: Military contracting in silicon valley. *Enterprise & Society*, 3(2), 247–284.

IDC digital universe study: Big data, bigger digital shadows and biggest growth in the far east. (2012). EMC. https://www.cs.princeton.edu/courses/archive/spring13/cos598C/idc-the-digital-universe-in-2020.pdf

Insurgencies and countering insurgencies (FM 3-24/MCWP 3-33.5, C1). (2014). Headquarters Department of the Army. https://irp.fas.org/doddir/army/fm3-24.pdf

International strategy for cyberspace prosperity, security, and openness in a networked world. (2011). The White House. https://obamawhitehouse.archives.gov/sites/default/files/rss_viewer/international_strategy_for_cyberspace.pdf

Jifa, G., & Lingling, Z. (2014). Data, DIKW, big data and data science. *Procedia Computer Science, 31*, 814–821. https://doi.org/10.1016/j.procs.2014.05.332

Kaldor, M. (2012). *New & old wars organized violence in a global era* (Vol. 3). Polity Press.

Kinney, S. (2019, January 24). Is big data more like oil or sunshine? *RCR Wireless News.* https://www.rcrwireless.com/20190124/data-analytics/big-data-oil-sunshine

Lane, J. (2020). *Democratizing our data: A manifesto.* MIT Press.

Marwala, T. (2024). *Militarization of AI has severe implications for global security and warfare.* United Nations University. https://unu.edu/article/militarization-ai-has-severe-implications-global-security-and-warfare

Meadow, C. T., & Yuan, W. (1997). Measuring the impact of information: Defining the concepts. *Information Processing & Management, 33*(6), 697–714. https://doi.org/10.1016/S0306-4573(97)00042-3

Medhora, R. P. (2018). *Data governence in the digital age.* Centre for International Governance Innovation.

Metz, S., & Johnson, D. V. (2011). *Asymmetry and U.S. Military strategy: Definition, background, and strategic concepts.* Strategic Studies Institute, US Army War College.

Musekiwa, R. (2023). *It is still all about the data.* LinkedIn. https://www.linkedin.com/pulse/still-all-data-robert-musekiwa/

Nake, F. (2001, July 23). *Data information and knowledge, organizational semiotics: Evolving a science of information systems.* IFIP TC8/WG8.1 Working Conference.

Nolin, J. M. (2019). Data as oil, infrastructure or asset? Three metaphors of data as economic value. *Journal of Information, Communication and Ethics in Society, 18*(1), 28–43. https://doi.org/10.1108/JICES-04-2019-0044

Nye, J. S. Jr. (2011). *The future of power.* Public Affairs.

Olson, M., & Davis, G. (1985). *Management information systems conceptual foundations, structure, and development* (2nd ed.). McGraw Hill.

Palmer, M. (2006, November 3). *Data is the new oil.* ANA Marketing. https://ana.blogs.com/maestros/2006/11/data_is_the_new.html

Peters, M. A., Jandrić, P., & Green, B. J. (2024). The DIKW model in the age of artificial intelligence. *Postdigital Science and Education.* https://doi.org/10.1007/s42438-024-00462-8

Reinsel, D., Gantz, J., & Rydning, J. (2020). *The digitization of the world from edge to core* (US44413318). IDC. https://www.seagate.com/files/www-content/our-story/trends/files/dataage-idc-report-final.pdf

Rickli, J.-M., & Mantellassi, F. (2023). Artificial intelligence in warfare military uses of AI and their international security implications. In *The AI wave in defence innovation_ assessing military artificial intelligence strategies, capabilities, and trajectories.* Routledge.

Rigol, X. G. (2021, July 8). *Data as a product vs data products. What are the differences?* Medium. https://towardsdatascience.com/data-as-a-product-vs-data-products-what-are-the-differences-b43ddbb0f123

Ronquillo, C., Currie, L. M., & Rodney, P. (2016). The evolution of data-information-knowledge-wisdom in nursing informatics. *Advances in Nursing Science, 39*(1), E1–E18.

Sayler, K. (2020). *Artificial intelligence and national security* (R45178). Congressional Research Service. https://crsreports.congress.gov/product/pdf/R/R45178/10

Senne, K. D. (2007). Integrated sensing and decision support. *Lincoln Laboratory Journal, 16*(2), 237–243.

Sezal, M. A., & Giumelli, F. (2022). Technology transfer and defence sector dynamics: The case of the Netherlands. *European Security, 31*(4), 558–575. https://doi.org/10.1080/09662839.2022.2028277

Shelton, H. H. (2000). *Joint vision 2020: America's military preparing for tomorrow.* Joint Cheifs of Staff. https://www.hsdl.org/?view&did=446826

Shultz, R. H., & Clarke, R. D. (2020, August 25). *Big data at war: Special operation forces, project maven, and twenty-first century warfare.* Modern War Institute. https://mwi.westpoint.edu/big-data-at-war-special-operations-forces-project-maven-and-twenty-first-century-warfare

Stenmark, D. (2002). *Information vs. knowledge: The role of intranets in knowledge management. Proceedings of the 35th Annual Hawaii International Conference on System Sciences,* 928–937. https://doi.org/10.1109/HICSS.2002.994043

Summary of the 2018 Department of Defense Artificial Intelligence Strategy Harnessing AI to Advance Our Security and Prosperity. (2018). U.S. Department of Defense. https://media.defense.gov/2019/Feb/12/2002088963/-1/-1/1/SUMMARY-OF-DOD-AI-STRATEGY.PDF

The Targeting Process (Field Manual FM 3-60 (FM 6-20-10)). (2010). Headquarters, Department of the Army.

The United States Air Force Artificial Intelligence Annex to The Department of Defense Artificial Intelligence Strategy. (2019). Department of the Air Force.

Trent, R. (2023). *Must learn AI security.* Independently published.

Ulu, C. (2023). *Siber Güç ve Asimetrik Strateji: Amerika Birleşik Devletleri'nin İnsansız Otonom Hava Sistemleri* [Doktora Tezi]. Ege Üniversitesi.

Wakabayashi, D., & Shane, S. (2018, June 1). *Google will not renew Pentagon Contract that upset employees*. The New York Times. https://www.nytimes.com/2018/06/01/technology/google-pentagon-project-maven.html

Wiig, K. M. (1993). *Knowledge management foundations, thinking about thinking, how people and organizations create, represent and use knowledge*. Schema Press.

Wittgenstein, L. (1968). II: Notes for lectures on "private experience" and "sense data". *The Philosophical Review, 77*(3), 275–320. https://doi.org/10.2307/2183568

Zeleny, M. (1987). Management support systems: Towards integrated knowledge management. *Human Systems Management, 7*(1), 59–70.

Zins, C. (2007). Conceptual approaches for defining data, information, and knowledge. *Journal of the American Society for Information Science and Technology, 58*(4), 479–493. https://doi.org/10.1002/asi.20508

8 The Ethical and Legal Challenges of AI in Wars

Kıvılcım Romya Bilgin

8.1 INTRODUCTION

The nature of war has an unchanging essence, on the other side, the character of war refers to the changing way in which wars are waged based on a various of factors, such as policy, technology, culture, economy, geography, and more. The developments in artificial intelligence (AI) technologies have a potential to cause radical changes in the character of war. Many scholars in war studies discuss the potential for AI's influence on the character of war to bring about a revolution in warfare (De Spiegeleire et al., 2017; Horowitz, 2018; Horowitz et al., 2018; Johnson, 2019; Lonsdale, 2004), because AI has a transformative influence across all dimensions of war. This influence is not solely into the realm of technology but also into implications for ethical and legal aspects of wars. In particular, the ethical and legal impact of AI technologies on war is intertwined with the impact of AI technologies in the field of warfare.

The relationship between war and technology has been characterized by inconsistency throughout history. Periods of wars have been marked by prolonged technological stagnation. Revolutionary changes in technology have changed the paradigm in warfare and then evolving threat environments, driven by the need for new destructive technologies, have resulted in technological advancements leading to revolutionary periods (Macgregor & Williamson, 2001). On the other hand, every technological revolution, from the discovery of gunpowder to the development of nuclear weapons, has consistently confronted humanity with new ethical dilemmas. AI technologies, as a significant technology, have placed humanity face-to-face with similar ethical dilemmas. The similarities between past technological revolutions and the ethical dilemmas posed by AI demonstrate that lasting nature of these concerns. Moreover, for now humanity is confronted with much more complex political, social, economic, and cultural conditions under a much more intricate technology that creates new ethical and legal dilemmas. International legal frameworks such as Article 36 of Additional Protocol I of the 1949 Geneva Conventions include efforts to address these ethical dilemmas by referring legal compliance. Article 36 provides the necessary groundwork for evaluating whether the use of new technologies in weapon systems complies with international law. This also necessarily brings the issue of ethical discussions to the forefront. However, the human factor remains a focal point irrespective of historical context and technological progress. Consequently, any ethical or legal consideration concerning AI in wars cannot exclude the presence of humans.

Rapid developments in military applications of AI technologies exceedingly challenge in determining the boundaries of the discussion on the ethical and legal evaluations of AI in wars. Additionally, every new development in AI technologies and every new military application has a potential to change the dimensions of the discourse within the scope of ethics, legality, and warfare. Nevertheless, it is still possible to establish fundamental concepts and outlines that need to be emphasized. In this context, three main sections have been designed for this chapter. First, the discussion regarding the relationship between war, ethics, and law will be conducted basing it on the ontology of each phenomenon. Second, international initiatives will be presented. Third, the focus will be on the ethical challenges arising from the rapid advancement of AI-based technologies in warfare. Finally, the international framework will be presented to explore the potential legal regulations that should be implemented on AI, considering the components of the just war theory and principles of International Humanitarian Law (IHL).

DOI: 10.1201/9781003441700-8

8.2 THE INTERSECTION AMONG ETHIC, AI, AND WAR

There has been a proliferation of scholarly works delving into the intricate relationship between AI, robots, machines, and ethics since the early 2000s (Allen et al., 2005; Anderson & Anderson, 2011; Cervantes et al., 2020; Deng, 2015; Formosa & Ryan, 2021; Tolmeijer et al., 2021; Vishwanath et al., 2023). However, despite the ongoing discussions, it is still imperative to define how ethics intersects with and relates to AI and war at the first stage. Furthermore, the conceptual relationship between them should be systematically structured. This stage is indispensable for creating a proper environment for implementation of precise political measures in advanced stages and the formulation and application of laws. Nevertheless, the primary and fundamental question that must be addressed is why ethics is important in the advancement and use of AI technologies. When answering this question, it is essential to keep in mind that AI is widely used in various aspects of daily life, ranging from banking applications to healthcare and even autonomous vehicles. This is because the use of AI is progressively expanding beyond military applications to encompass social, economic, political, and cultural domains. The relationship between AI and ethics should be comprehensively addressed, primarily covering all these areas and then how ethics, AI and war intersect with each other, in a complex way, should be considered. This complex relationship arises from the ambiguity in defining these concepts on a factual level and their intricate interconnections with other concepts, such as politics, violence, and morality.

The question of why ethics is important in the advancement and use of AI technologies can be answered based on four fundamental principles. First, AI applications can be influenced by the datasets used in their training. If societal biases exist in the training data, AI systems may learn and reproduce these biases. AI technologies designed with biases can exacerbate social inequalities or create new forms of inequality. Therefore, any AI technology is designed free from biases is crucial to ensuring fairness and equality in technology. The second principle is on the protection of personal data. AI often analyzes large datasets, potentially using personal information in the process. Ethical principles emphasize the necessity of preserving the privacy of this data, limiting its usage, and preventing misuse. The third principle is about transparency. AI makes decisions primarily through complex algorithms and deep learning methods. The decision-making processes of these algorithms frequently are not comprehended by humans. That's why transparency and comprehensibility in these algorithms are demanded as ethical rules in understanding the factors influencing decisions. The fourth principle acknowledges that AI systems' decisions often impact the lives, jobs, and daily activities of individuals. Therefore, it is crucial that the use of AI should prioritize a human-centered approach. Ethical rules should strive to guarantee that AI systems are in harmony with human needs and values. However, the ongoing debates about how the similarity between AI and human intelligence can be achieved and also how AI processes human emotions is highlighting the complexity of the relationship between AI, ethics, and humanity. Clear answers for these questions are yet to be found. Ultimately, ethical rules are essential to instill trust in AI between individuals and society. These ethical rules also play a crucial role in promoting the trustworthy development and use of AI. A trustworthy AI identifies acceptable risks beforehand and provides users with clear information.

The debates concerning ethics originated with the efforts of philosophers such as Socrates and Plato, who endeavored to establish the purpose of human life, the nature of a virtuous life, and the existence of morally universal values. Ethics fundamentally engages in discourse that concerns itself with the moral elements that arise in both individual and societal dimensions of human life, and it is essentially the activity of reflecting upon moral issues, and initially attempts to make sense of the relationship between the concepts of morality and ethics. Despite interchangeability of the two concepts in daily language, there should be a criterion to differentiate between them. While morality is an empirical and historically experienced phenomenon, ethics refers to the philosophical discipline that addresses this phenomenon. Therefore, when discussing a 'moral problem' in daily language, it should be understood as a problem of ethics (Ozlem, 2004, pp. 21–23). While ethics investigates what is considered valuable or seeks to understand the summum bonum, it raises several questions to determine the possibility of identifying universal and objective moral principles.

Various ethical theories, such as existentialist ethics, virtue ethics, feminist ethics, modeling ethics, or contractualism, can be considered to deepen the theoretical discussion between AI and ethics. However, when specifically addressing the relationship among AI, ethics, and war, particular attention should be given to deontological and utilitarian ethics. Deontological ethics, which focuses on the nature and content of actions without concerning itself with the consequences, emphasizes Kantian ethics' categorical imperatives and universal moral obligations. In this context, it advocates adherence to specific moral rules independently of outcomes. Kantian ethics formulates deontological and non-consequentialist principles that are inherently good to follow, independent of expediency or the pursuit of a greater public good (Wood, 2012). Kant formulates the categorical imperative, which is the main component of his ethical approach and deontological ethics, in the following manner: 'Act only according to that maxim by which you can at the same time will that it should become a universal law' (Kant, 1993, p. 30). Ülgen provides a comprehensive discussion on the compatibility and conflict between AI and ethics from the deontological perspective based on Kantian ethics. Kantian ethics, centered around human attributes and capacities, pose challenges for the implementation of AI and robotics. Kantian ethics also emphasize human self-determining capacity for rulemaking, including practical reasoning, judgment, self-reflection, and deliberation. These human attributes, essential for forming universally applicable moral rules, are lacking in AI and robotics. While a limited rational thinking capacity can be programmed into machines, it does not encompass the self-reflective and deliberative capacities of humans as conceptualized in Kantian ethics. Especially, the categorical imperative, a significant component of Kantian ethics, asserts that actions should be based on principles that can transform into universal laws. However, due to the lack of human-like reasoning and free will in AI, it seems challenging to comprehend and develop rules that are internally desired and universally applicable (Ülgen, 2017). It is obvious that there are challenges arising from the application of Kantian ethics to AI. The reason for that there is a mismatch between human attributes emphasized by Kant and the capabilities of AI and robotics. This might pose difficulties in adapting Kant's principles to machines lacking free will.

Utilitarianism, as a consequentialist ethical theory, posits that morally right action is determined by achieving the best consequences, operationalized as the maximization of utility. In other words, the ethical value of an action is defined by the results it produces or its consequences. For utilitarians, rooted in the principle of maximizing overall happiness or utility, the focus is on the consequences of actions. Utilitarianism includes the principle of balancing positive and negative consequences when assessing the net benefit of an action (Shaw, 1999, pp. 12–13). Utilitarian ethics in the AI context evaluates the ethical consequences of AI actions based on their overall impact on individual and society by maximizing happiness or utility, considering the general welfare, and balancing positive and negative outcomes. It also requires assessing issues such as privacy, prevention of bias, and fairness, with a long-term perspective to ensure sustainable benefits for society. Creating ethical frameworks is crucial for aligning AI applications with utilitarian principles. Utilitarianism has gained favor in the AI community due to its incorporation of ethical decisions, involving (1) formulating a utility function that expresses preferences concerning states of the world and (2) employing an optimization algorithm to determine the action (or rule) that maximizes expected utility (Kuipers, 2020, p. 426). From a utilitarian standpoint, military robots indeed offer significant advantages over humans in tasks that are dangerous such as the use of drones, reconnaissance missions, or bomb disposal robots in handling explosive devices. AI technologies have shown promising advancements not only in military contexts but also in the field of healthcare. For example, AI-assisted surgical robots enable surgeons to perform precise procedures with greater accuracy and precision. However, it should be remembered that AI design is not value-neutral, as the artifacts created inherently reflect the values of the designers and end users. Whether an action is deemed suitable for the public interest, from a utilitarian perspective, will depend on the magnitude of the benefits and happiness it provides to society. If an action increases general happiness and serves the general interests of the community, it can be considered an ethical action from a utilitarian standpoint. That is the reason why Gotterbarn references the manual 'Software Engineering Code of Ethics and Professional Practice (SE Code)', asserting that a computing professional is expected to address the broader needs of users and stakeholders, not just meet contract specifications (2001). In this context, one of the eight principles in the SE Code contributed by Gotterbarn, namely, 'Software engineers shall act consistently with the public interest' takes precedence (ACM Ethics, 1997). It can be observed that a utilitarian ethical perspective emphasizes the importance of assessing the consequences of AI actions on both individuals and society.

In the relationship between AI and ethics, another noteworthy issue is the connection between purpose and ethics. If the purpose of AI significantly influences the direction of ethical discussions, then an emphasis should also be placed on the purpose of developing and using AI technologies. The diverse goals of developing and using AI in military, social, economic, political, and cultural fields can make it challenging to provide a single answer to the purpose of AI. Taking this into account, Stahl attempts to answer this question from a non-sectoral inclusive perspective. According to Stahl, there should be three primary purposes of AI. The first is the use of AI for efficiency, emphasizing its potential to enhance productivity, leading to economic benefits and improved lives. The second purpose introduces a contrasting perspective, viewing technology, including AI, as a means of exerting social control. The collection of data and AI's role in detecting patterns for controlling human behavior, both subtly through nudging and more assertively, as seen in systems like China's social credit scoring. The third purpose is to enhance human flourishing. This involves the creation and implementation of AI in ways that contribute to the well-being of individuals. AI serves as a tool facilitating the identification of meaningful goals for individuals and groups, offering support in the pursuit of excellence to achieve these goals (Stahl, 2021, p. 23). Although Stahl's response is a good starting point when evaluating the relationship between AI and war from an ethical standpoint, it is not sufficient. Because at this stage, the purpose of AI must be also considered in terms of the intention behind it.

The responsibility of human actions is often associated with their intentions. However, an autonomous system can make decisions and function but cannot intend. This complicates the relationship between intention, responsibility, and the actions when considering both humans and autonomous systems. Therefore, while the concept of intention is a factor in determining human responsibility, it is a void for autonomous systems. Nonetheless, the intention of designing and developing an AI technology is generally determined by a series of factors and can vary according to various application areas. Intentions such as creating applications that facilitate human life, supporting education, and enhancing healthcare services can be considered within this scope. However, while categorizing these well-intentioned purposes; malicious intentions such as cyberattacks, manipulation, privacy violations, discrimination may also come into play. Particularly, the design of AI technologies for warfare with the intention to cause harm constitutes the focal point of the ethics and intention debate. While there is reference to the operator's intention, the existence of different levels of intention, such as the intention of the programmer, the patent holder, or the military user, prompts us to pose the question of 'intention of whom'. This situation is commonly denoted to as the 'problem of many hands', and it challenges traditional notions of responsibility and accountability, particularly in contexts where advanced technologies with high degrees of autonomy are involved.

The complexity of warfare environments and the innate risks of activities contribute to the increased uncertainty of effects, complicating the relationship between intention and impact. Goetze asserts that the designers' intentions become evident through their control over various aspects, such as deciding when the training of the machine learning system is deemed successful, selecting the dataset, formulating the reward function, adjusting the hyperparameters, and so on (Goetze, 2022, p. 399). Yet, with the developments in AI technology, one must consider the increasing complexity of the relationship between tools and intentions. Borgmann scrutinizes how technological advancements in contemporary society mold the intentions and values of individuals.

In his perspective, technology is intricately linked not only to the use of technological tools but also to how this usage is integrated with the intentions and lifestyles of individuals. Hence, while intention serves as the impetus for the development and usage of technology, technology has the potential to reshape the form of intention in turn (Borgmann, 1984, pp. 98–101). The ethical complexity of intentions in the context of AI and war makes it difficult to clearly define the boundaries between AI and ethics. In this context, AI for Social Good (AI4SG), as a rising and popular approach, provides an insight in AI and ethic relations. It refers to the use of AI technology for the aim of providing societal and environmental benefits without causing new harm or worsening new ones. AI4SG advocates the use of AI technologies in various domains such as improving people's lives, increasing social justice, promoting environmental sustainability, enhancing education, improving healthcare services, and enhancing the overall welfare of communities (Cowls et al., 2021; Floridi, 2023, pp. 142–157). In this framework, AI4SG aims to utilize the power of AI to benefit various sectors of society. AI4SG reminds the importance of using, developing, and deploying AI for benevolent intentions.

Examining the relationship between war and ethics through the lens of AI reveals that another focal point should be intelligence. AI systems are undeniably good in swiftly executing certain tasks, analyzing vast amounts of data, and possessing the capability to learn from extensive datasets. However, the human brain differentiates with the abilities to generate meaning, establish empathy, develop emotional connections, form an independent consciousness, and exhibit willpower. Durt places a particular emphasis on the potential misleading nature of any comparison between human intelligence and AI. Durt states that notable figures in the field of AI have been conscious of the fundamental difference in simulating processes like thought, experience, and understanding alongside AI. However, a significant portion of AI research is not only focused on mimicking human behavior as assumed but also on emphasizing the substantial data processing capacity of AI. In this context, Durt emphasizes that the determining factor in the evolution of AI is not its ability to imitate human behavior, but its high capacity in data processing. Therefore, Durt concludes that making a simple comparison between human intelligence and AI can be misleading (Durt, 2022, pp. 71–72). Larson takes a more skeptical stance about artificial and human intelligence comparison and highlights that early mistake in comprehending intelligence have gradually led to a theoretical deadlock within the field of AI. The intricate nuances of human intelligence were inevitably oversimplified at the inception of AI research, adopting a narrowed perspective. From that standpoint, Larson emphasizes the distinct differences between AI and human intelligence and points out that the AI systems encountered predominantly engaged in inductive reasoning and excelled at numerical computations for outcome predictions. In contrast, humans possess the ability to make informed conjectures based on context and experience, performing tasks that machines cannot replicate (Larson, 2021, pp. 119–121). Moreover, the human intelligence could determine war strategies using common tactics such as bluffing, setting traps, or deceiving. However, as of the current level of AI technology, it does not possess these complex abilities. These tactics are a product of cognitive evolution evolving in a specific environment and social competition with other humans. In the discussion on the relationship between AI and ethics, compliance with ethical rules in the design and use of AI should be considered, as well as compliance with ethical rules of cognitive abilities of human intelligence, such as bluffing, trapping or deceiving.

AI can provide an advantage in specific contexts due to its lack of emotions. At this point, it is crucial for AI systems to prioritize human safety and avoid undesired outcomes or biases. While the lack of emotions in AI can be advantageous in some situations, it should be noted that people make decisions by considering various contextual factors. Contextual intelligence also involves evaluating ethical responsibilities and societal impacts. However, AI technologies, even though they attempt to understand and implement this contextual intelligence, are not sufficient in this regard, at least for now. This situation poses a significant challenge in the intersection of AI technologies and ethics.

8.3 INITIATIVES IN THE REALM OF ETHIC OF AI: A COMPREHENSIVE REVIEW

In the past decade, it is imperative to accord particular attention to the multifarious initiatives launched by think tanks, academic institutions, and international organizations, which are aimed at the surveillance and assessment of progressions within the realm of AI. These initiatives seek to formulate a comprehensive set of the ethics of AI. The general objective of such endeavors is to ascertain that AI engenders a constructive impact on society, concurrently laying the foundation for the establishment of robust ethical standards. The first crucial initiative is the 'One Hundred Year Study on Artificial Intelligence', launched at Stanford University in 2014. The fundamental objective of it is to continuously monitor and comprehend developments, impacts, and potential risks in the field of AI. Serving the purpose of understanding the long-term effects of AI and guiding the community in ethical governance, this study aims to establish an interdisciplinary platform for the assessment of long-term impacts (AI100, 2021). The other initiative in 2015 is the 'Research Priorities for Robust and Beneficial Artificial Intelligence: An Open Letter', which constitutes an effort aimed at fostering positive impacts on the advancement of AI. Simultaneously, it underscores specific priorities for the ethical and reliable use of this technology (Russell et al., 2015). Ethical use of AI is a central focus for both initiatives that bring attention to the ethical challenges it may engender. By the way, 'The Asilomar AI Principles', introduced by the Future of Life Institute in 2017, stands as one of the most impactful frameworks for AI governance. It stresses the importance of guaranteeing the safety and security of AI systems throughout their operational lifetime and transparently understanding the reasons behind these systems. A total of 12 principles among 23 are specifically related to ethics and emphasize safety, transparency, responsibility, and alignment with human values. According to principles, designers and builders of AI systems are recognized as stakeholders with a moral responsibility to shape the implications of AI use. Additionally, the principles emphasize the importance of AI systems aligning with human values, respecting human dignity, rights, freedoms, and cultural diversity. The shared benefits and prosperity resulting from AI technologies should be widely distributed human control over AI decisions should be prioritized, and the power of advanced AI systems should enhance societal processes. Last, the principles call for avoiding an arms race in lethal autonomous weapons (Future of Life, 2017). Overall, the principles advocate for the ethical development of AI for the benefit of humanity.

Universities have also made significant contributions to the ongoing discussion about the ethical use of AI. 'The Montreal Declaration', conceived in 2017 under the auspices of the Université de Montréal, serves as a pivotal reference point in elucidating principles and guidelines for the ethical deployment of AI. It was created because of discussions and deliberations that took place during 'the Forum on the Socially Responsible Development of Artificial Intelligence' in November 2017. The declaration serves as frameworks to steer the responsible development and use of AI technologies, emphasizing considerations beyond technical aspects, such as privacy, fairness, accountability, and transparency (Montreal University, 2017). Meanwhile, universities and civil initiatives' initial efforts in the field of ethical AI have started to rapidly transform into more inclusive initiatives which are involving not only leading companies in the AI sector but a broader collaboration. In this context, the first initiative has been 'The Partnership on AI (PAI)'. PAI formed in 2016 by major tech companies like Google, Facebook, Amazon, and Microsoft. Its primary objective is to promote the responsible and ethical development of AI technologies, emphasizing transparency, fairness, accountability, privacy, and public engagement (Partnership on AI, 2023). Another illustrative initiative is AI4People, also known as 'Artificial Intelligence for People'. This initiative launched in 2018 that focuses on addressing the challenges and opportunities presented by AI and shapes the debate on the ethics of AI. AI4People Institute unites prominent AI companies, academic institutions, civil society groups, and government entities (Atomium European Institute, 2023). In 2018, AI4People presented 'AI4People's Ethical Framework for a Good AI Society' and influenced the European Commission's identification of the '7 Key Requirements for Trustworthy AI' (European Commission, 2019). All these initiatives specifically stress human empowerment, safety, data privacy, transparency, non-discrimination, environmental sustainability, and mechanisms for accountability for ethics of AI.

In comparison to the ethical concerns voiced by civil society and universities regarding the ethics of AI, the responses of states in terms of ethics are noticeably inadequate. Indeed, states prioritize the strategic, political, and economical advantages of AI and it causes ethical concerns on AI to be put in the background. Nevertheless, efforts are ongoing by the European Union (EU) and the United Nations (UN) to enhance the ethics of AI and states generally concentrate on the nexus of AI and ethics under the umbrella of these international organizations. EU has started a series of initiatives through its various bodies with the aim of bringing together member countries about ethics of AI. One of the most important initiatives of the EU was carried out by the European Council (EC) in 2017. The Council emphasized the need for urgency in addressing emerging trends, including the ethics of AI (European Council, 2017, p. 8). 'The European Commission's European Group on Ethics in Science and New Technologies (EGE)' in March 2018 has published in 'the Statement on Artificial Intelligence, Robotics and Autonomous Systems'. The statement advocates for the initiation of a comprehensive international process to establish an ethical and legal framework for the advancement and use of AI, robotics, and autonomous systems (European Commission, 2018). The European Commission has published '7 Key Requirements for Trustworthy AI' influenced from AI4People. The guidelines cover aspects like human agency, privacy, societal well-being, technical robustness, transparency, diversity, fairness, and accountability. These requirements emphasize human empowerment, safety, data privacy, transparency, non-discrimination, environmental sustainability, and mechanisms for accountability. The Guidelines also include a detailed assessment list and a definition of AI for clarification (European Commission, 2019). In April 2021, the European Commission proposed the initial regulatory framework for AI within EU (Council of the European Union, 2020). The proposal has included the examination and categorization of AI systems according to the varying levels of risk they present to users. Meanwhile, the European Parliament has actively engaged in AI-related efforts, and in October 2020, it passed several resolutions including those focused on ethics. On December 2023, the Parliament and the Council provisionally agreed on the AI Act. The approved text is set to undergo formal adoption by both the Parliament and the Council for it to be enacted as EU law (European Parliament, 2023). The EU has successfully fostered collaboration among member countries in addressing the ethical considerations of AI through diverse bodies. The recent parliamentary action further demonstrates the EU's achievement in establishing a unified legal framework among member states.

The efforts of the EU are pioneering and exemplary in the global system. Under the UN, there is a growing effort to establish a common ground for an ethical approach to the use of AI, and member states are invited to contribute to this endeavor. In November 2021, the initial worldwide normative instrument on the ethics of AI, titled the 'Recommendation on the Ethics of Artificial Intelligence', was endorsed by UNESCO. This instrument directly tackles the issues articulated in Human Rights Council Resolution 47/23, particularly within the realm of AI (UNESCO, 2021). In September 2022, 'the Principles for the Ethical Use of Artificial Intelligence in the United Nations System' was formulated by 'the High-level Committee on Programmes (HLCP)' and endorsed at an intersessional meeting in July 2022 and then received approval from 'the United Nations System Chief Executives Board for Coordination' (CEB, 2022). On the other hand, efforts under the UN umbrella continue a broader scale. The UN Secretary-General launched an AI Advisory Body to address risks, opportunities, and international governance related to AI. In 2023, the Interim Report titled 'Governing AI for Humanity' has been released by the UN Secretary-General's AI Advisory Body. The report emphasizes the necessity for a more harmonious alignment between international norms and the development and implementation of AI (UN, 2023). Whether initiated by research institutions, universities, civil society organizations, firms, or international bodies and states, the prominent themes in initiatives for creating ethics of AI include human dignity, human privacy, fairness, transparency, accountability, predictability, non-discrimination, responsibility, and human control. Together, these principles form a foundation for ethically navigating the evolving landscape of AI.

However, the traditional design of ethics that has been in place for millennia will be challenged by AI. Overcoming substantial barriers to the formalization of ethical principles will be essential for furthering the integration of ethics into AI (Powers & Ganascia, 2020, p. 28). That's why innovative approaches must be developed to tackle the ethical challeng associated with AI in the future. Simultaneously, AI technologies are advancing rapidly and there is a necessity for a concurrent and globally harmonized approach to ethical considerations. Any failure in achieving a harmonization may result in a growing misalignment the ethics of AI. This situation poses a significant challenge for humanity to the interface between AI and ethics.

8.4 AI IN WARS: ETHICAL CHALLENGES

There is an ongoing integration of AI into existing technologies, in addition to the development of novel technologies relying on AI. AI technologies in wars could be used in the domain of autonomous weapons, drone swarms, data processing and research, cyber security, threat monitoring, decision support systems, combat simulation, target recognition, biometric recognition, transportation, and casualty care and evacuation. In all of these areas, the use of AI naturally brings up the agenda on the topic of ethics of AI. However, in the context of military applications, AI technologies specifically serve two key purposes. First, it can enhance the performance of traditional and existing weapon systems. Second, AI can play a role in decision-making processes by providing assistance, facilitating, or even making autonomous decisions (CACDA-China Arms Control and Disarmament Association, 2019, p. 20). This dual focus also forms the focal point of the discussions on the ethical use of AI in wars. Autonomous Weapon Systems (AWS) constitute the epicenter of debates regarding the ethical use of AI in wars. On the other hand, another issue, albeit less discussed but ethically significant, is the use of AI in decision support systems during wars where strategic, tactical, and operational decisions play a crucial role, and the utilization of big data is progressively intensifying. In this context, it is observed that AI Decision Support Systems (AI-DSS), which influence human decisions, pose ethical challenges in decision-making comparable to the ethical dilemmas posed by AWS.

Ethical challenge discussion should begin with the definition of AWS. Because the definition of AWS is a sort of evidence the characteristics of them, such as autonomy, lack of human control, and lethality capacity and it places AWS at the center of ethical discussions. In this respect Horowitz and Scharre first define autonomy as the capacity of a machine to operate autonomously without human involvement. In this regard, autonomous systems are systems that perform certain tasks or functions autonomously, whether through software or hardware (Horowitz & Scharre, 2015, p. 5). According to them, machines can be classified into three categories based on their level of autonomy. 'Human in the loop' (semi-autonomous) machines operate for a period, then pause for human input before continuing. 'Human on the loop' (human-supervised autonomous) machines can function independently but have human monitoring for intervention in case of failure. 'Human out of the loop' (fully autonomous) machines can operate entirely on their own without human intervention. In this context, the term 'autonomy' refers to the machine's relationship with a human controller rather than its intelligence. While Horowitz and Scharre offer a comprehensive framework for understanding autonomy in AI systems, Taddeo and Blanchard conduct a comparative analysis of official definitions from states and also international organizations, including the North Atlantic Treaty Organization (NATO) and International Committee of the Red Cross (ICRC). They propose a value-neutral definition derived from the comparative analysis that aims to serve as a common ground for addressing ethical and legal challenges related to AWS. Taddeo and Blanchard address that at a minimum, an AWS is an artificial agent that can alter its internal states to achieve goals in its dynamic environment without direct intervention and may possess abilities to change its own transition rules. Deployed to apply kinetic force against a physical target, an AWS possesses the capability to autonomously identify, select, and engage targets without the need for external intervention. Subsequently, after deployment, an AWS has the capability to operate with or without human control (in, on, or out of the loop). A lethal AWS (LAWS) is a subset focused on applying kinetic force against human beings (Taddeo & Blanchard, 2022, pp. 37–39). Both definitions address autonomy within relation to human control which should be one of the central axes when discussing responsibility and accountability. Besides all, the fundamental characteristics that distinguishes AWS from other weapon systems and places it at the center of ethical debates is autonomy which is measured by five criteria defined by Christen et al. These are level of autarchy, independence from human control, interactive capabilities with the environment, learning capacity, and mobility level (Christen et al., 2017, p. 5). Christen et al.'s criteria highlight the unique characteristics of AWS, emphasizing the importance of responsibility and accountability in ethical discussions. The lethal capacity of AWS should be considered along with autonomy and the absence of human control. Lethality is considered one of the components AWS and not an absolute component, but it radically shapes the direction of the ethical debate surrounding AWS. LAWS are a specific subset of AWS designed with a distinct purpose, namely, the deployment of lethal force. This sets them apart from the broader range of applications associated with AWS, such as anti-material actions, causing damage and facilitating destruction (Taddeo & Blanchard, 2022, pp. 37–39). Especially when it comes to LAWS, the key issue is not only how they are defined but also how we will delineate their boundaries ethically and legally. In this context, the most significant effort to establish ethical rules on a global scale for providing a legal framework is carried out by the open-ended 'Group of Governmental Experts (GGE)', established under 'the United Nations Convention on Certain Weapons (CCW)', as mandated by the UN Office for Disarmament Affairs (UNODA). The purpose of this group is 'to explore and agree on possible recommendations regarding options related to emerging technologies in the field of LAWS' (UNDOA, 2016). Certain Conventional Weapons (CCW), while formulating recommendations to be adhered to by all states, also endeavors to contribute to the development of IHL with ethical emphases such as transparency, accountability, and non-discrimination.

Autonomy and lethality derive their power from AI technologies, enabling systems to operate independently and make lethal decisions with enhanced efficiency and precision. The functionality of AWS does not hinge on the inclusion of AI, yet the integration of AI has the potential to enhance these systems. These systems often incorporate AI features in decision-making processes, target detection and tracking, improving shooting accuracy, decision-making, and adapting to environmental conditions. Therefore, the issue of AWS and specifically LAWS cannot be separated from AI technologies, both technically and ethically. In discussions surrounding AI, autonomy, and lethality in wars, it is observed that the concept of human control is often presented as the answer. It is thought that by positioning human control against algorithms in AI-DSS and AWS, ethical questions and issues can be handled. Nevertheless, AI-DSS themselves do not 'make' decisions. AI-DSS just exert a direct and frequently substantial influence on human decisions, particularly as a result of the cognitive limitations and inclinations of humans when engaging with machines (Stewart & Hinds, 2023). However, as the use of AI systems in warfare increases and the systems themselves become more complex, there is a growing reliance on AI systems while human control weakens and decision-making capabilities of humans. This indicates that prioritizing the ethical framework by focusing on responsibility as a moral code before making any legal regulations. In this context, Responsible AI (RAI) emerges as a new conceptualization in the ethics of AI. Adopting a comprehensive approach that includes novel concepts such as human responsibility and RAI is a prerequisite for a meaningful examination of the legal challenges surrounded by the relationship between AI and ethics. RAI aims to implement transparent and fair AI models while ensuring ethical and accountable use of technology. Key principles in RAI include fairness, explainability, privacy, and security. Specifically, fairness being defined as equity for stakeholders is foundational and has a negative correlation with bias. Facilitating the removal of bias and discrimination is enhanced through the principles of explainability, fairness, and transparency (Fjeld et al., 2020; Lee & Floridi, 2021). Additional principles cover organizational commitment, including leadership, accountability, governance, diversity, culture, humanity, training, etc. (Ratzan & Rahman, 2023, p. 1).

RAI may be a good approach; however, it is not sufficient in addressing the potential ethical challenges that could arise from the use of AI technologies in wars. This is because it does not eliminate the responsibility of the human who serves the designer, programmer, developer, and implementer of any AI system. In this context, if meaningful human control (MHC) is to be discussed, the issue of the responsibility gap should also be addressed. Indeed, when assessing the ethic of AI in wars, the discourse generally tends to focus on the concept of responsibility gap. This primarily brings the initial question especially in the use of AWS and the question revolves around the issue of who will assume or share responsibility when an AI-based system is used, or a decision is made relying on this system. This raises other questions about whether the responsibility will fall on the programmer, the manufacturer, or the institutional owner of the technology, and whether the responsibility will be distributed among relevant parties, or if there will be an environment where no one can be held accountable. However, if no one can be held responsible for the outcomes arising from the use of AI technologies, the basis of ethical discussions collapses. Yet, responsibility has a direct relationship with ethics and humanity, although how this relationship should be structured remains unclear. Amoroso and Giordano adopt a broader perspective in addressing the responsibility gap for AWS. In fact, their discussions provide a more comprehensive insight regarding responsibility gap on AI technologies. According to them, while individual criminal responsibility is often emphasized, the responsibilities of states and institutions should also be considered. However, states and institutions may not fully fill the responsibility gap. Even the adoption of a regime of no-fault liability may not offer a completely satisfactory solution. Therefore, the issue of responsibility for the use of AI-based technologies, particularly AWS, in warfare, is a complex problem that requires a complementary approach among responsibility regimes (Amoroso & Giordano, 2019). Matthias drew attention to these discussions by introducing the responsibility gap, which specifically focuses on the human aspect of the human-machine relationship (Matthias, 2004). Following this, the discussion about the responsibility gap has grown more complex, delving into moral and legal viewpoints to understand how a responsibility gap may emerge in the absence of human responsibility.

The issue of the responsibility gap only points out one aspect of the discussion. Nevertheless, emphasis should also be placed on accountability because responsibility necessitates being held accountable. Accountability primarily functions as a deterrent by employing credible threats of punitive measures against those engaging in harmful actions. Second, it establishes a clear responsibility for a specific actor to undertake measures ensuring compliance with pertinent legal considerations associated with the action. Third, accountability assumes a crucial moral dimension by assigning moral responsibility for an action. This involves identifying appropriate moral consequences invoking relevant moral emotions such as shame and guilt and recognizing the responsibility to make moral amends or redress. As a result, accountability plays a pivotal role as a deterrent, a legal framework, and a moral imperative, extending its significance beyond the context of warfare to encompass broader aspects of social life (Morgan et al., 2020, p. 33). In the context of operations supported by AI systems, the inability to determine responsibility indicates a deficiency in ethical understanding regarding the use of AI in wars. Failure to take responsibility can complicate the prevention and punishment of crimes occurring in wars. If individuals or institutions are not held accountable for these crimes, it can reduce the deterrent effect in wars and create a conducive environment for the recurrence of similar offenses. This situation can also lead to serious human rights violations and undermine the legality of wars under international law. Therefore, in situations involving crimes are committed, the determination of responsibility and accountability of these crimes within the use of AI should eliminate the responsibility gap that may arise. Otherwise, situations may arise where war crimes cannot be defined.

The concept of MHC is proposed as a solution in discussions on how to address the responsibility gap and accountability issue. Roff and Moyes highlight the importance of integrating MHC throughout the entire life cycle of AWS, not just during deployment. This includes design, development, and training phases before conflicts occur, along with maintaining human control during attacks and establishing accountability structures in the post-conflict period. They assert that the fundamental elements of MHC should form the basis for evaluating new technologies related to AWS (Roff & Moyes, 2016). Horowitz and Scharre outline three criteria to establish MHC, which encompasses three pivotal elements: (1) informed decision-making by human operators regarding weapon use, (2) provision of sufficient information to human operators for the lawful execution of actions, and (3) the utilization of well-designed, tested weapons by properly trained human operators to ensure effective control. These standards are designed to guarantee that commanders make deliberate decisions, have the necessary information for legal accountability, and exercise responsible use of weapons. Additionally, proper design, testing, and training contribute to effective weapon control and help mitigate unacceptable risks (Horowitz & Scharre, 2015, p. 4). Nevertheless, it should always be remembered that concept of the MHC is subject to interpretation, and there are challenges in achieving a consensus, especially considering the divergent perspectives of states possessing autonomous weapons and those without (Burri, 2018, p. 100). The interpretability of MCH can create a new gap and may give rise to new ethical debates not only in the development of AI-based systems but also in every decision during any war. Therefore, the focus should be on discussions concerning MHC, which determine the legitimization or delegitimization of future warfare. While some endorse the use of AI-based systems, notably AWS in warfare, others reject it due to moral and ethical concerns. However, one thing is clear that the legitimization of certain forms of algorithmic violence is already in progress (Ferl, 2024).

The lack of emotional capacities in AI technologies may signify a rational perspective over human judgments. Rationally made decisions in wars can foster an environment where warring parties can easily reconcile and contribute to peaceful endeavors. When considering the notion that AI systems are rational, it can be assumed that the actions of AWS and AI-DSS are predictable. Predictability refers to the extent to which the actions of a machine can be foreseen. According to Holland, autonomous systems exhibit three distinct facets of predictability or unpredictability. First two are 'technical predictability' and 'operational predictability'. Technical predictability is about the machine features and operational predictability is about the environment in which the machine operates. Both technical and operational predictability have potential to being unpredictable that leads to a third one. Third one is about meaning of unpredictability, and it matters on the extent to which one can predict the outcomes or effects of a system's utilization. Despite the predictability stemming from technology, every form of AWS introduces a certain level of 'operational unpredictability' (Holland Michel, 2020). While machines are immune to physical, cognitive, and psychological fatigue, they also pose risks associated with malfunctions and errors due to hardware failures. Insufficient data or algorithmic regulations also raise questions about the rationality of a decision. Therefore, there is a gap regarding the predictability of AI systems in wars. Moreover, evaluating the decisions and consequences of attacks in wars raises the inherent problem of the concept of rational decision-making. An action may seem rational and consistent due to its rationality but may not necessarily be ethical. Conversely, an ethical action is not obliged to be rational.

Currently, AI technologies provide a low level of predictability in complex and chaotic warfare environments, and there is no conclusive evidence that they are sufficiently secure. This situation renders the use of a weapon that is, to some extent, unpredictable and operates beyond human control ethically controversial, even if technological limitations are overcome in the future. In this case, the approach of AI4SG toward the use of AI in warfare can be considered a new topic for discussion. This discussion could lead to the development of interim formulas, such as deploying AI weapons only in areas devoid of civilian populations, such as deep sea, deserts, and space, to ensure predictability and reliability of AI weapons. One of the prominent views in the discussion of ethics and rationality is the idea that the occurrence of war crimes could be minimized by deploying killer robots. This is based on the rationale that robots operate on a rational decision basis, and unlike soldiers, they cannot lose control over their emotions, thereby reducing the likelihood of committing war crimes (Arkin, 2010). This argument raises ethical concerns, as it assumes that removing emotional elements from decision-making inherently leads to morally superior outcomes and supports the rational outputs. On the other hand, Schwarz emphasizes the distribution of morally complex decisions via technological interfaces, nodes, and system components and underscores the challenges in assuming responsibility and fully understanding the gravity of these decisions. Despite the complexity, Schwarz contends that taking a life should be a morally challenging undertaking, emphasizing the need to view it beyond mere technological choices (Schwarz, 2021, p. 69). At this juncture, the fundamental question it should be posed is whether MHC is feasible. This is since it has not been distinctly determined where the boundaries of human control on AI should commence. If one considers that the challenges within the realm of ethics of AI can be surmounted through the implementation of MHC, it is imperative to also account for the human touch and human sight. Both hold an exceptionally significant role in the emergence of an ethical understanding in people. The understanding gained through human touch and human sight is decisive in human beings' ability to evaluate the fairness or injustice of certain situations resulting from the individual and social consequences of behaviors. The inadequacy of AI technologies regarding human touch and human sight raises different debates in the ethics of AI. Within the realm of ethics of AI there exist a plethora of urgent ethical concerns, including but not limited to privacy, consent, bias, transparency, discrimination, accountability, fairness, reliability, safety, neutrality, privacy, security, and inclusivity. The moral concerns in question should be regarded as essential ethical lenses to be applied when utilizing AI technologies that have the potential to alter the character of war. At this point, the critical question is how these ethical lenses should be applied.

There's a risk that warfare might move far away from human-centered values and think more like machines. In this situation, people's ability to make ethical decisions may weaken, emphasizing the importance of understanding the ethical impact of evolving technology in modern wars. In addressing the expanding impact of AI technologies, there arises a question: how can ethical principles be maintained in warfare environments where human influence is diminishing? Leveringhaus highlights that in the ethical discussion of AI-based systems, spanning from healthcare to military applications, it is appropriate to find specific ethical solutions for certain technologies. At the same time, Leveringhaus notes that a universal ethical solution may not be possible for all robots (Leveringhaus, 2018, p. 41). Consequently, Leveringhaus advocates for a case-by-case ethical examination, considering the unique characteristics of each technology. Otherwise, there is a risk of getting stuck in the unresolved nature of ethical debates confined within patterns.

8.5 AI IN WARFARE: LEGAL CHALLENGES

The surrounding the relationship among war, ethics, and law is not as simple as it may initially seem. Ethical debates and legal debates are intricately linked, and these two spheres cannot be dissociated from one another. Ethics plays a significant role in shaping new and potential legal norms, especially in situations where new technologies pose challenges in implementation. In the context of international law, particularly the Martens Clause, reflects this important intersection between ethics and law. Those who recognize the inevitability of war often attempt to address the complex issue between war and ethics through the framework of the just war theory. The reason of that ethics leads to the establishment of formal codes of conduct in warfare, exemplified by the Hague and Geneva Conventions. These codes serve as guidelines for the drafting and enforcement of rules of engagement for states, armies, military personnel, as well as for the prosecution of individuals involved in war crimes, including soldiers and non-combatants. Certainly, the increasing role of AI in wars has brought the issue to a different dimension. At this point, the just war tradition and the law of war should continue to provide critical guiding considerations in wars. The just war theory asserts that war can be considered morally justifiable in cases of legitimate defense and protection of civilian populations as long as it adheres to established rules.

The foundation of the just war theory is based on humane and ethical grounds. In this context, just war establishes a broad moral framework by addressing the ethical aspects of the justice of war (jus ad bellum), the justice in war (jus in bello), and the ethical principles for going to war (jus ante bellum). However, in discussions of just war, especially when AWS are involved, the jus in bello holds particular significance. This is because jus in bello determines the ethical rules to be followed during war, and AI pose a serious challenge to these ethical rules and the fundamental three principles: the principle of necessity, the principle of proportionality, and the principle of discrimination. However, AI technology has challenged the fundamental principles of traditional just war theory, the law of war, and also human right law. Because AI-driven weapons can especially pose distinct risks of violations of jus ad bellum and jus in bello, which are inherent in the development of any weapon (Regan & Davidovic, 2023, pp. 1–2). Besides all, the principles of jus ante bellum which covers the exhaustion of diplomatic efforts and the legitimacy of the authority play a crucial role in determining whether resorting to war is ethically and legally justified by using AI. For this reason, comprehensive regulations regarding AI technologies need to be enacted in national and international legal frameworks which should encompass humanitarian law as well as human rights law.

Burri presents five arguments concerning significant aspects of AI and their impact on international law. The focal points include automation, personhood, weapon systems, control, and standardization. The first argument is that the surge of AI is transforming the legal landscape by automating routine tasks such as contract drafting. While machine learning demonstrates efficacy in legal assessments, its application to international law faces limitations due to the distinctive features of the field. The overarching conclusion is that international law and the tasks of international lawyers are improbable to be entirely automated. The second argument is about that artificially intelligent entities may be acknowledged as legal persons within the confines of national law; however, uncertainty shrouds the identification of who will shoulder the international legal obligations related to these entities in the global legal context. The third argument concerns the term MHC which encounters challenges in interpretation and consensus due to varying perspectives among states. Its inherent vagueness permits diverse interpretations. Despite addressing concerns about automated warfare, the true challenge lies in the deployment of AI and algorithms in warfare, characterized by a data-driven arms race. The regulation of MHC may divert attention from broader challenges posed by AI in warfare, encompassing its civilian applications. The fourth argument claims that artificially intelligent entities are deemed lawful when subject to human control, whereas AWS might be deemed unlawful unless under human control. This underscores the significance of contemplating legal precedents and fundamental rights in shaping the legal framework for AI, emphasizing the necessity of a comprehensive approach to tackle the challenges arising from AI across various legal domains. Finally, the swift advancement of AI, primarily led by companies and occasionally by states, has given rise to ethical concerns. Initiatives, like 'the IEEE's Global Initiative for Ethical Considerations in AI', strive to establish universal ethical standards for human-machine interactions. However, this standardization is unfolding beyond the traditional arenas of international law-making (Burri, 2018). Through this emphasis, Burri provides scholarly insights, raise awareness, and discourse concerning the intricate issues arising at the intersection of AI and international law.

The discussions on the development of legal regulations within the framework of MHC are generally focused on LAWS. In this context, several organizations, such as the ICRC and Article 36, have put forth a set of criteria that serves as a comprehensive guide for the factors that should be taken into consideration in determining MHC. The most recent sessions addressing LAWS at the Convention on CCW took place from May 15 to 19, 2023. Despite a decade of deliberations, the conclusion reached by the 'UN Convention on Certain Conventional Weapons (CCW) Group of Governmental Experts on Emerging Technologies in the Area of Lethal Autonomous Weapons Systems' was that IHL is entirely applicable to the emerging technologies associated with LAWS. However, the role of AI in warfare is more complex, and therefore, broader and more comprehensive discussions should take place within the legal framework.

International law has not overlooked the emergence of AI, and ethical principles have played a role in establishing legal norms. However, the legal norms do not directly confront the primary issues related to AI, such as bias and opacity. Instead, it takes a subtle procedural approach by indirectly tackling bias through transparency and data governance and while dealing with opacity requires enhancing interpretability (Christen et al., 2017, p. 118). Nations are duty-bound to adhere to the stipulations delineated in IHL, reflecting a commitment to the principles and regulations that govern the treatment of individuals during armed conflicts. This obligation underscores the imperative for states to conduct their military operations in accordance with the established norms of IHL, aiming to mitigate the impact of armed conflicts on civilians and vulnerable populations. The inherent responsibility to uphold IHL signifies a broader commitment to the protection of human rights and the preservation of human dignity within the complex landscape of international relations.

The distinction, the proportionality, and the necessity are three principles that form the core tenets of IHL and guide the ethical conduct of parties engaged in armed conflicts. These principles underscore a steadfast commitment to upholding human rights and safeguarding human dignity under all circumstances. The principle of proportionality in the law of targeting mandates a military commander to refrain from executing or to halt an attack if it is anticipated to cause excessive harm to civilians and critical infrastructures. The proportionality of means is codified in Article 51(5)(b) of the Additional Protocol I (ICRC, 2023b). This principle serves as the final assessment before carrying out a planned operation and requires ongoing monitoring during an ongoing attack. If, at any point, the commander determines that the attack violates the proportionality rule, they are obligated to cancel or suspend the attack. The principle of proportionality is generally discussed in the context of AWS and LAWS, but it should also be considered from the perspective of AI-DSS. Gunneflo and Noll explore the role of proportionality in AI-supported decision-making within IHL. Proportionality reasoning is becoming a central norm in IHL, influencing the rules, and allowing more discretion. It also highlights how proportionality facilitates the incorporation of emerging technologies, especially digital decision support with AI. The discussion underscores the shift from traditional legal distinctions to a more comprehensive compromise-seeking approach marked by law (Gunneflo & Noll, 2023, pp. 95–98). The design and use of AI technologies should adhere to this principle. This way, the level of force applied, and decision taken by these systems can be aligned with the intended military targets.

The principle of distinction is closely linked to the principle of proportionality. The principle of distinction is worded in Articles 48, 51(2), 52(2) of the Additional Protocol I, and also Article 13(2) of the Additional Protocol II (ICRC, 2023a) and emphasizes the necessity to differentiate between combatants and civilians, as well as between military objectives and civilian objects, during armed conflicts. This cardinal principle prohibits directing attacks against civilians and civilian objects by ensuring that only legitimate military targets are subject to attack. The principle of distinction contributes to the protection of civilians and the humane conduct of hostilities, seeking to uphold ethical standards during wars (ICRC, 2023b). AI technologies become a significant ethical and legal concern when employed in areas such as target identification and operational decision-making, raising questions about how the principle is applied and how the distinction between civilians and military targets is preserved. Especially, automatic target recognition (ATR) systems used by AWSs make AI technologies contentious within the framework of the principle of distinction. For Sharkey, assigning such critical decisions to a nonhuman agent constitutes a violation of the breach of distinction, given that current AI capabilities, and most likely future ones in the foreseeable future, do not align with the requirements of this principle (Sharkey, 2010, p. 378). This principle ensures that only legitimate military targets are subject to attack, which is crucial for preserving ethical codes in AI during warfare.

Just war theory places greater emphasis on the intended purpose of the means used rather than the technical characteristics. Therefore, the principle of necessity emphasizes using force only when absolutely needed for a legitimate military goal. It aims to minimize harm to civilians and maintain a balance between military requirements and humanitarian concerns during armed conflicts. The principle of necessity aligns most closely with the requirement for 'justified use' of AI by emphasizing the requirement to justify the purpose behind utilizing AI and mitigating potential harmful effects (Blanchard & Taddeo, 2022, p. 289). Considering these three principals at every stage, from the design to the use of AI, is essential for mitigating the challenges between AI and IHL.

All these principles are complemented by the Martens Clause which introduced for the first time in the Preamble of the 1899 Hague Convention. The Martens Clause serves as a guide in the application of IHL and allows recourse to the general principles of civilized nations, the dictates of public conscience, and the principles of humanity in situations where existing rules of humanitarian law are inadequate or insufficient. This aims to make the laws of war more comprehensive and flexible to protect civilian populations and other vulnerable groups (ICRC, 1997). The legal framework might fall behind ethical standards and it might be necessitated occasional adjustments to align with evolving ethical considerations. In these situations, the Martens Clause offers a mechanism to facilitate this adaptation (Sparrow, 2017). Therefore, the Martens Clause provides a robust framework for the ethical and legal use of AI in the field of warfare, despite the absence of explicit legal regulations made by IHL. In the context of new technologies such as robot soldiers or unmanned ground vehicles (UGV), the Martens Clause implies that developers, states, and individuals still carry responsibility of fundamental ethical principles such as human dignity, protection of civilian populations, and prevention of unnecessary suffering in armed conflicts, even in the absence of explicit regulations or treaties governing their use.

As can be seen, the legal use of AI in warfare is developing two main axes. First, the advancement of AI technologies conforming to just war theory and IHL principles for use in the battlefield may not be possible. This assertion is based on the unpredictability of algorithms. That's why attributing responsibility to the algorithm is difficult and thereby creating a responsibility gap. All of this leads to uncertainty regarding the legal framework for the use of AI in warfare. On the other hand, the ethical development of AI technologies in line with ethical rules may not only strengthen the just war theory but also facilitate the implementation of IHL by potentially reducing civilian casualties. In determining which of these two axes will take precedence, the path traversed by AI technologies will be decisive.

8.6　CONCLUSION

Establishing a robust conceptual relationship among ethics, AI, and war is crucial for the development of effective policy measures and the implementation of laws in warfare. Therefore, each new discussion addressing the emerging ethical and legal challenges resulting from the advancement of AI technologies provides an opportunity to perceive the relationships among these factors more clearly and allows for the conceptual restructuring of their interconnections. However, the relationship is becoming increasingly complex and challenging as AI technologies in wars continue to advance and their usage becomes more prevalent by adding layers of difficulty to the intersections of war, ethics, law, and technology. In this complex relationship, human dignity, privacy, justice, transparency, accountability, predictability, non-discrimination, responsibility, and human control are fundamental ethical principles that must be considered at every stage of the design, development, and use of AI technologies. There is a broad consensus on these principles. Nevertheless, there is a lack of clarity on how these ethical principles should be implemented within a legal framework. This constitutes the fundamental challenge in the ethical and legal dimensions of AI. To address these challenges, it is essential to establish a robust ethical foundation for legal codes and develop transparent policy measures that define the limits of AI technology usage. This way, the use of AI in future wars can be aligned with human rights and legal standards.

Indeed, there is a consensus on the necessity of establishing regulations regarding AI-based new wars; however, uncertainty persists regarding how these regulations will be formulated. Currently, LAWS take precedence in the regulations concerning the use of AI technologies in wars. However, regulations are required not only for LAWS but also for unarmed AI technologies such as decision aids, detention or military human enhancement technologies such as exoskeletons, neuro-enhancement devices, and augmented reality systems. In this context, international discussions and negotiations have intensified, particularly among countries such as the USA and China. Even though these countries have made some national arrangements, these arrangements remain limited. By the way, the negotiations for regulations that started within the UN, EU, and NATO are important in the military use of AI technologies. However, harmonization between national regulations and international regulations should be considered a separate issue. MHC and RAI are among the prominent topics in the negotiations on the regulations regarding the ethics of AI.

The fundamental issue arises not from needs assessment but rather from the emphasis on ethical considerations and the incorporation and application of these ethical issues within the legal framework. Therefore, there is still a long way to go in in defining the structure of the relationship between AI, ethics, and law in the context of war. In summary, the advent of AI technologies has the potential to bring about significant changes in the character of war. These changes have ethical and legal consequences that must be carefully considered.

REFERENCES

ACM Ethics. (1997). *Software engineering code*. The Committee on Professional Ethics. https://ethics.acm.org/code-of-ethics/software-engineering-code/

AI100. (2021). *One hundred year study on artificial intelligence (AI100)*. Stanford University. https://ai100.stanford.edu/

Allen, C., Smit, I., & Wallach, W. (2005). Artificial morality: Top-down, bottom-up, and hybrid approaches. *Ethics and Information Technology, 7*(3), 149–155. https://doi.org/10.1007/s10676-006-0004-4

Amoroso, D., & Giordano, B. (2019). Who is to blame for autonomous weapons systems' misdoings? In E. Carpanelli & N. Lazzerini (Eds.), *Use and misuse of new technologies* (pp. 211–232). Springer International Publishing. https://doi.org/10.1007/978-3-030-05648-3_11

Anderson, M., & Anderson, S. L. (2011). *Machine ethics*. Cambridge University Press. https://www.google.com/books?hl=tr&lr=&id=N4IF2p4w7uwC&oi=fnd&pg=PP1&dq=Machine+ethics&ots=5ZUZsmfUMq&sig=drh0bj4xSyoLkrytsnYdOCmw5oI

Arkin, R. C. (2010). The case for ethical autonomy in unmanned systems. *Journal of Military Ethics, 9*(4), 332–341. https://doi.org/10.1080/15027570.2010.536402

Atomium European Institute. (2023). *AI4People*. Atomium-EISMD. https://eismd.eu/ai4people/

Blanchard, A., & Taddeo, M. (2022). Jus in bello, necessity, the requirement of minimal force, and autonomous weapons systems. *Journal of Military Ethics, 21*(3–4), 286–303. https://doi.org/10.1080/15027570.2022.2157952

Borgmann, A. (1984). *Technology and the character of contemporary life: A philosophical inquiry*. University of Chicago Press. https://www.google.com/books?hl=tr&lr=&id=qS3pJL_BcdkC&oi=fnd&pg=PA1&dq=Technology+and+the+character+of+contemporary+life:+A+philosophical+inquiry&ots=TGg15jCHTK&sig=bKxhLlOUSwG5J9zq78V2IvFc0cM

Burri, T. (2018). International law and artificial intelligence. *German Yearbook of International Law, 60*(1), 91–108. http://dx.doi.org/10.2139/ssrn.3060191

CACDA-China Arms Control and Disarmament Association. (2019). *The militarization of artificial intelligence*. UNODA. https://digitallibrary.un.org/record/3972613

CEB. (2022). *Chief executives board for coordination*. UN System Chief Executives Board for Coordination. https://unsceb.org/sites/default/files/2022-09/Principles%20for%20the%20Ethical%20Use%20of%20AI%20in%20the%20UN%20System_1.pdf

Cervantes, J.-A., López, S., Rodríguez, L.-F., Cervantes, S., Cervantes, F., & Ramos, F. (2020). Artificial moral agents: A survey of the current status. *Science and Engineering Ethics, 26*(2), 501–532. https://doi.org/10.1007/s11948-019-00151-x

Christen, M., Burri, T., Chapa, J., Salvi, R., Santoni de Sio, F., & Sullins, J. (2017). *An evaluation schema for the ethical use of autonomous robotic systems in security applications*. University of Zurich Digital Society Initiative White Paper Series, 1. https://papers.ssrn.com/sol3/papers.cfm?abstract_id=3063617

Council of the European Union. (2020). *Presidency conclusions—The charter of fundamental rights in the context of artificial intelligence and digital change. 11481/20*. European Council. https://www.consilium.europa.eu/media/46496/st11481-en20.pdf

Cowls, J., Tsamados, A., Taddeo, M., & Floridi, L. (2021). A definition, benchmark and database of AI for social good initiatives. *Nature Machine Intelligence, 3*(2), 111–115. http://dx.doi.org/10.2139/ssrn.3826465

De Spiegeleire, S., Maas, M., & Sweijs, T. (2017). *Artificial intelligence and the future of defense: Strategic implications for small and medium sized providers* (pp. 2–5). The Hague Centre for Strategic Studies.

Deng, B. (2015). Machine ethics: The robot's dilemma. *Nature, 523*(7558), 24–26. https://doi.org/10.1038/523024a

Durt, C. (2022). Artificial intelligence and its integration into the human lifeworld. In W. Burkhard, P. Kellmeyer, O. Müller, & S. Vöneky (Eds.), *The Cambridge handbook of responsible artificial intelligence: Interdisciplinary perspectives* (pp. 67–82). Cambridge University Press. https://www.durt.de/publications/ai-lifeworld/

European Commission. (2018). *Statement on artificial intelligence, robotics and "autonomous" systems*. Brussels, 9 March 2018. Publications Office of the European Union. https://op.europa.eu/en/publication-detail/-/publication/dfebe62e-4ce9-11e8-be1d-01aa75ed71a1

European Commission. (2019). *Ethics guidelines for trustworthy AI | Shaping Europe's digital future*. European Commission. https://digital-strategy.ec.europa.eu/en/library/ethics-guidelines-trustworthy-ai

European Council. (2017). *European Council meeting (19 October 2017) – Conclusion. EUCO 14/17*. European Council. https://www.consilium.europa.eu/media/21620/19-euco-final-conclusions-en.pdf

European Parliament. (2023). *EU AI Act: First regulation on artificial intelligence*. European Parliament News. https://www.europarl.europa.eu/topics/en/article/20230601STO93804/eu-ai-act-first-regulation-on-artificial-intelligence

Ferl, A.-K. (2024). Imagining meaningful human control: Autonomous weapons and the (de-) legitimisation of future warfare. *Global Society, 38*(1), 139–155. https://doi.org/10.1080/13600826.2023.2233004

Fjeld, J., Achten, N., Hilligoss, H., Nagy, A., & Srikumar, M. (2020). *Principled artificial intelligence: Mapping consensus in ethical and rights-based approaches to principles for AI*. Berkman Klein Center Research Publication, 2020–1. https://papers.ssrn.com/sol3/papers.cfm?abstract_id=3518482

Floridi, L. (2023). *The ethics of artificial intelligence*. Oxford University Press.

Formosa, P., & Ryan, M. (2021). Making moral machines: Why we need artificial moral agents. *AI & Society, 36*, 839–851. https://doi.org/10.1007/s00146-020-01089-6

Future of Life. (2017). *Asilomar AI principles*. Future of Life Project. https://futureoflife.org/open-letter/ai-principles/

Goetze, T. S. (2022). *Mind the gap: Autonomous systems, the responsibility gap, and moral entanglement*. 2022 ACM Conference on Fairness, Accountability, and Transparency, 390–400. https://doi.org/10.1145/3531146.3533106

Gunneflo, M., & Noll, G. (2023). Technologies of decision support and proportionality in international humanitarian law. *Nordic Journal of International Law, 92*(1), 93–118. https://doi.org/10.1163/15718107-bja10055

Holland Michel, A. (2020). *The black box, unlocked: Predictability and understandability in military AI*. United Nations Institute for Disarmament Research. https://unidir.org/sites/default/files/2020-09/BlackBoxUnlocked.pdf

Horowitz, M. C. (2018). *Artificial intelligence, international competition, and the balance of power*. 2018, 22. https://www.google.com/books?hl=tr&lr=&id=b646EAAAQBAJ&oi=fnd&pg=PA372&dq=Artificial+intelligence,+international+competition,+and+the+balance+of+power&ots=WD_IeLbZ7I&sig=oRCmonjYlFqpSvBoXfW0Ao_yLr8

Horowitz, M. C., Allen, G. C., Kania, E. B., & Scharre, P. (2018). *Strategic competition in an era of artificial intelligence center for a new American security*. https://www.cnas.org/publications/reports/strategic-competition-in-an-era-of-artificial-intelligence

Horowitz, M. C., & Scharre, P. (2015). *Meaningful human control in weapon systems*. Center for a New American Working Paper. https://www.files.ethz.ch/isn/189786/Ethical_Autonomy_Working_Paper_031315.pdf

ICRC. (1997). *The martens clause and the laws of armed conflict*. International Review of the Red Cross; 1. https://www.icrc.org/en/doc/resources/documents/article/other/57jnhy.htm

ICRC. (2023a). *International humanitarian law databases*. International Committee of the Red Cross. https://ihl-databases.icrc.org/en/customary-ihl/v1/rule1

ICRC. (2023b). *The principle of proportionality*. International Committee of the Red Cross. https://www.icrc.org/sites/default/files/wysiwyg/war-and-law/04_proportionality-0.pdf

Johnson, J. (2019). Artificial intelligence & future warfare: Implications for international security. *Defense & Security Analysis, 35*(2), 147–169. https://doi.org/10.1080/14751798.2019.1600800

Kant, I. (1993). *Groundwork of the metaphysic of morals*. Cambridge University Press. https://www.taylorfrancis.com/chapters/edit/10.4324/9780203714805-2/groundwork-metaphysic-morals-immanuel-kant

Kuipers, B. (2020). Perspectives on ethics of AI. In M. D. Dubber, F. Pasquale, & S. Das (Eds.), *The Oxford handbook of ethics of AI* (pp. 409–429). Oxford University Press. https://www.google.com/books?hl=tr&lr=&id=9PQTEAAAQBAJ&oi=fnd&pg=PA421&dq=Perspectives+on+Ethics+of+AI&ots=yRlTJsPLGK&sig=Z2EugSGIbfh9nwzvRKevwIt5FIw

Larson, E. J. (2021). *The myth of artificial intelligence: Why computers can't think the way we do.* Harvard University Press. https://doi.org/10.4159/9780674259935

Lee, M. S. A., & Floridi, L. (2021). Algorithmic fairness in mortgage lending: From absolute conditions to relational trade-offs. *Minds and Machines, 31*(1), 165–191. https://doi.org/10.1007/s11023-020-09529-4

Leveringhaus, A. (2018). Developing robots: The need for an ethical framework. *European View, 17*(1), 37–43. https://doi.org/10.1177/1781685818761016

Lonsdale, D. J. (2004). *The nature of war in the information age: Clausewitzian future.* Routledge.

Macgregor, K., & Williamson, M. (2001). Thinking about revolutions of warfare. In M. Knox & W. Murray (Eds.), *The dynamics of military revolution 1300–2050* (pp. 1–14). Cambridge University Press.

Matthias, A. (2004). The responsibility gap: Ascribing responsibility for the actions of learning automata. *Ethics and Information Technology, 6*(3), 175–183. https://doi.org/10.1007/s10676-004-3422-1

Montreal University. (2017). Montreal declaration for a responsible development of artificial intelligence. Montréal Declaration on Responsible AI: An Initiative of Université de Montréal. https://recherche.umontreal.ca/english/strategic-initiatives/montreal-declaration-for-a-responsible-ai/

Morgan, F. E., Boudreaux, B., Lohn, A. J., Ashby, M., Curriden, C., Klima, K., & Grossman, D. (2020). *Military applications of artificial intelligence.* RAND Corporation. https://www.rand.org/content/dam/rand/pubs/research_reports/RR3100/RR3139-1/RAND_RR3139-1.pdf

Ozlem, D. (2004). *Etik.* İnkilap Kitapevi.

Partnership on AI. (2023). *Partnership on AI – Home.* Partnership on AI. https://partnershiponai.org/

Powers, T. M., & Ganascia, J.-G. (2020). The ethics of the ethics of AI. In M. D. Dubber, F. Pasquale, & S. Das (Eds.), *The Oxford handbook of ethics of AI* (pp. 25–51). Oxford University Press. https://www.google.com/books?hl=tr&lr=&id=9PQTEAAAQBAJ&oi=fnd&pg=PA27&dq=The+Ethics+of+ethics+of+AI&ots=yRlTJsLPGT&sig=fCtpkaPCx2gJmOpSv7_cTbftctM

Ratzan, J., & Rahman, N. (2023). Measuring responsible artificial intelligence (RAI) in banking: A valid and reliable instrument. *AI and Ethics,* 1–19. https://doi.org/10.1007/s43681-023-00321-5

Regan, M., & Davidovic, J. (2023). Just preparation for war and AI-enabled weapons. *Frontiers in Big Data, 6,* 1020107. https://doi.org/10.3389/fdata.2023.1020107

Roff, H. M., & Moyes, R. (2016). *Meaningful human control, artificial intelligence and autonomous weapons.* Briefing Paper Prepared for the Informal Meeting of Experts on Lethal Au-Tonomous Weapons Systems, UN Convention on Certain Conventional Weapons. https://article36.org/wp-content/uploads/2016/04/MHC-AI-and-AWS-FINAL.pdf

Russell, S., Dewey, D., & Tegmark, M. (2015). Research priorities for robust and beneficial artificial intelligence. *AI Magazine, 36*(4), Article 4. https://doi.org/10.1609/aimag.v36i4.2577

Schwarz, E. (2021). Autonomous weapons systems, artificial intelligence, and the problem of meaningful human control. *Philosophical Journal of Conflict and Violence, 5*(1), 52–72. https://doi.org/10.22618/TP.PJCV.20215.1.139004

Sharkey, N. (2010). Saying 'No!' to lethal autonomous targeting. *Journal of Military Ethics, 9*(4), 369–383. https://doi.org/10.1080/15027570.2010.537903

Shaw, W. H. (1999). *Contemporary ethics: Taking account of utilitarianism* (1st ed.). Wiley-Blackwell.

Sparrow, R. (2017, November 17). *Ethics as a source of law: The martens clause and autonomous weapons.* Humanitarian Law & Policy Blog. https://blogs.icrc.org/law-and-policy/2017/11/14/ethics-source-law-martens-clause-autonomous-weapons/

Stahl, B. C. (2021). Concepts of ethics and their application to AI. In B. C. Stahl, *Artificial intelligence for a better future* (pp. 19–33). Springer International Publishing. https://doi.org/10.1007/978-3-030-69978-9_3

Stewart, R., & Hinds, G. (2023, October 24). *Algorithms of war: The use of artificial intelligence in decision making in armed conflict.* Humanitarian Law & Policy Blog. https://blogs.icrc.org/law-and-policy/2023/10/24/algorithms-of-war-use-of-artificial-intelligence-decision-making-armed-conflict/

Taddeo, M., & Blanchard, A. (2022). A comparative analysis of the definitions of autonomous weapons systems. *Science and Engineering Ethics, 28*(5), 37. https://doi.org/10.1007/s11948-022-00392-3

Tolmeijer, S., Kneer, M., Sarasua, C., Christen, M., & Bernstein, A. (2021). Implementations in machine ethics: A survey. *ACM Computing Surveys, 53*(6), 1–38. https://doi.org/10.1145/3419633

Ülgen, O. (2017). Kantian ethics in the age of artificial intelligence and robotics. *Questions of International Law, 43*(1), 59–83.

UN. (2023). *Interim report: Governing AI for humanity.* UN Secretary-General's AI Advisory Body; United Nations. https://www.un.org/en/ai-advisory-body

UNDOA. (2016). *CCW convention, fifth review conference, report of the 2016 informal meeting of experts on lethal autonomous weapons systems, CCW/CONF.V/2.* UNDOA. https://docs-library.unoda.org/Convention_on_Certain_Conventional_Weapons_-_Fifth_Review_Conference_(2016)/FinalDocument_FifthCCWRevCon.pdf

UNESCO. (2021). *Recommendation on the ethics of artificial intelligence.* UNESCO. https://www.unesco.org/en/articles/recommendation-ethics-artificial-intelligence

Vishwanath, A., Bøhn, E. D., Granmo, O.-C., Maree, C., & Omlin, C. (2023). Towards artificial virtuous agents: Games, dilemmas and machine learning. *AI and Ethics, 3*(3), 663–672. https://doi.org/10.1007/s43681-022-00251-8

Wood, A. W. (2012). *Kantian ethics.* Stanford University.

9 Artificial Intelligence in Strategic Planning and Military Operations

August Cole, Don Howard, Robert Latiff, George Lucas,
Gregory M. Reichberg, and Kaushik Roy

9.1 INTRODUCTION

There has been an enormous increase in the integration of artificial intelligence (AI) into advanced military systems, by some accounts at least doubling in the last five years. This, of course, follows and is a result of a correspondingly large growth in the so-called Internet of Things (IoT). It is a simple fact that when almost "everything is connected to everything," the overwhelming amount of data and enormous numbers of network connections are far too complex to be controlled conventionally. While integrating this ocean of information poses enormous practical challenges in terms of data management and policy, it is already clear that this complex situation lends itself well to mediation by AI systems.

In the military domain, recent breakthroughs in robotics, software, satellites, and autonomy present defense leaders with a new challenge to consider anew the technological possibilities as they relate to the conduct of military operations and the attendant ethical questions. There is a particularly strong operational focus today on areas such as robotic ground, air, and sea systems; largely autonomous anti-armor standoff weapons, and lower cost commercial geospatial targeting and intelligence data. Whichever nations – or other non-state groups – adopt these technologies will likely find that there are technological asymmetries that can be turned into advantages. With the framework that AI will impact every level of military operations, the current Ukrainian conflict shows that:

- Commercial space satellites and AI data processing present unprecedented access to communications/bandwidth and targeting information for military forces.
- Laws and norms, particularly enforceability, lag in shaping or restricting new weapons systems.
- Pervasive mobile devices and resilient mobile connectivity integrated with AI-powered apps offer visual, location, and other battlefield data to all parties in a conflict, while also drawing in civilian participation in new ways.
- Integrating AI, drones, and software into modern military operations is impossible without civilian technical expertise present close to military operations.

9.2 THE EMERGENCE OF ALGORITHMIC WARFARE

Dr. Alex Karp, CEO of Palantir Engineering, labels this new, AI-driven integrated and comprehensive form of warfare "algorithmic warfare." With strong support from the US Department of Defense and the Chairman of the US Joint Chiefs of Staff at the time, Army General Mark Milley, Palantir's "Meta-Constellation" uses complex and sophisticated AI algorithms to integrate otherwise incomprehensibly vast arrays of data from discrete sources, both public and private, in order to provide a comprehensive, real-time operational picture of discrete regions of military conflict in Ukraine directly to commanders in those battlefields, as well as to their supervisors in Kyiv. Sources

DOI: 10.1201/9781003441700-9

of vital strategic information include weather reports from NOAA, GPS positional data, photos of enemy positions and equipment taken by individual citizens in the vicinity of combat zones and uploaded to social media feeds from commercial satellites (such as Elon Musk's privately funded commercial "Starlink" system). "Meta-constellation" reportedly can assimilate and analyze all this information and combine it with data sets provided by more secure military systems (e.g., NATO intelligence services) to establish a comprehensive perspective of real-time events and enable a flexible and optimized response by commanders in the field. This genuinely new use of AI by Ukrainian armed forces has been credited as a key source of their remarkable success in confronting a much larger and more fully equipped invasion force (Ignatius, 2022).

US military operations are now using AI systems, as well, for not just imagery analysis but targeting. In September 2021, the Secretary of the US Air Force revealed publicly at a military and industry event that the service's targeting protocols already have incorporated AI capabilities to locate targets in a fusion of intelligence and military effects. This revelation offers a concrete confirmation of the potential that current systems have already and forfends their growing ubiquity (Amanda, 2021).

Such new developments in applied uses of AI invite military planners and strategists to consider a *single comprehensive operational picture* through networking all available and accessible data silos, repositories, and battlespace perspectives at their disposal in a military operation. In the US, for example, this vision is a key component of the US Army's "Vantage" program, a successor to its earlier "Project Maven" technology enterprise that employs data analytics platforms like Meta-Constellation and even OpenAI's "ChatGPT" to leverage data-driven strategic decision-making. Consider the prospects, promises, and potential pitfalls of the three areas singled out above.

9.2.1 Small Satellites

Small satellites such as those that make up the Starlink communications (a small satellite "mega-constellation") and others whose multi-spectral sensors gather information about Earth's surface have major implications for military operations. This imagery and data can be bought on the open market by everybody from governments to NGOs to individuals. As with other areas, more data can bring more problems in terms of integration, user access, and restrictions. Yet AI software is at the center of processing such data by sorting it or finding patterns, and it is also at the cutting edge of how it is being used. For the Ukrainian military, space-based surveillance afforded its units to have situational awareness enabling targeting and intelligence collection against Russian military forces. The growing demand for commercial space data and its AI processing will further change a crucial dynamic for the kinetic aspects of modern warfare: Military units trying to evade detection, and therefore targeting by new generations of weapons such as hypersonic missiles and drone swarms (Eftimiades, 2022). As high-speed and ultra-precise weapons become more commonplace while air-defense weapons become more effective in creating highly contested environments, finding targets from space will only grow in importance. Further, as the Starlink space-based internet connectivity platform showed, the military effectiveness of commercial space systems is of immediate benefit and can be quickly integrated into military information technology networks (Woody, 2022).

9.2.2 Increased Use of Robotic Systems

If there was any doubt that the 2020s would point toward the future of autonomy, AI, and warfare, the use of military robots starting in the 2020 Armenia-Azerbaijan conflict (Reed & Rife, 2022) put skeptics on notice. As we can see in Ukraine presently, even a rudimentary level of control and semi-autonomy exists. No longer the domain of large weapons systems from the world's best funded militaries, small and (relatively) inexpensive drones can now be sourced worldwide and equipped with AI-powered target-recognition systems. Ukraine fielded TB2 armed drones from Turkey and Russian forces continue to fly Iranian Shahed-136 and other drones supplied by Tehran. Russia's experience in Ukraine fielding imported Iranian drones shows a willingness to regularly employ them in warfare for surveillance and strike missions, the latter of which is effectively a loitering munition (Trofimov & Nissenbaum, 2022). Though AI is not yet a central capability, Russia's use of Iranian drones points to how robotic systems can be rapidly adopted and integrated into military operations. Further onboard targeting capabilities from onboard AI-powered image analysis would likely make such systems even more desirable – and effective.

For Western nations, the same holds true in terms of operational necessity when potential competitors advance robotic warfare even further. In a sign of the centrality of such systems, NATO's Data and Artificial Intelligence Review Board is developing rules and guidelines for the responsible use of autonomy that should be completed by the end of 2023 (NATO, 2023).

9.2.3 SMARTPHONES AND AI

The smartphone may be the most exponentially relevant frontier in AI and warfare due to its ubiquity, affordability, and the almost horizonless potential of regularly updated software. The prevalence of handheld devices in conflict zones means they can be overlooked for their tactical and strategic implications compared to, say, loitering attack drones or hypersonic missiles. Yet as Ukraine shows, there is an AI-powered fusion of image and text processing and generation, data collection, and information-warfare-style cognitive campaigns using personalized social media and other vectors, and electronic targeting capabilities exploiting the use of mobile devices by military personnel and civilians alike. As with the small satellite and AI nexus, it is harder and harder for military units to hide on battlefields where operational transparency is the norm. As Ukraine's software developers showed by creating apps to warn civilians of impending air raids by Russian forces, mobile devices raise urgent questions about the development of software and data that can directly limit harm to vulnerable non-combatants. Yet there are clear offensive uses too, even in the hands of civilians – all with the same device that in the span of a few minutes can be used to save a family from an incoming missile and target a nearby Russian tank unit. The Ukrainian eVorog app helps users target Russian equipment or personnel and this in turn is screened by AI-powered systems such as Palantir's "Meta-Constellation" (Bajraktari, 2022; Harwell, 2022).

The duality of such devices and their software is further evident when it comes to the new crop of generative AI software capable of creating text and images that are interchangeable with human-created content. This goes right to the heart of the battle over information – and disinformation in conflict zones around the world. Large language models or LLMs such as Open AI's ChatGPT or Google's Bard and image generation software like Stable Diffusion or Dall-E give an individual the tools to wage a campaign quite easily for attention and influence that, enabled by social media or other software employing these generative tools, is unprecedented. Indeed, whether by individuals or nations, these AI tools foretell an era of cognitive warfare that will see the precision targeting of individuals at scale using new AI systems (Cole & Le Guyader, 2023). Not all systems are the product of major companies. Open-source generative AI tools are readily available, such as Bloom, an LLM text model from Hugging Face. Early in the conflict in Ukraine, a spoofed video of Ukrainian President Volodymyr Zelensky ordering Ukrainian forces to stand down emerged as a harbinger for what is to come (Allyn, 2022).

All this is to caution that AI's promising developments are not without serious downsides and reservations, especially when pursued by reckless and irresponsible state and non-state actors who may not be constrained by the guardrails specified by, for example, ChatGPT itself. The longstanding debate over whether to deploy autonomous lethal weapons becomes even more urgent when paired with the continued advances in AI. Recently, however, concern has also turned to the expanding use of AI in the aforementioned decision-making support systems and command and control (C2) systems for military senior leaders.

On the one hand, as per the foregoing accounts from Ukraine, the *in-bello* use of AI and other advanced technology in military systems is understandable. The nation wants to equip its fighting forces with the best tools available to prevail in conflict. In the past, new technologies for weapons fell mostly in the physical realm – faster jets, more accurate missiles, more lethal explosives, and greater survivability. While new technologies today include these advances, the most significant breakthroughs are increasingly based in the digital or non-physical realm. Whereas before, the effects of new technologies were more deterministic, today they are incredibly complex, adaptive, and far more unpredictable in their behavior and outcomes.

The rapidly increasing use of AI in weapons and C2 systems to stay ahead of competitors pursuing the same edge carries with it risks. It is predicated on assumptions that require careful ethical examination, especially when this rush to expand the use of military AI up the chain of command either assumes such systems and their data sets will behave as intended or brushes aside the known shortcomings and vulnerabilities. More importantly, the use of AI in military systems in many cases usurps – or could usurp – human authority and decision-making in tasks that demand human agency. The assessment of conditions for the application of lethal force, the decision to put lives at stake in going to war, and the behavior of personnel and weapons in war are all human activities, still optimally controlled and carried out by human commanders.

9.3 CURRENT AND FUTURE AI USAGE FOR MILITARY STRATEGY

9.3.1 PROSPECTIVE USES OF AI IN NATIONAL SECURITY SYSTEMS

In the conduct of war, activities are often referred to in relation to the so-called OODA loop, comprising *Observe, Orient, Decide, and Act* (McIntosh, 2011). The OODA concept was coined by iconoclastic military innovator John Boyd, a US Air Force officer whose influences shaped some of the most important 20th-century weapons systems, including the F-16 Fighting Falcon, which went on to become the most widely produced US fighter ever and is currently being supplied to Ukraine nearly 50 years after its first flight. With this framework, he sought to create a systemic way of understanding military decision-making for both operational insights and to develop novel tactics disrupting that cycle.

Each of these phases in a military's OODA loop requires and employs data and, as the earlier examples in Ukraine illustrate, is being significantly reshaped by AI.

War and violent conflicts are preceded by some form of intelligence preparation of the battlefield, the *observe* function. In orienting oneself to the realities of the battlefield, commanders will need to call upon past experience and training. They will act based upon the creation of several courses of action and simulation of outcomes based on probable enemy reactions. Finally, in the actual battle, data will be employed in various automated and possibly autonomous weapons. Each side attempts to collect as much data as possible on the area, terrain, population, and other external factors. Each side also tries to know all it can about the adversary's intentions and capabilities. AI will play an important role in gathering and analyzing large amounts of data.

Based on the data collected and the commander's assessment of the situation, courses of action are developed (*orient*), commander's intent is disseminated, commands are issued (*decide*), and forces are controlled (*act*). As adversaries engage on the battlefield, weapons (both physical and cyber) are employed to close with, and defeat, the enemy. We are witnessing the OODA process integrated with AI unfold in real time on the battlefields in Ukraine.

As military decision-making and analysis AI systems proliferate, it is worth reexamining the OODA concept, which itself dates back to 1986. In the current and near-future of human decision-making amidst rapid AI innovation and implementation, it is worth reconsidering where a human will fit into the OODA loop for strategic, as well as tactical, decision-making. This is another way of asking which of the OODA elements are to be effectively automated due to ubiquitous sensors and data, as well as ever-faster software processing speeds. Rather than the OODA, it is likely we are arriving at an O-A loop, with human inputs bound to Orient-Act, while autonomous systems handle the Observe (what a system is "seeing") and Decide (narrowing or selecting a sequence or unitary action). Human direction, much like today's prompts of a generative AI system, remains important with military decision-making, as does the ultimate "act," for reasons of ethics, values, and law – for now. It is also possible that strategic decision-making may see human inputs limited, necessarily, to the Orient, especially if adversary systems are unshackled to allow machine-speed military operations. Whether those will indeed be more effective than operations managed or executed by human commanders remains to be seen. But no military can afford to rest upon the assumption that a tentative approach will be a victorious one, especially at the strategic level. Even so, the ethical considerations of this moment need to be kept at the fore, just as they have with prior periods of military technological breakthrough.

Former Secretary of Defense Ashton Carter who helped usher in the US regulations and rules on AI that postulated a strong human hand in overseeing expanding autonomy, wrote, "While some advancements in AI are breathtakingly new, their novelty should not be exaggerated. Right and wrong are certainly not new" (Carter, 2022).

9.3.2 AI-Enhanced Situational Awareness

As with influencing a tightening of the OODA loop through automating data collection and decision-making, AI has the potential to significantly enhance military planners' capabilities in intelligence analysis, situation assessment, threat detection, reconnaissance, and target recognition. The following taxonomy underscores the near-term ways that AI will transform military decision-making through evolved perception and situational awareness.

1. **Intelligence Analysis:** AI can analyze vast amounts of data, including satellite imagery, signals intelligence, social media feeds, and other open-source intelligence, to identify patterns, detect anomalies, and extract relevant information. Machine learning algorithms can automatically process and categorize this data, assisting military planners in identifying potential threats, predicting enemy activities, and assessing the overall intelligence picture.
2. **Satellite Data and Global-Scale Monitoring:** AI algorithms can analyze satellite data, such as high-resolution imagery, to provide real-time monitoring of global activities. By combining AI with computer vision techniques, it becomes possible to automatically detect changes on the ground, track movements of military assets, identify new constructions (such as Chinese airfields on artificial islands in the South China Sea), and monitor critical infrastructure. This enables military planners to gather valuable intelligence and keep track of potential threats and adversaries.
3. **Situation Assessment:** AI can aid in rapidly assessing the evolving battlefield situation by analyzing multiple streams of data in real time. This includes data from sensors, surveillance systems, and unmanned aerial vehicles (UAVs). AI algorithms can process this information, recognize patterns, and provide real-time updates on the movement of enemy forces, changes in terrain, and other critical factors. This assists military planners in making informed decisions and adapting their strategies accordingly.
4. **Threat Detection:** AI can help identify potential threats by analyzing sensor data, such as radar, sonar, and infrared systems. Machine learning algorithms can learn to recognize patterns associated with different types of threats, such as enemy vehicles, aircraft, or vessels. By continuously monitoring sensor data and employing AI-based threat detection systems, military planners can receive timely alerts about potential hostile activities and take appropriate countermeasures.
5. **Reconnaissance:** AI-powered drones and autonomous systems can be used for reconnaissance missions, reducing the risk to human personnel. These systems can employ computer vision and machine learning algorithms to analyze visual data captured by UAVs or other surveillance platforms. AI can assist in identifying enemy positions, recognizing

military assets, and mapping the battlefield terrain. This information enhances situational awareness and enables military planners to devise effective strategies.

6. **Target Recognition:** AI can play a crucial role in automatic target recognition (ATR) systems. By training machine learning models on vast datasets of known targets, AI algorithms can learn to identify and classify objects of interest, such as enemy vehicles, aircraft, or personnel. This assists military planners in distinguishing between friendly and hostile forces, aiding in target prioritization and minimizing the risk of collateral damage.

Overall, by leveraging AI technologies to augment perception, military planners can enhance their intelligence analysis capabilities, gain real-time situational awareness, detect potential threats, conduct efficient reconnaissance, and improve target recognition. These advancements empower decision-makers with timely and accurate information, enabling them to make more informed choices and improve the efficiency of military operations.

9.3.3 Use of AI in Intelligence Systems

The intelligence cycle itself is heavily dependent on data and automated systems. From large geospatial databases to weapons data, to shipping and transportation, to large-scale simulations of collection platforms, to the ultimate processing and dissemination of information, automated systems play a crucial role. AI can play a significant role in enhancing the following types of national security intelligence systems:

1. **Image Analysis:** AI can be employed for advanced image analysis tasks, such as object recognition, facial recognition, and anomaly detection. It can help in identifying objects, individuals, or activities of interest in images or videos, aiding in surveillance and threat detection.
2. **Signal Analysis:** AI algorithms can assist in signal analysis by automatically processing and interpreting large volumes of data from diverse sources, such as communications intercepts, radar signals, or sensor networks. AI can help identify patterns, anomalies, and potential threats hidden within complex signal data.
3. **Document Analysis:** AI-powered techniques like natural language processing (NLP) can be used to analyze and extract information from various types of documents, including text-based intelligence reports, social media posts, and open-source materials. NLP algorithms can enable automated document categorization, sentiment analysis, named entity recognition, and summarization, facilitating efficient analysis of vast amounts of textual data.
4. **Financial Flow Analysis:** AI can aid in identifying suspicious financial activities and detecting money laundering or terrorist financing. By leveraging machine learning algorithms, AI systems can analyze financial transactions, detect patterns of behavior, and flag potentially illicit activities for further investigation.[1]
5. **Structured and Unstructured Data Analysis:** AI can assist in handling structured and unstructured data relevant to national security. Machine learning models can be trained to classify and categorize data, detect trends and anomalies, and perform predictive analytics. AI can analyze data from various sources, such as social media, news articles, sensor networks, and intelligence databases, providing valuable insights to intelligence analysts.

In all these applications, AI systems rely on large datasets for training and continuous learning. They can adapt and improve over time, increasing their accuracy and efficiency in performing intelligence analysis tasks. Once again, however, it is important to ensure ethical considerations, data privacy, and human oversight in the development and deployment of AI systems for national security purposes.

9.3.4 Principal Benefits from Use of AI in National Intelligence Operations

1. **Data Analysis and Pattern Recognition:** AI can process and analyze vast amounts of data quickly and efficiently. In intelligence and security operations, this capability allows for the identification of patterns, anomalies, and correlations that human analysts might miss. AI algorithms can sift through large volumes of structured and unstructured data, including texts, images, videos, and social media feeds, to identify relevant information and extract actionable insights.

2. **Enhanced Threat Detection:** AI can improve the detection and prediction of security threats. Machine learning algorithms can learn from historical data and real-time inputs to identify potential risks and suspicious activities. This can be applied to various security domains, such as cyber security, counterterrorism, border security, and surveillance systems. AI can identify emerging threats, detect unusual behaviors, and provide early warnings, enabling proactive response measures.

3. **Automation and Efficiency:** AI can automate repetitive tasks and streamline intelligence and security workflows. This allows human analysts to focus on higher level analysis and decision-making, leveraging the speed and accuracy of AI systems. By automating processes like data collection, data fusion, and report generation, AI can significantly reduce the time and effort required for routine tasks, improving operational efficiency.

4. **Predictive Analytics:** AI algorithms can utilize historical data and real-time inputs to generate predictive models. These models can help anticipate potential security threats, identify vulnerabilities, and forecast future trends. For example, AI-powered predictive analytics can assist intelligence agencies in identifying potential hotspots for criminal activities, predicting cyberattacks, or forecasting geopolitical developments.

5. **Risk Assessment and Mitigation:** AI can support risk assessment and mitigation strategies by analyzing complex scenarios and evaluating multiple factors simultaneously. AI algorithms can assess the likelihood and impact of security incidents, identify vulnerabilities, and recommend mitigation measures. This can aid in resource allocation, strategic planning, and optimizing security operations.

6. **Decision Support:** AI can provide decision support to intelligence and security professionals by presenting relevant information, alternative scenarios, and recommendations. By leveraging AI's analytical capabilities, decision-makers can have access to comprehensive insights and assessments, aiding them in making informed and timely decisions.

7. **Adaptive and Continuous Learning:** AI systems can continuously learn and adapt from new data, experiences, and feedback. This allows them to improve over time and enhance their performance. In intelligence and security operations, this adaptability enables AI systems to evolve alongside evolving threats and challenges.

9.3.5 Operational Enhancements

AI has the potential to significantly enhance military operations across domains and the entire operational cycle, from training to casualty evacuation.

1. **Simulation and Training:** AI can be used to create realistic virtual environments for military training, allowing personnel to practice complex scenarios without real-world risks. AI algorithms can generate intelligent adversaries and adaptive environments and simulate various combat situations to improve training effectiveness and decision-making skills.

2. **Logistical Support:** AI can optimize and automate logistics processes, including supply chain management, maintenance scheduling, and resource allocation. Machine learning algorithms can analyze historical data to predict equipment failure, optimize inventory levels, and plan efficient transportation routes, thereby improving overall logistical efficiency.

3. **Transportation:** AI can enhance transportation capabilities through autonomous vehicles, drones, and unmanned aerial systems. These technologies can improve the speed, safety, and precision of logistics and troop movement, reducing the risk to human personnel.
4. **Casualty Evacuation/Treatment:** AI-powered robots and drones can assist in casualty evacuation by swiftly navigating dangerous or inaccessible areas and providing immediate medical assistance. AI algorithms can also analyze medical data, assist in diagnosing injuries, and recommend appropriate treatment options, aiding healthcare providers in delivering effective care.
5. **Strategic and Tactical Planning:** AI can analyze vast amounts of data and provide real-time intelligence to support strategic and tactical decision-making. It can assist in identifying patterns, predicting enemy actions, and optimizing military operations, leading to more effective planning and execution.
6. **Battlespace Management:** AI can help manage complex battlespaces by processing and integrating data from various sensors, surveillance systems, and intelligence sources. This can enhance situational awareness, enable real-time threat assessment, and facilitate faster response times, improving overall operational effectiveness.
7. **Cyber Operations:** AI can bolster cyber security efforts by detecting and responding to cyber threats in real time. Machine learning algorithms can identify patterns of malicious activities, recognize anomalies, and autonomously respond to cyber-attacks, thereby fortifying military networks and systems (Helkala et al., 2023).
8. **Psychological Operations:** AI can support psychological operations by analyzing social media data, news, and other sources of information to understand public sentiment, identify influential actors, and tailor messaging strategies. This can aid in shaping narratives, conducting information campaigns, and achieving psychological effects on adversaries or populations.
9. **Ethical Use of Semi-Autonomous and Autonomous Weapons:** AI can enable the effective and ethical deployment and control of semi-autonomous and autonomous weapons systems. It can enhance target identification, reduce the risk of civilian casualties through advanced recognition algorithms, and improve the coordination and decision-making capabilities of these systems, while still ensuring human oversight and adherence to ethical and legal frameworks.

Once again, advocates of "responsible" AI (like ChatGPT itself) remind us that it is crucial to ensure responsible and ethical use of AI technologies in military contexts, maintaining human control and oversight to mitigate potential risks.

9.3.6 Downsides and Vulnerabilities

There are, however, accompanying downsides and specific vulnerabilities also associated with using AI in intelligence and operations. Specifically:

1. **Data Privacy and Security:** AI systems rely heavily on vast amounts of data, including personal and sensitive information. The collection, storage, and processing of such data can pose significant risks to privacy if not properly safeguarded. Breaches or unauthorized access to AI systems could lead to the exposure of classified information or compromise the privacy of individuals.
2. **Bias and Discrimination:** AI systems are only as good as the data they are trained on. If the training data is biased or incomplete, the AI algorithms can perpetuate or amplify those biases, leading to discriminatory outcomes. In national intelligence operations, biased AI algorithms could result in unjust targeting, profiling, or surveillance of specific individuals or groups.

3. **Adversarial Attacks:** AI systems can be vulnerable to adversarial attacks, where malicious actors intentionally manipulate the input data to deceive or mislead AI algorithms. For example, by subtly modifying an image or altering text, an attacker may be able to fool AI-based recognition systems or manipulate the outcomes of intelligence analysis, leading to inaccurate or misleading conclusions.

4. **Lack of Human Oversight:** Relying too heavily on AI systems without adequate human oversight can be problematic. AI algorithms can make mistakes or produce false positives/negatives, especially in complex intelligence analysis scenarios. Human judgment, critical thinking, and contextual understanding are still essential in evaluating intelligence data and making informed decisions.

5. **Dependence on Unreliable or Incomplete Data:** AI systems require large amounts of high-quality data for effective training and decision-making. However, in intelligence operations, relevant data may be scarce or incomplete, especially in rapidly evolving situations or in covert operations. Relying solely on AI could result in incomplete or inaccurate assessments due to the lack of necessary data.

6. **Other General Ethical Concerns:** The use of AI in national intelligence raises ethical concerns regarding surveillance, privacy invasion, and potential misuse of technology. The responsible and transparent use of AI in intelligence operations is crucial to ensure adherence to legal frameworks and ethical principles.

7. **Technical Limitations:** Finally, it is well-documented that AI systems have certain limitations. They may struggle to handle ambiguous or contradictory information, lack common-sense reasoning abilities, or be susceptible to misinterpretation in complex intelligence contexts. Overreliance on AI without acknowledging its limitations can lead to erroneous conclusions or missed opportunities.

Important advantages accrue with the use of automated and autonomous systems, not the least of which is that they free humans from doing boring, repetitive, and dangerous jobs and are very efficient in what they do. However, these systems, still somewhat in their infancy, are plagued by some important issues, including one of the most important: explainability.

An AI system cannot explain in any comprehensible way why it makes the decisions it does. This has already been born out in the civilian sector: AI-powered language translation software from Google made significant improvements in accuracy without the software's designers understanding how that happened. This lack of transparency in the process is antithetical to the military mindset, where it is critical to be able to explain recommendations and to be able to understand, and trust, the system. To be fair, laboratories around the globe are working to improve AI in this regard, but to date, there is no known AI system that can, in human terms, explain the rationale for its decisions. With AI systems, explanations are themselves often highly complex. The counterargument is that human decision-making may not be logical or correct either. Yet correct or incorrect, the human cognitive process is at least explainable in human terms with a commensurate military legal, moral, and ethical framework.

Another vulnerability with potentially catastrophic consequences for military operations is the risk of spoofing and false data. AI systems have been reportedly shown to fall prey to adversarial inputs and stray information, ranging from out-of-place pixels on a stop sign that can trick a driverless car into accelerating instead of braking to deliberately misclassifying database inputs. Data manipulation, hacking, and spoofing are problems that must be mitigated, even if solving them may remain elusive as AI adoption increases. AI systems are only as good as their training data and a small amount of corrupted training data could have huge impacts on the predictive ability of such systems. Many AI systems operate with what are known as neural networks in which the computer emulates the neurons of the brain. Operating on billions of pieces of data, the system learns from what it is seeing. The learning accomplished, however, is only as good as the data provided. Biased, incomplete, or just plain incorrect data will most likely result in the machine learning the wrong lesson.

There are difficulties in adequate testing. How can you test a system if its operation cannot be fully explained or is, by definition, unpredictable? This issue then becomes how such systems can be trusted, a foundational principle of military operations. Military testing cannot realistically anticipate, much less replicate, what an enemy will do. AI will inevitably be faced with some enemy behavior for the first time in battle, with unpredictable results. Perversely, it is possible that some nations or groups may seek to employ their untrained AI systems in battle in order to refine them.

Even more concerning, recent research has shown that, if they are not properly designed to prevent such reactions, AI systems can exhibit unexpected learned "aggressive" behavior. Add to this the fact that our national defense strategy itself has become more aggressive, with keywords like increased lethality and defending forward, and you have the potential for bad behavior by the machines that have to implement in computer code this very strategy. Computer code codifies the algorithms that are, themselves, reflections of the values of the engineers and designers. This raises additional concerns about bias in AI systems and about the ability of a machine to even consider the measured and proportional responses so critical to the laws of armed conflict. Systems designed to win at all costs may not know how to even consider a stalemate or a truce. There is a growing recognition by senior military and intelligence leaders that the training of machine learning systems could introduce unintended biases, with unintended negative consequences.

There is a risk that an adversary's exploitation of our system vulnerabilities might occur faster than humans could react, or might be so subtle as to be undetectable, with the result that control could be ceded nearly completely to the systems, with the human being completely unaware. If the process is so automated as to eliminate the need for human involvement, we must either accept a high level of risk of manipulation by adversaries or ensure that systems are trustworthy and that the possibility of manipulation is mitigated. Completely verifying and securing such complex systems would be nearly impossible, and it would be difficult to build such absolutely trustworthy AI systems. In light of such vulnerabilities, military and civilian war planners must consider the downsides and unknowns before rushing to deploy such technologies.

Another problem is that automated systems, including AI systems, tend to be brittle. They are great at doing what they were designed to do, but they fail miserably when asked to do something different. In a highly complex battlefield, the data may be like nothing ever seen before, and machine learning systems that depend on proven data sets may be useless.

To address these downsides and vulnerabilities, it is important to have robust governance frameworks, rigorous data privacy and security measures, regular audits and testing of AI systems, continuous human oversight, and ongoing research and development to mitigate the risks associated with AI in national intelligence operations.

9.4 ETHICAL CONSIDERATIONS OF AI IN MILITARY STRATEGY

9.4.1 The Arrival of Large Language Models

During the course of writing this chapter, LLMs such as ChatGPT began to emerge and have been rapidly adopted by industry, academia, and other fields. Despite some serious shortcomings, these models have demonstrated numerous practical benefits in their ability to ingest and summarize enormous amounts of data. The merging of strategic and tactical decision-making through AI programs like Meta-Constellation not only flattens the command hierarchy and distributes decision-making authority more widely across the battlefield. It also shifts ethical considerations closer to the point of action.

For the purposes of this chapter, consider, for example, the following question:

> How might a generative AI Large Language model be utilized to augment and improve the capacity for strategic military planning?

Then ponder this answer, provided by ChatGPT 3.5 on 28 June 2023: *A generative AI large language model like ChatGPT can potentially be utilized to augment and improve the capacity for strategic military planning in several ways:*

> **Data Analysis and Synthesis:** The model can process vast amounts of textual data, including military reports, historical records, intelligence assessments, and academic research. It can analyze this information and synthesize key insights, trends, and patterns to assist military planners in making informed decisions.
>
> **Scenario Generation:** By providing relevant inputs and parameters, the AI model can generate hypothetical scenarios for military planning exercises. It can simulate various geopolitical, operational, or tactical situations, allowing planners to explore different strategies and responses without real-world consequences. This can help identify potential risks, vulnerabilities, or opportunities.
>
> **Course of Action Evaluation:** The AI model can evaluate different courses of action suggested by military planners. It can analyze the potential outcomes, risks, and benefits of each option based on historical data, simulations, and expert knowledge. This can provide decision-makers with a more comprehensive understanding of the implications of different strategies.
>
> **Red Teaming and Wargaming:** The AI model can act as a virtual adversary or partner during red teaming exercises and wargaming simulations. It can provide realistic and dynamic responses based on its understanding of military doctrines, historical precedents, and strategic thinking. This can enhance the realism and complexity of training scenarios and help identify potential vulnerabilities.
>
> **Decision Support:** When faced with complex and rapidly evolving situations, military planners can leverage the AI model as a decision support tool. They can interact with the model through conversational interfaces, posing questions, seeking alternative perspectives, and exploring different what-if scenarios. This can aid in information gathering, hypothesis testing, and generating novel ideas.
>
> **Knowledge Dissemination:** The AI model can serve as a repository of military knowledge and best practices. It can provide instant access to a wide range of information, doctrines, and historical lessons, supporting military planners in their research and analysis. Additionally, it can facilitate knowledge sharing and collaboration among military personnel by acting as a virtual assistant during discussions and brainstorming.

The content of these ChatGPT descriptions of potential AI benefits closely tracks the summary opinions of AI experts and advocates in the current literature. Without doubt the kinds of contributions to planning and operations outlined briefly here would prove extensive, highly useful, and beneficial as enhancements to the process of strategic military planning, as well as to fine-tuning resulting military operations. These computer-generated summaries are admittedly a bit general, lacking specifically in the details that might need to be provided to flesh them out and operationalize promised features such as "decision support" and "knowledge dissemination." These suggestions are far more detailed and robust, however, than were ChatGPT's initial responses to virtually identical questions put to earlier versions of this LLM four to six months earlier. And they come with a sensible, if broad, disclaimer from the tool itself, which concludes by warning its interrogators:

> However, it's important to note that while generative AI language models can offer valuable assistance, strategic military planning involves complex geopolitical, ethical, and contextual considerations. The AI model should be treated as a tool to aid human decision-making rather than a substitute for human judgment and expertise. Human oversight and validation are crucial to ensure the accuracy, relevance, and ethical implications of the AI model's outputs.

This is as much as to say that the very real Devil is in the details, which remain for users to comprehend perceptively and flesh out responsibly. For example, how accurate, comprehensive, and (importantly) impartial are the "vast amounts of data" the program promises to synthesize, ranging from "military reports, historical records, intelligence assessments, [to] academic research." Can this synthesis distinguish between reputable sources (such as refereed publications in leading academic journals) and the copious amounts of unfiltered nonsense and misinformation purveyed on web blogs and social media? Will the model have access to the requisite, highly classified information necessary for effective planning and operations – and will this access represent a security vulnerability that might be exploited by adversaries?

Here it must be emphasized how every effort must be made to cull data that is ethically sourced; i.e., selected, stored, and constantly monitored to eliminate the possibility of bias or the appearance of other vulnerabilities. The value or utility of any subsequent "scenario generation" and resulting "course of action evaluations" depend critically upon this ethical sourcing (Boyhun, 2021). Indeed, it will prove dangerous, not to mention unethical, to "scrape" materials indiscriminately from widely available sources without the corresponding ability to vet those sources accurately. The more data assimilated, moreover, the harder it will be for human users (who should be better at this critical cross-examination task) to keep up with the sheer volume of data gathered – meaning that we will have to establish that the assimilation process itself, even prior to any synthesis, is *trustworthy* and *reliable*.

9.4.2 Other General Concerns Arising from the Increased Use of AI

To date, effective incorporation of the benefits of AI in military strategy and operations has depended upon a robust public-private partnership. This has not been without controversy. Famously, Google engineers protested *en masse* to their company's involvement in the US Army's "Project Maven," intended to harness the advantages of big data and AI analysis. Their reluctance to involve themselves in military and defense contracting disrupted that program and resulted in a new project and partnership with Palantir, cited above. Defending his own and his company's collaboration with government and military services through the "Army Vantage" program, Palantir CEO Alex Karp stated: "Silicon Valley screaming at us for over a decade did not make the world any less dangerous. We built software products that made America and its allies stronger – and we are proud of that."

But such partnerships introduce new and unpredictable vulnerabilities. Following Russian attacks on Ukrainian infrastructure early in that conflict, Elon Musk agreed to allow Ukrainian citizens and the Ukrainian government to utilize his Starlink satellite network to provide internet and cell phone connectivity. Subsequently, months later, Musk began to balk at assuming the costs of this service unilaterally and threatened to terminate satellite service unless the US or another government stepped up to subsidize these expenses. It was precisely at this moment, however, that the Ukrainian military's use of Starlink data for its highly successful "meta-constellation" program came to light. It was not just video gaming or phone calls to relatives that were threatened by Mr. Musk's recalcitrance but, unbeknownst to him and to the wider work, his abrupt withdrawal of service threatened the very military successes the Ukrainians had achieved to that point.

That crisis was averted when Musk agreed to continue supporting the costs of satellite communication … for the time being. But this conundrum illustrates the danger of relying on the *in-bello* cooperation of private-sector leaders and employees, whose independence, ethics, and profit-driven motives do not always coordinate with an inherently governmental function (like waging a war) or fit neatly within a military sense of the chain of command, good order, and discipline. Unexpected disruptions of this kind of collaboration in an AI context could prove disastrous, particularly in light of the complexity and sensitivity of the technologies algorithms and data involved.

9.5   CONCLUSION

The US military, NATO countries, and other militaries around the world are racing to incorporate AI into their weapons systems, with varying degrees of commitment to the ethical use of this new technology. Military decision-makers with strategic planning responsibilities have myriad matters to sort out. Yet as other military technology breakthroughs of the past century have shown, focusing on the ethical dimensions can help them navigate the thorniest challenges amid this rush to field one of humanity's most consequential inventions. There is much to weigh for military leaders at this moment. Errors in individual military technology systems may have minimal or manageable operational risk, but when they are combined with broadly used AI capabilities, likely employing different algorithms, answers may be confusing and inconsistent at times when clarity and decisiveness are required for a pace of warfare that increasingly moves at machine speed. The greatest risks posed by military applications of AI – increasingly autonomous weapons, and algorithmic C2, among others – are that the interactions between the systems deployed will be extremely complex, impossible to predict or trust, and subject to catastrophic forms of failure that are hard to anticipate, much less mitigate. As a result, there is a serious risk of harm to civilians, operational miscalculations that cost the lives of service members, or even the accidental escalation of conflict if machine learning or algorithmic automation is not ethically and judiciously used in these kinds of military applications. As Professor Don Howard has concluded, "we cannot let unrealistic fears about a Terminator-AI apocalypse prevent our taking advantage of the opportunities for moral progress that properly designed and deployed autonomous weapons afford. We must, of course, ensure that such systems are being used for good, rather than malign purposes, as we must with any technology, and especially technologies of war. Indeed, with autonomous weapons we need to be more vigilant, still. But minimizing death and suffering in war is the ultimate goal. If autonomous weapons can contribute to progress toward that goal, then we must find a way to license their use in full compliance with what law and morality demand" (Howard, 2022).

NOTE

1. Alex Karp of Palantir defends the wider and successful use of meta-constellation not only in military operations but also to track down human traffickers and drug smugglers and even locate missing and exploited children.

REFERENCES

Allyn, B. (2022, March 16). Deepfake video of Zelenskyy could be "tip of the iceberg" in info war, experts warn. *NPR.* https://www.npr.org/2022/03/16/1087062648/deepfake-video-zelenskyy-experts-war-manipulation-ukraine-russia

Amanda, M. (2021, September 22). AI algorithms deployed in kill chain target recognition. *Air & Space Forces Magazine.* https://www.airandspaceforces.com/ai-algorithms-deployed-in-kill-chain-target-recognition/

Bajraktari, Y. (2022, September 13). The first networked war: Eric Schmidt's Ukraine trip report [Substack newsletter]. *Special competitive studies project.* https://scsp222.substack.com/p/the-first-networked-war-eric-schmidts

Boyhun, K. (2021). Ethics and big data. In D. C. Ellis & M. Grzegorzewski (Eds.), *Big data for generals … and everyone else over 40* (pp. 119–128). Joint Special Operations University Press.

Carter, A. (2022). The moral dimension of AI-assisted decision-making: Some practical perspectives from the front lines. *Daedalus, 151*(2), 299–308. https://doi.org/10.1162/daed_a_01917

Cole, A., & Le Guyader, H. (2023, December 19). *Cognitive warfare – Innovation hub.* https://www.project-censored.org/wp-content/uploads/2023/01/Cognitive-a-6th-Domain-of-Operations.pdf

Eftimiades, N. (2022, May 5). Small satellites: The implications for national security. *Atlantic Council.* https://www.atlanticcouncil.org/in-depth-research-reports/report/small-satellites-the-implications-for-national-security/

Harwell, D. (2022, March 24). Instead of consumer software, Ukraine's tech workers build apps of war. *Washington Post.* https://www.washingtonpost.com/technology/2022/03/24/ukraine-war-apps-russian-invasion/

Helkala, K. M., Cook, J., Lucas, G., Pasquale, F., Reichberg, G., & Syse, H. (2023). AI in cyber operations: Ethical and legal considerations for end-users. In T. Sipola, T. Kokkonen, & M. Karjalainen (Eds.), *Artificial intelligence and cybersecurity: Theory and applications* (pp. 185–206). Springer International Publishing. https://doi.org/10.1007/978-3-031-15030-2_9

Howard, D. (2022). In defense of (virtuous) autonomous weapons. *Notre Dame Journal of Emerging Technologies*, *3*(2), 230–260.

Ignatius, D. (2022, December 19). Opinion | How the algorithm tipped the balance in Ukraine. *Washington Post*. https://www.washingtonpost.com/opinions/2022/12/19/palantir-algorithm-data-ukraine-war/

McIntosh, S. E. (2011). The wingman-philosopher of MiG alley: John Boyd and the OODA loop. *Air Power History*, *58*(4), 24–33.

NATO. (2023, February 7). *NATO starts work on artificial intelligence certification standard*. NATO. https://www.nato.int/cps/en/natohq/news_211498.htm

Reed, C. A., & Rife, J. P. (2022, January). *New wrinkles to drone warfare*. U.S. Naval Institute. https://www.usni.org/magazines/proceedings/2022/january/new-wrinkles-drone-warfare

Trofimov, Y., & Nissenbaum, D. (2022, September 17). Russia's use of Iranian kamikaze drones creates new dangers for Ukrainian troops. *Wall Street Journal*. https://www.wsj.com/articles/russias-use-of-iranian-kamikaze-drones-creates-new-dangers-for-ukrainian-troops-11663415140

Woody, C. (2022, August 8). *The US Air Force is signing up for Starlink after watching it help Ukraine stay online amid Russia's ongoing attacks*. Business Insider. https://www.businessinsider.com/us-air-force-contracts-with-starlink-as-it-helps-ukraine-2022-8

10 Autonomous Weapons Systems and the Future of Warfare

Andrew N. Liaropoulos

10.1 INTRODUCTION

The application of artificial intelligence (AI) in the military domain is an issue that has troubled the scientific and military communities for decades. The military applications of AI range from image recognition for surveillance to situational awareness on the battlefield and from logistics for battle management to a compressed decision-making loop (Horowitz, 2018). AI should not be understood as a weapon, but rather as an enabler and force multiplier of a wide range of military capabilities, both kinetic and non-kinetic (Johnson, 2019, p. 150). One potential application of AI is the use of lethal autonomous weapons (LAWs) systems. Algorithms program such weapons systems to select and engage targets based on predetermined criteria; thus, they can operate without direct human intervention. Along with the diffusion of dual-use emerging technologies in warfare, autonomous weapons systems (AWS) consist of an ongoing revolution in military affairs (Raska, 2021) that is expected to shape the way war is conducted in the future. Technology is challenging humanity's interaction with war. Uncertainty over what AI weapons can do in the future has triggered a security dilemma among great powers and thereby an arms race is already underway (Payne, 2021, p. 217). These systems are increasingly becoming more autonomous and the question that inevitably arises is what sort of authority-autonomy should machines be given over life-or-death decisions (Scharre, 2018, p. 10).

Proponents of autonomous weapons argue that such systems could potentially enhance military capabilities by providing faster, more precise, and more efficient responses in combat situations. Furthermore, they claim that these weapons may reduce the risk to human soldiers, by assigning certain tasks to machines. One such example is the case of autonomous drones that are used for either surveillance or targeted killing and thus minimize the number of human pilots that are in danger. On the other hand, the use of such weapons raises ethical, legal, and strategic issues. The main concern is the potential loss of human control over lethal decision-making. Allowing machines to make life-or-death decisions raises moral and accountability issues. Machines, after all, cannot comprehend complex ethical matters and thus cannot be held responsible for errors or violations of international laws. Machines cannot be moral actors. When a warbot makes calculations balancing between mission objectives and potential casualties, it is not using its *own* judgment, but the judgment of its programmers (Payne, 2021, p. 229). Adding to the above, there are also practical concerns about the operational security of these weapons systems. AWS are exposed to cyber security risks, such as hacking, spoofing, or other malicious actions. Last, but certainly not least, there is a legitimate fear that the extensive use of AWS might lead to an arms race and that the proliferation of these systems will enable non-state actors or rogue states to acquire and misuse them (Etzioni & Etzioni, 2017).

The chapter focuses on how the use of AWS affects the conduct of the war in tactical, operational, and strategic terms. The purpose is to identify and properly frame the implications of such weapons systems not only on the battlefield but also about broader issues like the arms race, deterrence, and stability in the international order. In terms of structure, the chapter first tackles the definitional debate and approaches the types of autonomy in modern weapons systems. A common

DOI: 10.1201/9781003441700-10

understanding of AWS is still lacking in both the scientific and policy circles. The following two sections review the advantages and challenges that such weapons systems pose to military practitioners and policymakers. Building on the above, the final section discusses the potential impact of AWS on the international system and stresses the likelihood of unintended escalation and the need for arms control or international regulations in the development and deployment of AWS. The chapter ends with the conclusions that summarize the main points discussed and highlight the need for policy development in this area. The research is based on a thorough literature review of scholarly articles, books, reports, and relevant publications that explore the strategic implications of autonomous weapons. Additionally, real-world examples and case studies where autonomous weapons have been deployed or tested are used to support the arguments and provide practical insights.

10.2 DEFINING AUTONOMY IN WEAPONS SYSTEMS

Autonomy is a tricky and poorly understood concept. There are numerous approaches and competing views in the literature concerning the differences between autonomy, automated, and autonomous (Wood, 2023). In general, autonomy refers to the capacity of a machine to execute independently a task or function (Scharre, 2018, p. 34).

Automatic systems, such as landmines, respond mechanically to external inputs and are unable to discriminate the inputs. These systems have no complex decision-making process and the relationship between environmental inputs and behavioral outputs is linear (Limata, 2023, p. 2). Automated systems are those that can execute certain commands/tasks in a pre-programmed way. They are rule-based systems, and their reasoning relies on rules and procedures like checklists. Automated functions, mainly found in applications that include calculations and computations, are constrained by algorithms that determine their modus operandi and are thus predictable. Autonomous systems, on the other hand, possess the capability to select from various action options by processing sensory input, enabling them to accomplish goals by optimizing across a predefined set of parameters. Even though they operate within a pre-programmed range of actions, they can exercise independent judgment and choose an alternative course of action that complies with the mission statement (Kunertova, 2021, p. 2). Autonomous systems can engage in probabilistic reasoning using a given set of inputs, formulate diverse courses of action, and independently choose and execute the optimal option without requiring human intervention at any stage of the process (Cummings, 2017, p. 3). These systems are goal-oriented. The human selects a goal and the machine acts to achieve that goal. The user is aware of the goal, but not of the process-reasoning that the machine will apply to accomplish this task. The reason is that autonomous systems use algorithms to accomplish the task and algorithms are not neutral data extractors of processors (Limata, 2023, p. 2). Thus, autonomous systems are far more complex and unpredictable compared to automated ones.

Autonomy refers to three distinct concepts (Scharre, 2018, pp. 34–40; Scharre & Horowitz, 2015, pp. 5–7). First, autonomy has to do with the type of task that is being automated. Tasks vary in complexity, and the repercussions of machine failure in executing them also differ. The task might involve automatic breaking and avoiding a collision or an early warning indication that might lead to the launch of a nuclear missile. Second, autonomy is related to the complexity of the machine. In cases of self-learning machines, the cognitive processes of the code are very complex and while the human operator comprehends the task intended for the machine, the process is not transparent to the operator. This might lead to unpredictable behavior. Third, the level of autonomy is defined by the human-machine command-and-control relationship. Concerning the military Observe, Orient, Decide and Act (OODA) loop, developed by John Boyd (Johnson, 2023), one can identify three degrees of autonomy, semiautonomous, supervised autonomous, and fully autonomous. In semiautonomous systems, the machine executes a task and subsequently pauses, awaiting the user's action before proceeding. In other words, there is a *human in the loop* and the system cannot carry out a certain action without human approval. One such example is Israel's Iron Dome system which detects incoming rockets and predicts their trajectory. It sends the information to the human operator who decides on whether to launch an interceptor rocket (Etzioni & Etzioni, 2017, pp. 78–79). In supervised autonomous systems, machines can perform the loop, but the human user can observe and intervene in any phase, to stop it. In this case, the *human is on the loop.* Such a case is the Phalanx Close-In Weapon System (CIWS), which has been used on US Navy ships, to detect, track, engage, and use force against missiles and high-speed aircraft threats (Etzioni & Etzioni, 2017, p. 78). Finally, in fully autonomous systems, the *human is out of the loop.* Machines, once activated, perform their tasks without any further communication with the user (Scharre, 2018, pp. 34–40; Scharre & Horowitz, 2015, pp. 5–7).

While the above typology aids in distinguishing between various weapons systems, it is crucial to recognize that it indicates policy preferences rather than inherent limitations or capabilities of the weapons. The technology used in a semiautonomous system may be precisely the same as that in a fully autonomous system, the key distinction lying in whether the human operator selects to control or supervise the decision-making processes. Such a case is the Patriot missile systems, which can be operated manually by a human or set to operate fully automatically, depending on the operational environment and the urgency of the perceived threat (Gardner, 2021, p. 89). The degree of autonomy can vary. Precision-guided munitions (PGMs), also labeled smart missiles or smart bombs, are indicative of the above. There are cases, where the human user does not have direct control over the munition's movements but does have real-time control over the weapon's aimpoint. This capability enables the human controller to potentially cancel the attack. In other cases, the weapon once fired cannot be canceled, yet the destination is preset (Scharre, 2018, p. 47). In semi AWS, automation is used to search for and detect targets and to carry out the operation, but it is the human who decides to engage with a specific target. Supervised AWS, once activated, can search for, detect, decide to engage, and finally engage a target on their own, but the human controller is still present and has the authority to intervene (Scharre, 2018, pp. 51–52). Critical to protection from PGM attacks are automated defense systems. These include systems like the US Aegis, the US Patriot, the German MANTIS, the Israeli Trophy, and the Russian Arena system (Scharre, 2018, p. 52).

Autonomous systems are primarily used for non-critical functions like mobility, interoperability, intelligence, surveillance, reconnaissance, logistics, and maintenance. The integration of autonomous functions in machines is favored in areas where human cognitive capabilities are not essential. Concerning mobility, machines demonstrate autonomy in tasks like take-off, landing, and navigation, relying on pre-programmed maneuvers. This autonomy becomes crucial in situations where communication is challenging due to factors such as weather conditions, complex topography, or deliberate denial. In the targeting process, humans typically remain involved, either directly or indirectly. Automated target recognition systems may have humans in the loop, especially when dealing with fast or distant targets. Even when a system is capable of autonomously detecting, tracking, prioritizing, and selecting targets, the ultimate decision to engage the target is made by a human operator. While some systems can operate without the direct involvement of the human operator, this capability is limited in cases with extremely short response windows, such as for the protection of ships or troops against incoming projectiles (Kunertova, 2021, p. 2). As AWS advance in complexity, speed, and sophistication, there is a growing concern that human supervisors may increasingly rely on and trust these systems, becoming less inclined to override their recommendations. This shift in trust could result in a transfer of decision-making authority over matters of life-and-death from human supervisors to the autonomous weapons themselves (Schott & Lang, 2017, p. 2). Fully AWS, capable of engaging targets with humans out of the loop, are not yet operational.

As argued above, autonomy is an elusive concept. Distinguishing between automated and autonomous systems in practical terms can be a challenging task since some systems exist within a substantial gray area. The progression of autonomous functionality in weapons systems is gradual, and some highly advanced automatic systems already exhibit behaviors that could be characterized as autonomous. This is the case with automated defense systems that may autonomously target the origin of incoming fire. These systems also introduce ambiguity in categorizing actions as either defensive or offensive (Altmann & Sauer, 2017, p. 118). The fact that autonomy can range from automatic piloting to the autonomy of engaging with a target explains the lack of a commonly agreed definition of lethal AWS. For the purposes of the present analysis, lethal AWS are considered fully autonomous ones that can decide about selecting and engaging targets without any human supervision (Lewis, 2015, p. 1311). Since 2018, United Nations Secretary-General António Guterres has consistently expressed the view that lethal AWS are politically and morally unacceptable. In the 2023 New Agenda for Peace, the Secretary-General reaffirmed this stance and urged states to establish, by 2026, a legally binding instrument. This instrument should prohibit lethal AWS that lack human control or oversight and are unable to comply with international humanitarian law. Additionally, it should aim to address the regulation of other forms of AWS (Guterres, 2023, p. 27).

10.3 ADVANTAGES OF AUTONOMOUS WEAPONS SYSTEMS

Militaries around the world are increasingly investing in the development of AWS, which can have a wide range of applications in the modern battlefield. In general, the arguments in support of using such systems fall into two categories. The first refers to the military advantages, especially at the tactical level, and the second category of arguments relates to moral justifications regarding their usage.

The first advantage is speed and faster reaction times. AWS can operate at high speed and process vast amounts of data much more quickly than human operators. War, among other things, is also a race against time. The rapid response offered by such weapons systems can be crucial in dynamic and fast-paced situations. Thus, the military that integrates better these systems will gain a tactical advantage. In a fluid and data-intensive environment, the advantage goes to the party that can execute the OODA loop more quickly and effectively (Ploumis, 2022, p. 3). Indicative of this is that in a virtual dogfight simulation in August 2020, an AI algorithm defeated a human F-16 pilot. The simulation was part of the Defense Advanced Research Project Agency's (DARPA) Air Combat Evolution program, which explored automation in air-to-air combat and aimed to improve human trust in AI systems (Eversden, 2020).

The second advantage is precision and accuracy. AWS incorporate cutting-edge technologies, advanced sensors, and computing capabilities that enable them to achieve high levels of precision and accuracy in targeting. A network of such weapons platforms could enhance the capabilities of a battalion commander to coordinate fire and maneuver tactics against an enemy position. This can minimize collateral damage and increase the effectiveness of military operations (Ayoub & Payne, 2016, p. 806).

The third advantage is that such weapons can operate continuously without fatigue. Unlike human operators, AWS do not experience emotions and cognitive stress. In contrast to human limitations regarding endurance and emotions, machines can contribute to consistent decision-making operational performance. Whereas a human sniper is emotionally and physically constrained after some hours or days operating on the battlefield, the autonomous sniper would not get tired or hesitate to take a shot (Horowitz, 2019, p. 770). Likewise, in the case of pilots, the physical burden produced by demanding maneuvers and the cognitive focus during a dogfight will inevitably cause physical and mental exhaustion. Again, robot pilots are not subject to these constraints (DeSon, 2014).

In general, AWS can act as force multipliers in a variety of ways. They are increasing the capabilities of a military force by providing additional firepower, surveillance, and support without proportionally increasing the number of human personnel. They are expanding the battlefield, projecting firepower and combat capabilities, to zones that were until then unreachable. For example, such systems are more suitable than human soldiers in a mission that involves harmful radiological material (Etzioni & Etzioni, 2017, p. 72). The experience from the military operations in Iraq and Afghanistan has demonstrated the utility of unmanned vehicles in clearing roads and fields from mines and detecting weapons at checkpoints (Kunertova, 2021, p. 2). Furthermore, some autonomous systems are designed to adapt to changing circumstances and learn from their experiences. These learning capabilities can enhance the interaction with the human operator and improve coordination. This adaptability can enhance their effectiveness in data-intensive operational environments and unpredictable situations.

From an ethical point of view, there is the argument that the wide use of AWS reduces risk to human life. Reducing the need for human operators to be physically present on the battlefield and replacing them with AWS like killer-robots does not only involve operational parameters, as demonstrated above, but also moral ones. It can be argued that the proliferation of such systems is not only morally acceptable but also ethically preferable to human fighters (Etzioni & Etzioni, 2017, p. 74). Professor Ronald Arkin envisions that future autonomous robots could exhibit humane behavior on the battlefield, simply because a machine is not constrained by self-preservation instincts or biased by emotions and preconceptions that blur its judgment (Arkin, 2010, p. 333). Robots do not need to protect themselves. That allows them to take up more risks to be certain that a particular target is indeed a legitimate one. In case of a mission failure, a commanding officer can even sacrifice them, to eliminate further human casualties on the opponent's side (Wood, 2020, p. 220). Fear, stress, and anger are inherent characteristics of war and distort the judgment of the human soldier. AWS on the other hand can be designed in a way that incoming data that potentially leads to life-or-death decisions is not distorted by preconceived notions or emotions (Arkin, 2010, p. 333). Going a step further, these systems could also be used to objectively monitor and report any form of unethical behavior on the battlefield (Arkin, 2010, p. 334). Paradoxically enough, if we follow this line of argument, the so-called dehumanization of warfare – replacing humans on the battlefield with machines – might make warfare more humane. Some scholars even argue that there is a moral imperative to develop moral machines on the battlefield. In case AWS overcome the technological shortcomings of differentiating between targets and demonstrate targeting and judgment systems that are equal or even superior to the abilities of humans, then it is fair to explore the possibility of replacing humans with such ethical weapons systems (Umbrello et al., 2020).

Finally, the integration of military and moral perspectives on the utilization of AWS underlines the following proposition: if these systems genuinely reduce the human cost of war, opting for military operations with minimal casualties emerges as a favorable political decision. Investigating this hypothesis is difficult in the case of fully AWS since they have not been deployed yet. On the other hand, testing this hypothesis in the case of drones provides us with some useful insight. The evidence from the deployment of drones over the last two decades, as part of the counterterrorism campaigns conducted by the United States and Israel, demonstrates that policymakers and to an extent societies are more willing to support the use of force when it involves drone strikes (Lin et al., 2012; Walsh & Schulzke, 2015). According to Professor Amy Zegart, the fact that drones entail lower human and financial costs makes targeted killing or even continuation of war a more feasible option (Zegart, 2020). Drawing on this analogy, it can be anticipated that if lethal AWS demonstrate relative effectiveness, the political implications of taking action will once more experience a significant reduction. Nevertheless, the wisdom of resorting to military violence, even in the context of a seemingly costless war, is a subject of intense debate.

10.4   THE CHALLENGES OF INTEGRATING AUTONOMOUS WEAPONS SYSTEMS

As with many cases in the past, the introduction of new weapons systems on the battlefield entails not only potential advantages but also disadvantages. AWS do not escape this reality. In common with the previous section, one can discern two groups of arguments against the use of such weapons. The first group refers to operational concerns regarding the potential for unintended consequences. These concerns refer to inherent limitations like complexity, security, bias, and friction. The second group stresses the ethical challenges of applying these weapons systems.

The first challenge is the complexity of the systems. Autonomous weapons exhibit a high level of complexity, necessitating extensive code to generate algorithms for the application of force.

Making these weapons less complex means that their methods of operation are easily understood by the enemy and thereby they become predictable and less useful. Making them more complex implies that the risk of mistakes and failures increases (Alloui-Cros, 2022, p. 6). When these systems operate in environments that surpass the limitations of their programming, there is a potential for malfunctions and failures. Even with commanders possessing a thorough understanding of how the systems are intended to function and deploying them appropriately, the sophistication of the programming introduces the risk of unintended behaviors and accidents (Horowitz, 2019, p. 771). Daniel Trusilo demonstrates the complexity of such weapons systems in two case studies. The first refers to a swarm system designed for maritime intelligence and the second involves a next-generation humanitarian notification system. In both scenarios, the AWS exhibited a dual nature, demonstrating both unpredictability and enhanced reliability, depending upon the specific level at which the system is analyzed (Trusilo, 2023).

The second challenge, amplified by the previous one, is the cyber security of these systems. AWS heavily rely on advanced technologies, including AI and communication networks. This dependence exposes them to cyber security risks, such as hacking, spoofing, or other malicious actions. Compromised AWS could be turned against their operators or be used to carry out unauthorized attacks. A commander will be hesitant to deploy a weapons system that might break the chain of command or execute an operation based on distorted data (Kunertova, 2021, p. 3). Let's imagine a scenario where an adversary cuts communications links between the human operator and the weapons platforms, thus forcing them to act without any oversight. Depending on the context of the confrontation, this could lead to civilian casualties or unintended escalation risks (Johnson, 2020).

The third challenge relates to trust in the systems and automation bias. The human user may over-rely on an autonomous system that is in general reliable, even though the user does not fully understand how the system functions. For example, by over-trusting the sensors, the human operator outsources judgment to algorithms. This is called automation bias (Horowitz, 2019, p. 773) and appeared during the invasion of Iraq in 2003, when two friendly aircraft were shot down by the Patriot missile system that identified them as hostile targets (Hawley, 2017). Bearing in mind the complexity of these weapons systems, it is only natural that the risk of automation bias is a crucial concern (Gardner, 2021, p. 92). Adding to that, trusting the algorithm too much may decrease the confidence in the ability of the human to solve problems (Kunertova, 2021, p. 3).

Fourth, there is the friction paradox (Gardner, 2021). It is only natural to assume that since the human constraints that relate to human behavior like fatigue, stress, and physical and cognitive limitations do not appear in the AWS, the latter will not suffer from friction. Nothing is further from the truth. Apart from the complexity, the automation bias, and the cyber security concerns that are present and hinder the decision-making process, there is also the inability of these weapons systems to grasp a wider understanding of reality and make judgments. These weapons systems are as good as the parameters they are set with and the data they are provided with. They can comprehend large amounts of data and build complex statistical models, based on probabilistic reasoning, but they cannot understand context. Based on narrow AI applications, machines can outperform human capabilities in terms of pattern identification but can do that only in relation to what they are trained for. Lacking an understanding of context, thereby the causes and consequences of an action, AWS would find it difficult to discriminate between combatants and civilians (Scharre, 2018, p. 256) or might misunderstand statements or actions that aim to deescalate a conflict, like the case of the naval blockade during the Cuban Missile Crisis (Gardner, 2021, p. 91; Horowitz, 2019, p. 765).

In a hypothetical scenario, only machines with a general AI would be able to think for themselves and take action. Till now, general AI-based weapons systems have not been developed (Kunertova, 2021, p. 3). Despite the advanced capabilities of the machines, friction is inherent in AWS as they rely on algorithms, essentially mathematical formulas crafted by humans, making them susceptible to bias (Limata, 2023, p. 9). Thus, even in the case of fully AWS, where the human is out of the loop, the human element, channeled through the algorithm, is present and holds the potential to introduce flaws and biases. According to Gardner, we are facing a paradox regarding friction in

AWS. Replacing certain human functions with machines may reduce friction by easing stress on the human decision-makers, but on the other hand, it will also generate friction due to the perceptual limitations and complexity of such systems, the inherent human constraints, and the interaction between humans and machines (2021).

Furthermore, there is the complex issue of ethics and the dehumanization of warfare. The deployment of lethal AWS fundamentally reshapes the dynamics between humanity and technology. Entrusting life-or-death decisions to machines signifies a deliberate move toward dehumanizing warfare, attributing casualties to algorithms. Surrendering significant human control distances weapon users from the full awareness of the repercussions of their actions, thereby diminishing the humane aspect and leading to moral de-skilling (Galliott, 2017, p. 337; Guha & Galliott, 2023). Adding to that, allowing dehumanization in one aspect of our lives sets a precedent, leaving us vulnerable to machine decision-making intruding into other spheres. Military leaders must possess the ability to assess the necessity and proportionality of an attack, differentiating between civilians and legitimate military targets. This requires not only comprehending increasingly complex weapons systems, particularly autonomous ones, but also understanding the context in which they may be employed. Machines lack the capacity for ethical decision-making and fail to grasp the intrinsic value of human life, remaining oblivious to contextual details and the consequences of their actions.

Finally, the challenges inherent in current methods of remote warfare would be exacerbated with advanced autonomy, further disconnecting humans from the application of force. While states understandably seek to minimize risks to their own troops by substituting humans with machines, this approach may render military action more politically acceptable in the domestic context and thus lower the threshold for engaging in conflicts. Additionally, it may shift the burden of harm onto civilian populations and increase the possibility of targeted killing campaigns (Haas & Fischer, 2017). For instance, a tyrannical leader could deploy emotionless robots to sow fear and execute killings without concern for soldiers who might sympathize with the victims and potentially turn against the dictator (Etzioni & Etzioni, 2017, p. 78).

10.5 STRATEGIC STABILITY, INTERNATIONAL ORDER, AND THE FUTURE OF WARFARE

Having demonstrated the advantages and disadvantages of employing AWS in the future battlefield, it is now necessary to explore the implications of these weapons systems in relation to issues like proliferation, deterrence, escalation, and strategic stability.

To begin with, and as stated above, AWS are complex and lack understanding of context. These characteristics make such systems unpredictable, and in a nonlinear environment, this can lead to escalation. Due to the benefits of speed, precision, and low risk to humans, AWS benefit first-strike strategies. Thus, in a scenario with such weapons systems, the incentive is to strike first, to launch a preemptive attack (Gardner, 2021, p. 94; Johnson, 2020, p. 34). AI and AWS may have a strong impact on how deterrence and escalation are carried out. After all, up until recently, deterrence involved humans signaling each other to avoid certain courses of action. However, what happens when the decision-making process is no longer exclusively human? What happens when automatization is involved and decisions are taken at machine speeds? A wargame conducted by RAND in 2020 simulated the above parameters and demonstrated the potential escalation risks associated with AWS. The scenario portrayed China as the dominant global power, facing opposition from a military alliance comprising the United States, Japan, and South Korea in East Asia. As tensions heightened between the two sides, China sought support from North Korea, a non-formal military ally that received Chinese assistance. In an act of solidarity, North Korea launched a missile over Japan. In response, automated defense systems from the United States, Japan, and South Korea automatically initiated counterbattery fire directed at North Korea. This autonomous action had the potential to draw North Korea into the conflict, highlighting a situation where AI-driven weaponry took a potentially escalatory step (Wong et al., 2020). It is only natural to assume that in a nuclear deterrent scenario, AWS might create additional friction and challenge well-established norms regarding the use of nuclear weapons. Swarms of AWS can be used to target command-and-control capabilities, mobile missile launchers, and ballistic missile submarines. The existence of new weapons that undermine the deterrent capabilities of a nuclear power might trigger the incentive for a preemptive nuclear strike, thus destabilizing the international order (Gardner, 2021, p. 95; Horowitz, 2019, p. 782).

Another great concern regarding the strategic implications of AWS on global stability is proliferation and the arms race. History teaches us that restriction of weapons is not an easy process. When a state attains pioneering capabilities in a particular domain, other states typically strive to obtain similar capabilities. In cases of bilateral antagonism and tension, it can escalate into arms races. Ultimately, even states reluctant to engage in arms race may find themselves compelled to do so for the sake of deterrence and self-defense (Antebi, 2013, p. 74). In contrast to nuclear weapons that had several constraints regarding their development, autonomous weapons are relatively easy and cheap to make. They don't necessitate costly, highly regulated, or challenging-to-obtain raw materials. Additionally, the widespread availability and swiftly decreasing unit costs of drones imply that these capabilities will progressively advance, becoming more autonomous and easier to manufacture on a larger scale (Johnson, 2019, p. 153). The reason is the dual-use nature of these weapons systems. Whereas stealth technology is applied only in the military domain, the same cannot be argued for the underlying technology of AWS. Drones, robots, and machine learning capabilities are designed by the private sector and used for commercial purposes (Altmann & Sauer, 2017, pp. 124–125; Horowitz, 2019, p. 778). In sharp contrast to previous technological developments, like the atomic weapons, the progress regarding the military applications of AI is mainly driven by the industry and not the governments. Concisely, future AWS will be the adaptations of commercial algorithms. The various national drone programs indicate that we are in the early phase of a robotic revolution that depends on corporate research. Unlike armored vehicles and combat aircraft, AWS can be purchased at low cost, they are relatively easy to use, and they can produce the same if not greater lethal effect (Araya, 2022, p. 10).

Recent conflicts have vividly demonstrated not only the diffusion of such weapons systems but also their potential to offer a decisive advantage (e.g. the extended and sophisticated use of drones by Azerbaijan against Armenia and by the Houthis against Saudi Arabia). Based on the above, it is only natural to assume that they are also going to be deployed by terrorist organizations and that these weapons systems are perceived as an asymmetric option to erode a militarily superior enemy (Johnson, 2020, p. 36). States and non-state actors with a disregard for human rights and a military disadvantage may not be deterred by the unpredictability of these weapons systems, which could result in harm to civilians. Instead, they might perceive the development of such systems as a means to bridge their capability gap (Horowitz, 2019, p. 775).

In addition, evaluating the adversary's capabilities with relative accuracy is a difficult task, due to the nature of these systems. Although sensors and munitions determine the capabilities of weapons systems, autonomy is not about hardware, but software. The latter can be easily copied and stolen (Altmann & Sauer, 2017, p. 126). Autonomy does not involve counting missiles, but understanding algorithms. It is hard to imagine any state allowing an international body to inspect the software that runs its weapons systems. Even if that happened, it is obvious that the software could be altered, to hide the capabilities of such systems (Gardner, 2021, p. 92). Thus, the proliferation of the AWS will only increase ambiguity and strengthen the security dilemmas. Unlike bombs and missiles that can be inspected and restricted, certifying the AI capabilities of AWS is challenging. Lacking verification mechanisms increases uncertainty, thus making it harder for states to reach an agreement on how to control such weapons systems.

In light of all the above, it is critical to stress the necessity of regulating the use of AWS. The latter is a complex task, involving ethical, legal, and technical parameters. In this direction, the most notable regulation developed so far is the report by the UN's Group of Governmental Experts – GGE, published in May 2023. The GGE concluded that when characterizing AWS, it is crucial to consider the potential future developments of these technologies. The group underscored that states must adhere strictly to international humanitarian law throughout the entire life cycle of these weapons systems. Thus, states are urged to restrict the categories of targets and the extent and duration of operations in which the weapons systems are involved, ensuring comprehensive training for human operators. If a weapons system is unable to conform to international legal standards, its deployment should be avoided (Group of Governmental Experts [GGE], 2023).

10.6   CONCLUSION

In the same way that the proliferation of machines led to mechanized warfare, AI is enabling the automatization of machines and empowering them to exercise conscious intellectual activities, among them the choice to kill. As these systems move from a popular science fiction concept to reality, military practitioners, robotic designers, ethicists, lawyers, and policymakers discuss the constraints of using them on the battlefield. These systems integrate advanced technologies, including facial recognition, computer vision, autonomous navigation, and the coordination of swarms. Such weapons systems can be deployed from land, sea, and air platforms and can operate on the battlefield, without human intervention to terminate their functions. They accelerate the sensor-to-shooter cycle and partly remove the human cost of war. On the other hand, the power of automation makes these weapons systems unpredictable since they can break the chain of command and thus escape any liability and accountability.

The fact that these weapons systems combine advanced capabilities with lower cost and that due to the absence of a normative and legal framework that regulates their use, they will operate with ambiguous rules of engagement, makes them an ideal option for a variety of actors and types of operations. Thus, military organizations must weigh the balance between the potential risks linked to AWS and the prospect of granting adversaries wielding fully autonomous weaponry a distinct asymmetric advantage.

Based on the above analysis, it is safe to argue that the future of warfare depends among other things, also, on how governments, military organizations, and the international community navigate the ethical, legal, and strategic challenges associated with AWS. Striking a balance between technological advancements and ensuring human control and oversight will be crucial in shaping the role of these weapons systems in future conflicts. Denying the digital dehumanization of warfare and ensuring meaningful human control over the use of force is crucial in shaping our relationship with technology. Sophisticated AWS will be developed and deployed soon. Future work should expand the discussion on the challenges that such weapons systems pose not only to the commander and the policymaker but to humanity in general.

REFERENCES

Alloui-Cros, B. (2022). Does artificial intelligence change the nature of war? *Military Strategy Magazine*, *8*(3), 4–8.

Altmann, J., & Sauer, F. (2017). Autonomous weapon systems and strategic stability. *Survival*, *59*(5), 117–142. https://doi.org/10.1080/00396338.2017.1375263

Antebi, L. (2013). Who will stop the robots? *Military and Strategic Affairs*, *5*(2), 61–77.

Araya, D. (2022). Artificial intelligence for defence and security. *Waterloo, Centre for International Governance Innovation*. https://www.cigionline.org/static/documents/Araya_AI-for-Defence_SpecialReport_Q4fjNfp.pdf

Arkin, R. C. (2010). The case for ethical autonomy in unmanned systems. *Journal of Military Ethics*, *9*(4), 332–341. https://doi.org/10.1080/15027570.2010.536402

Ayoub, K., & Payne, K. (2016). Strategy in the age of artificial intelligence. *Journal of Strategic Studies*, *39*(5–6), 793–819. https://doi.org/10.1080/01402390.2015.1088838

Cummings, M. L. (2017). *Artificial intelligence and the future of warfare* (Chatham House Research Paper). Chatham House. https://www.chathamhouse.org/sites/default/files/publications/research/2017-01-26-artificial-intelligence-future-warfare-cummings-final.pdf

DeSon, J. S. (2014). Automating the right stuff? The hidden ramifications of ensuring autonomous aerial weapon systems comply with international humanitarian law. *Air Force Law Review, 72*, 85–123.

Etzioni, A., & Etzioni, O. (2017). Pros and cons of autonomous weapons systems. *Military Review, 97*(3), 72–81.

Eversden, A. (2020, August 20). *AI algorithm defeats human fighter pilot in simulated dogfight.* C4ISRNet. https://www.c4isrnet.com/artificial-intelligence/2020/08/21/ai-algorithm-defeats-human-fighter-pilot-in-simulated-dogfight/

Galliott, J. (2017). The limits of robotic solutions to human challenges in the land domain. *Defence Studies, 17*(4), 327–345. https://doi.org/10.1080/14702436.2017.1333890

Gardner, N. (2021). Clausewitzian friction and autonomous weapon systems. *Comparative Strategy, 40*(1), 86–98. https://doi.org/10.1080/01495933.2021.1853442

Group of Governmental Experts (GGE). (2023, May 24). *Report of the 2023 session of the Group of Governmental Experts on emerging technologies in the area of lethal autonomous weapons systems, GGE.1/2023/2.* https://dig.watch/updates/gge-on-lethal-autonomous-weapons-systems-adopts-2023-report

Guha, M., & Galliott, J. (2023). Autonomous systems and moral de-skilling: Beyond good and evil in the emergent battlespaces of the twenty-first century. *Journal of Military Ethics, 22*(1), 51–71. https://doi.org/10.1080/15027570.2023.2232623

Guterres, A. (2023). A New Agenda for Peace. Our Common Agenda Policy Brief 9. United Nations. https://dppa.un.org/en/a-new-agenda-for-peace

Haas, M. C., & Fischer, S.-C. (2017). The evolution of targeted killing practices: Autonomous weapons, future conflict, and the international order. *Contemporary Security Policy, 38*(2), 281–306. https://doi.org/10.1080/13523260.2017.1336407

Hawley, J. (2017). *Patriot wars, automation and the patriot air and missile defence system.* Center for a New American Century. https://www.cnas.org/publications/reports/patriot-wars

Horowitz, M. C. (2018). Artificial intelligence, international competition, and the balance of power. *2018, 22.* https://www.google.com/books?hl=tr&lr=&id=b646EAAAQBAJ&oi=fnd&pg=PA372&dq=Artificial+intelligence,+international+competition,+and+the+balance+of+power&ots=WD_IeLbZ7I&sig=oRCmonjYlFqpSvBoXfW0Ao_yLr8

Horowitz, M. C. (2019). When speed kills: Lethal autonomous weapon systems, deterrence and stability. *Journal of Strategic Studies, 42*(6), 764–788. https://doi.org/10.1080/01402390.2019.1621174

Johnson, J. (2019). Artificial intelligence & future warfare: Implications for international security. *Defense & Security Analysis, 35*(2), 147–169. https://doi.org/10.1080/14751798.2019.1600800

Johnson, J. (2020). Artificial intelligence, drone swarming and escalation risks in future warfare. *The RUSI Journal, 165*(2), 26–36. https://doi.org/10.1080/03071847.2020.1752026

Johnson, J. (2023). Automating the OODA loop in the age of intelligent machines: Reaffirming the role of humans in command-and-control decision-making in the digital age. *Defence Studies, 23*(1), 43–67. https://doi.org/10.1080/14702436.2022.2102486

Kunertova, D. (2021). From robots to warbots: Reality meets science fiction [Application/pdf]. *CSS Analyses in Security Policy, 292.* https://doi.org/10.3929/ETHZ B 000508358

Lewis, J. (2015). The case for regulating fully autonomous weapons. *The Yale Law Journal, 124*(4), 1309–1325.

Limata, T. (2023). Decision making in killer robots is not bias free. *Journal of Military Ethics, 22*(2), 118–128. https://doi.org/10.1080/15027570.2023.2286044

Lin, P., Abney, K., & Bekey, G. A. (2012). Killing made easy: From joysticks to politics. In *Robot ethics: The ethical and social implications of robotics* (pp. 111–128). MIT Press. https://ieeexplore.ieee.org/document/6733961

Payne, K. (2021). *I, warbot: The dawn of artificially intelligent conflict.* Oxford University Press.

Ploumis, M. (2022). AI weapon systems in future war operations; strategy, operations and tactics. *Comparative Strategy, 41*(1), 1–18. https://doi.org/10.1080/01495933.2021.2017739

Raska, M. (2021). The sixth RMA wave: Disruption in military affairs? *Journal of Strategic Studies, 44*(4), 456–479. https://doi.org/10.1080/01402390.2020.1848818

Scharre, P. (2018). *Army of none: Autonomous weapons and the future of war* (1st ed.). W. W. Norton & Company.

Scharre, P., & Horowitz, M. C. (2015). Autonomy in weapon systems. *Center for a New American Security Working Paper.* https://s3.us-east-1.amazonaws.com/files.cnas.org/hero/documents/Ethical-Autonomy-Working-Paper_021015_v02.pdf

Schott, R. M., & Lang, J. (2017). *Killer robots: The future of war?* [DIIS Policy Brief]. The Danish Institute for International Studies. https://pure.diis.dk/ws/files/817702/Killer_robots_WEB.pdf

Trusilo, D. (2023). Autonomous AI systems in conflict: Emergent behavior and its impact on predictability and reliability. *Journal of Military Ethics, 22*(1), 2–17. https://doi.org/10.1080/15027570.2023.2213985

Umbrello, S., Torres, P., & De Bellis, A. F. (2020). The future of war: Could lethal autonomous weapons make conflict more ethical? *AI & SOCIETY, 35*(1), 273–282. https://doi.org/10.1007/s00146-019-00879-x

Walsh, J., & Schulzke, M. (2015). *The ethics of drone strikes: Does reducing the cost of conflict encourage war?* (ADA621793). US Army War College Press. https://apps.dtic.mil/sti/citations/ADA621793

Wong, Y. H., Yurchak, J., Button, R. W., Frank, A. B., Laird, B., Osoba, O. A., Steeb, R., Harris, B. N., & Bae, S. J. (2020). *Deterrence in the age of thinking machines.* RAND Corporation. https://www.rand.org/pubs/research_reports/RR2797.html

Wood, G. N. (2020). The problem with killer robots. *Journal of Military Ethics, 19*(3), 220–240. https://doi.org/10.1080/15027570.2020.1849966

Wood, N. G. (2023). Autonomous weapon systems: A clarification. *Journal of Military Ethics, 22*(1), 18–32. https://doi.org/10.1080/15027570.2023.2214402

Zegart, A. (2020). Cheap fights, credible threats: The future of armed drones and coercion. *Journal of Strategic Studies, 43*(1), 6–46. https://doi.org/10.1080/01402390.2018.1439747

11 Autonomous Weapons and the Need for an Evolution of the Clausewitzian Trinity toward an Emphasis on Tactical Positioning

Kyu-hyun Jo

11.1 INTRODUCTION

The Clausewitzian Trinity has long served as a classic topic among scholars of *On War*. Much of the discussion has centered around how each element of the Trinity functions independently and cooperatively with each other to form the Trinity that gives the original term its name. The current scholarly literature has emphasized the roles and functions of Clausewitz's emphases on strategy, tactics, and politics, or more specifically, how generals utilize strategies and tactics and coordinate with politicians, and ultimately, the public to cooperate in fighting against enemies (Herberg-Rothe et al., 2011, pp. 48–73). Most recently, Kyu-hyun Jo examined how the management of territories and resources found inside territories are strategically mobilized for war, thereby expanding the classic Trinity's application to the political organization of "military territoriality" in wars (Jo, 2021). However, as much as these scholarly endeavors have shed much light on how promising the Clausewitzian Trinity is in contemporary and general warfare, very few scholars have examined the limitations of the Clausewitzian Trinity in the advent of the usage and application of artificial intelligence and autonomous weapons. With regard to autonomous weapons and the Clausewitzian Trinity, scholarship remains bifurcated. On the one hand, there are several articles and books warning about the dangers of autonomous weapons; on the other hand, there are articles devoted to understanding how the Clausewitzian Trinity can be applied to modern warfare, with "modern" understood to mean warfare since the early 1980s.

On the former, topics have ranged from robotic warfare, artificial intelligence and its use in asymmetrical warfare, friction and autonomous weapons, human rights violations and autonomous weapons, autonomous weapons and international law (Carlini, 2020, pp. 21–29; Crootof, 2016; Davison, 2018; Gardner, 2021, pp. 86–98; Horowitz, 2016, pp. 25–36; Lewis, 2015, pp. 1309–1325; Petrenko, 2012, pp. 81–102; Rudischhauser, 2017; Simpson & Müller, 2016, pp. 302–322; Sparrow, 2007, pp. 62–77; Warren & Hillas, 2018, pp. 218–249). On the latter, topics are diverse such as the Clausewitzian paradigm in relation to East Asia, a comparison between Clausewitz and Sun Zi, Clausewitz and the contemporary politics of war, a Foucauldian interpretation of the relationship between war and power, Clausewitzian effectiveness, Clausewitz and contemporary strategies, and Clausewitz's position on small wars (Čajić, 2016; Daase & Davis, 2015; Dimitriu, 2020; Summers, 1983; Torkelson, 2008). In general, recent scholarship on Clausewitz features analyses about technology and its impact on warfare posing indirect warnings against various test cases which seek to demonstrate the continuing relevance of Clausewitz in the 21st century.

Yet, there is no attempt to link the two bodies of scholarship together to critically examine whether or not autonomous weapons expose the limited applicability of the Clausewitzian Trinity as a paradigm. There are just two articles rejecting the continued relevance of Clausewitz to technological warfare in the 21st century. Although both raise crucial questions about whether advanced

DOI: 10.1201/9781003441700-11

technology can break down the Clausewitzian Trinity, they only focus on disproving Clausewitz's relevance to the 21st century and do not substantiate their stances by precisely showing how autonomous weapons and warfare actually render Clausewitz irrelevant based on legal and moral grounds (Meilinger, 2007, pp. 116–145). There is a small body of literature addressing the legal and ethical implications of autonomous weapons, but it mostly concentrates on autonomous weapons without referencing Clausewitz. One book claims that 21st-century technology has proven Clausewitz wrong by showing that war is "not a game of Go," or as orderly as Clausewitz imagined wars to be, but the book does not show how it is the dissolution of the classic Trinity which provides the ultimate reason for such a diagnosis (Hadji-Janev & Hristovski, 2017, pp. 325–340; Heyns, 2016b, pp. 350–378; Saxon, 2014, pp. 100–107; Wallace, 2018).

While the classic model is deeply analyzed in its own terms, the primary problem is that there is no comparative lens with which to judge the merits and disadvantages of the Trinity to offer a more balanced and critical evaluation of the concept. In contrast to the literature on the Clausewitzian Trinity, while this literature is more relevant to contemporary concerns about war from the perspective of technological advancements, it ignores possibilities for considering the potential and limitations of the Trinity, which leaves an objective assessment of the Trinity's weaknesses to be desired. The literature on autonomous warfare suffers from the exact opposite problem. While there are various warnings about the dangers of autonomous warfare and how technology can destroy a sense of order on the battlefield, there is no serious attempt to link autonomous weapons and warfare's function in destroying the Clausewitzian Trinity which is actually why war has become more unpredictable and dangerous and *thereby* requires the Clausewitzian Trinity to undergo an evolution with an emphasis on tactical positioning of soldiers and weapons from the perspective of technological warfare in the 21st century.

The mobilization of the three elements of the Trinity requires precise coordination and a common will to support the coexistence of the three tenets of the Trinity in the name of national defense and national honor; time is a critical ingredient. However, because technology exists to win the race against time, and autonomous weapons expedite victory by killing many people indiscriminately, coordination between the three elements of the Trinity is very difficult if not impossible to achieve before autonomous weapons release indiscriminate and extensive destruction. The prevention of such an outcome through an evolution focused on tactical positioning is important because bureaucratic coordination is both the central feature and weakness of the Clausewitzian Trinity and the essential purpose for which autonomous weapons have their strategic importance in modern warfare. The clash between these two identities is what precisely describes the antagonistic complementation between the Clausewitzian Trinity and autonomous weapons.

Therefore, I will make two arguments. First, after analyzing each tenet of the Clausewitzian Trinity, I will argue that the Trinity's inherent weakness lies in technological warfare's capability to destroy bureaucratic efficiency within militaries and the coordination between a military and citizens, which could also be an enemy's objective behind formulating their own Trinity. Conceptually, the Clausewitzian Trinity must accordingly shift its concentration on tactical positioning of soldiers and weaponry to maintain pace with technological advancements. I will bridge the two currently disjointed bodies of scholarship together by arguing that because autonomous weapons have the fundamental capacity to kill large numbers of people indiscriminately, the precise coordination between generals, the government, and the public, for which strategy, tactics, and politics have essential meaning and aim—to defeat an enemy—becomes limited in scope. Although friction may still be visible if two opposing nations use similar levels of autonomous warfare, the existence of the friction is independent from the Clausewitzian Trinity because uncertainties about autonomous weapons are largely due to a flooding of information and a naive belief that technology would get rid of the traditional sense of friction that arose from within the Trinity, whereas the Clausewitzian Trinity had constant friction because of uncertainties about how enemy soldiers would maneuver under the assumption that there was a technological balance of power between two opposing sides (Murray, 1997, p. 63). Since autonomous weapons only require a few highly trained specialists to give out commands, and the degree of technical sophistication directly translates into a clear imbalance in technical capabilities between two opposing sides, there is always an incentive to develop more sophisticated weapons to outpace the enemy and destroy the Trinity using such technological superiority.

Thus, considering that it does not take many technicians to develop autonomous weapons and many commanders to decide whether or not to use them, and considering that a major objective behind deploying more sophisticated autonomous weapons than an enemy is to ensure a rapid and complete destruction of the enemy's Trinity before it could even form, the Trinity's full formation may not always be necessary or even possible, depending on the degree of a military's technological development. Furthermore, because speed is essential in determining the effective mobilization of autonomous weapons, the minimization of the Clausewitzian Trinity may be desirable for one nation, which, in turn, translates into a simplicity and rapidity with which the Trinity can be destroyed by another. Since these mutually contradictory features are going to be common and desirable for all nations, the Clausewitzian Trinity requires modification and adaptive evolution toward focusing on tactical positioning when confronted with autonomous warfare (Horowitz et al., 2018, pp. 139–171; McAdam, 1983, pp. 735–754).

My second argument is that autonomous warfare translates my first theoretical argument into a real phenomenological fact. Sociologist Ian Roxborough's recommendation that Clausewitz's Trinity needs to be updated to reflect technological warfare, F. Hoffman's warning that autonomous warfare represents a "seventh military revolution," and Phillip Meilinger's argument that the Clausewitzian Trinity can be rendered obsolete because of such developments holistically point to the capability of autonomous warfare to engineer an adaptive evolution of the Clausewitzian Trinity (Hoffman, 2017, p. 1931; Meilinger, 2007, pp. 116–145; Roxborough, 1994, pp. 619–634). Considering the indiscriminate and rapid destruction that autonomous weapons inflict, they require the Clausewitzian Trinity to adaptively evolve and systematically decapitate enemy troops before the coordination behind the functioning of the Trinity can occur. The real danger and devastation of autonomous weapons is not simply the ease and extent to which it can kill people. It is increasingly demonstrated using military drones in the mission to kill terrorists such as Osama Bin Laden and Qasem Soleimani, and the rapidly expanding drone market, the lethal potential to decapitate communication, and the formation of any military strategy based on numbers and mobilization of troops alone (Frantzman, 2021). Autonomous weapons signal the inevitable adaptive evolution of the Clausewitzian Trinity. Technological superiority can inspire the evolution of bureaucratic efficiency, which depends on the aforementioned coordination.

While it may be true that modern warfare involves a careful consideration of domestic and international politics, inherent human capabilities, environmental awareness, and tactical acumen to place weapons in the right places, technological superiority still largely determines the qualitative and quantitative capability of autonomous weapons. Since neither an individual nor an entire bureaucracy can foresee the consequences of technological developments with absolute certainty, no agency or individual can be held singularly responsible, thereby reducing the role of bureaucracy and individual liability. Human strength is not only utilized differently from manning weapons to developing drones or killer robots, but also thereby economized to the greatest degree, with less exposure to enemy fire. Since autonomous weapons economize human strength in killing people, the efficiency necessitates the evolution of the Clausewitzian Trinity to focus on tactical positioning of soldiers and weapons due to the rapidity with which autonomous weapons can destroy targets. It is the ultimate and definitive destruction of the Clausewitzian Trinity that autonomous weapons attain the highest strategic value in wars, and tactical positioning must be the next evolutionary step for the Clausewitzian Trinity to be an effective bulwark against autonomous weapons.

11.2 SPEED, EFFICIENCY, AND LIMITED BUREAUCRATIC COORDINATION IN AUTONOMOUS WARFARE AND THE NEED FOR THE CLAUSEWITZIAN TRINITY TO REFLECT A BUREAUCRATIC DEMAND FOR TECHNOLOGICAL PROGRESS

Clausewitz defines the Trinity as a coordination of passion, chance, and reason. Clausewitz emphasizes the need to coordinate generals, politicians, and citizens to transform war from a duel between two or more nations into an instrument of policy (von Clausewitz, 1982, p. 38). Since war is a violent activity aimed at compelling a defeated military to fulfill the victor's will, it necessarily features all energies and resources devoted to exerting violence to be invested to their full potential (Aron, 1974, p. 50; von Clausewitz, 1982, pp. 102–103). Although it could be possible that two opposing sides could engage in similar levels of violence, a complete stalemate is impossible since both sides will try to intensify violence to capture a favorable moment in a war (Aron, 1974, p. 51; von Clausewitz, 1982, p. 112). The destruction of military power, the subjugation of a country, and subsequent submission of an enemy's will describe the ultimate objective behind the investment of passion, chance, and reason (Torkelson, 2008, p. 12; von Clausewitz, 1982, p. 123). The Clausewitzian Trinity's central purpose is to ensure the fulfillment of a complete and an unquestionable victory which even the enemy can concur.

However, since any war's outcome is bound to be unpredictable, unexpected setbacks might occur to cause "insignificant irregularities or delays," or friction at all times (Monte, 2018, p. 216). Friction—any form of difficulty associated with maintaining control due to weather, terrain, time lags, or difficulty in logistics or supplying provisions—can be physically manifest through the existence of resistance from within a military bureaucracy or through a minor external resistance from an enemy or plans and commands are not often delivered precisely causing many misinterpretations and mistakes in executing the commands (Beyerchen, 1992, pp. 67–68; Torkelson, 2008, p. 13; von Clausewitz, 1982, p. 164; Wallace, 2018, p. 5). Strategies must be carefully and precisely used to attain the central objective in wars, which is to compel an enemy to fulfill a particular nation's will (von Clausewitz, 1982, p. 241), and to that end, a proper assembly of soldiers to overcome the enemy is highly desirable (von Clausewitz, 1982, p. 304, p. 388).

Insofar as diverse strategies, the mobilization of soldiers and civilians, and the use of diverse weapons and strategies to position militaries in appropriate times and positions all render wars to have the common end of seizing victory and using that victory to demand specific political and economic terms to clearly confirm an enemy's defeat, for Clausewitz, war is "nothing but a continuation of political discourse." War is a domain of policies, or more accurately, an instrument of policy, through which it attains "its absolute form" (Torkelson, 2008, p. 15; von Clausewitz, 1982, p. 403). War does not merely reflect or inherit partial features of a policy, but becomes a policy, for war is a political choice bearing the character of a policy and also its scale to the extent that the sword "replaces the pen" and preserves its independent nature thereafter (von Clausewitz, 1982, p. 410). War is politics by other means because victory is not an end but a means to compel an enemy to completely obey to every command that a victor wishes to enforce.

Clausewitz's interpretation of war as a science is still enduring because he was able to philosophically articulate the intrinsic nature of war and because he was able to situate the need for strategies by interpreting victory and the subjugation of the enemy as the ultimate end of war (Čajić, 2016, p. 77). Yet, the main weakness of Clausewitz's theory is that if technological superiority ensures the destruction of the Clausewitzian Trinity through a systematic decapitation of a military bureaucracy which provides the Trinity with reason and a rationale for winning a war, it would be in an enemy's interest to prevent the realization of technological superiority by generating a communication failure within the Clausewitzian Trinity using superior technology (Wallace, 2018, p. 33). Therefore, ironically, it is also true that there needs to be some coordination within a bureaucracy to formulate a strategy to defeat an enemy; it can be pointed out that technological superiority can become an objective for the formation of a Clausewitzian Trinity but also simultaneously the principal cause behind the Trinity's modification toward enhancing tactical positioning to keep pace with technological developments. The precision with which a military can adequately respond in real time with absolute technological superiority is never guaranteed insofar as technological superiority and a constant competition to attain that very condition render the non-linearity of modern warfare as its perpetual feature. War is never a game of Go in the real world if there is a constant and persistent pressure to pursue technological excellence and evolution (Neuneck, 2008, pp. 50–60; Wallace, 2018, pp. 40–43).

If we reinterpret the classic Clausewitzian Trinity and the transformation of war into policy to more precisely reflect the reality of most battles fought between militaries on land, the classic Trinity can be better approximately translated into what Jo (2021) calls "military territoriality," in which human morale connects with a new Trinity of strategy, plan, and battle to stress the importance of preparedness before a war. Autonomous weapons place the Trinity and "military territoriality" in a limbo, for while the rapid formation of the Trinity using advanced telecommunications also enables the appearance of "military territoriality," the rapidity behind the making of these two crucial elements can also backfire and be a prime cause for their elimination and disappearance. Moreover, because the success of the Clausewitzian Trinity and "military territoriality" all depend on the effective coordination within a military bureaucracy and the communication of policies to citizens during a war, the development of more sophisticated killer robots, automatic weapons, and even swifter space shuttles to enable militaries to maneuver and execute operations on land and in space can simultaneously have the objective of formulating a superior Clausewitzian Trinity and "military territoriality" which can be proven through the destruction of an enemy's formation of the same elements (Jo, 2021, pp. 395–409). The Clausewitzian Trinity is not a stagnant object, but a flexible program whose elements can and must change in accordance with the objective of rendering a complete military victory possible.

The proliferation of autonomous weapons is largely a product of the speed and efficiency with which the weapons themselves have been equipped to kill human targets and how these weapons have evolved to harm enemy targets' interests or their physical safety (Leveringhaus, 2016, pp. 37–39). These weapons do not require many personnel to handle or operate them, which means that soldiers can be spared from fighting against enemy forces themselves. However, on the other hand, war becomes more dehumanized because, as philosopher Alex Leveringhaus has argued, autonomous weapons are uninhabited, or they do not require an operator to physically operate the weapons, they are pre-programmed, or commands for a finite range of actions circumscribe the utility of an autonomous weapon, and finally, once programmed, autonomous weapons do not require human intervention, or they become independent agents because of the pre-planned programming (Leveringhaus, 2016, p. 4). Moreover, as Massimiliano Gaetano, Kenneth Scheve, and David Stasavage have meticulously shown through data analysis, a rapid increase in technological development decreased the need to field mass armies between the 19th and 20th centuries (Onorato et al., 2014, pp. 476–477). What this finding implies is that there was also a decrease in the need for the coordination of soldiers from military bureaucracies because an attention to technological precision, not the precise positioning of soldiers, became a decisive factor in clinching victories on the battlefield across the two centuries.

If autonomous weapons acquire the status of being the most advanced form of military technology in the 21st century and are expected to advance ever more in the future, it is not difficult to predict that autonomous weapons might be capable of forcing the territorial essence of passion and reason, or battle, strategy, and planning to lose their unique values. Although Robert Sparrow is right to point out that there is not much of a moral difference between an autonomous weapon and a semi-autonomous one insofar as it is human beings issuing directions and making specific programs to target enemies, the more important point is that there is no need for an elaborate bureaucracy to issue such directions (Sparrow, 2007, p. 65).

Furthermore, because the ease with which autonomous weapons kill selected targets, even if one nation arrives at a specific conclusion through bureaucratic meetings, is the source of these weapons' indiscriminate nature, a Clausewitzian Trinity cannot avoid being modified toward enhancing tactical positioning to reflect bureaucratic decisions' need to react to the pace of technological change (Monte, 2018, p. 22). If the conclusion of such meetings was to obliterate another enemy by destroying their bureaucratic foundations, it implies that on a theoretical level, a Clausewitzian Trinity is subject to committing suicide, for one Trinity is being driven by autonomous weapons and their indiscriminate nature and effectiveness to destroy another. Meaningful human control of autonomous weapons is difficult to define because the level of autonomy associated with autonomous weapons can wildly vary according to the level of sophistication required from human interactions with such weapons and the range of targets which the autonomous weapon can destroy (Bode & Huelss, 2022, pp. 27–33). Since the autonomy of autonomous weapons is ruthlessly simple to denote a capability to "fire and forget," there is always a possibility for autonomous weapons to be "liberated" from human control and render distinctions between enemies and allies meaningless. The proportionality of the damage an autonomous weapon can inflict or is allowed to inflict can differ based on contextual details (Bode & Huelss, 2022, p. 44).

This problem would be more serious if the precise identity of an "enemy soldier" is complicated by the existence of enemy civilians wielding weapons and resisting exactly as enemy soldiers would do, which would confuse autonomous weapons into firing indiscriminately at civilians, causing human rights violations (Horowitz, 2016, pp. 29–30). Since autonomous weapons rapidly and indiscriminately destroy human targets, and that is the only principal function which largely describes their worth on the battlefield, the need to translate the Clausewitzian Trinity into "military territoriality"—human morale and the more physical Trinity of battle, plan, and strategy—becomes more important and tactical positioning can absorb any rate of technological progress by always ensuring that the authority to decide where and when to position soldiers and weapons rests with the judgment and strategic thinking of a commander. The rationality to distinguish allies from enemies is not something which could be encoded precisely to account for every kind of "image" that an enemy might appear in front of autonomous weapons, but the more serious problem is that once a certain range of rational decisions is encoded in an autonomous weapon, the need for a military bureaucracy to congregate and communicate its decisions with civilians and soldiers becomes obsolete. As Warren and Hillas argue, the question of whether autonomous weapons can be intelligent and conscious as human beings, not merely approximating them, can never be clearly answered (Warren & Hillas, 2018, p. 248). Hence, the strategic deployment of such weapons along with the precise positioning of soldiers must always be within the domain of the human mind.

Furthermore, because autonomous weapons can only be autonomous insofar as human beings allow them to be, it is important to specify that "autonomy" cannot mean "acting for one's own reasons" (Leveringhaus, 2016, p. 47). If an autonomous weapon is encoded to target an enemy's military headquarters because such a decision arose from a military's headquarters, the autonomy of an autonomous weapon is in a limbo between the ironic situation of being borne from a Clausewitzian Trinity to destroy another. Humans can control autonomous weapons by specifying the range of rational choices, but it is also precisely due to this limited nature of control exerted from human beings that a Clausewitzian Trinity must also change by repositioning soldiers or weapons or luring enemy to unfamiliar terrain. Such adaptability allows the Clausewitzian Trinity to avoid being a target even though it was borne from a military bureaucracy and informed to a public.

Autonomous weapons rob the need for reason, psychology, or the rationality to devise meticulous strategies since the only command which instructs autonomous weapons is determining who to kill, which is programmed in advance by a human operator. No human operator can have the prescience to predict an autonomous weapon's decisions in advance, and even if an operator did have such prescience, there is no further mechanism to control an autonomous weapon to cease doing a harmful action (Leveringhaus, 2016, pp. 79–86). Due to this absence of reason, there is also no passion in autonomous warfare, since barely any human involvement on the battlefield is necessary (Bode & Huelss, 2022, p. 16; Leveringhaus, 2016, pp. 48–49). Yet, despite the highly probable absence of direct human involvement in conducting autonomous warfare, there still can exist an increased concern for chance and friction because autonomous weapons have yet to acquire the capability to distinguish military and civilian targets, and because nations with excellent technological capabilities will constantly try to maintain a balance against each other by developing ever more sophisticated weapons. The rapidity with which autonomous weapons eliminate targets might not be sufficient to address friction, because even if autonomous weapons have a fixed idea of who would be enemies, they cannot accurately assess context, which means that former enemy soldiers who surrendered to be allies would not be registered as allies if autonomous weapons cannot liberate themselves from the fixed idea (Gardner, 2021, pp. 86–98). Technological development, however sophisticated, always encounters an urge to maintain a balance of power, rendering the existence of friction unavoidable.

Yet, such a concern does not render the Clausewitzian Trinity as a concept to be more important than technology. Instead, what is at stake is whether the Clausewitzian Trinity can keep pace with technological developments because technological progress itself is always a relative concept. Even if territorial calculations and strategic placement of soldiers will still matter, directives which nations enter into autonomous machines will more likely destroy friction than preserve it. More specifically, the traditional Clausewitzian sense of friction, largely stemming from an uncertainty about what an enemy would do under the assumption that there is technological parity between two opposing sides, is less valid in autonomous warfare. War becomes ever more unpredictable with technological development and technological superiority will never become technological supremacy so long as the urge to develop ever more sophisticated autonomous weapons to destroy the traditional sense of friction by outpacing each other in terms of technological advancement remains. Although the flooding of information might be a source of friction given that one can never know how much information an enemy might have, since the flooding of information is necessary to provide a basis for developing more sophisticated autonomous weapon, the flooding of information is not always a source of friction but a source of stronger certainty than an opposing side, for it is with copious information that one gets the confidence to develop new weapons.

It is not the flooding of information, which is a source of friction, but the degree of the flooding which can be a source of friction, but since information is essential to know how new a technology is, the degree of the flooding of information cannot definitively be said to be a source of friction either, for more information is always better than having less information to test new technologies (Murray, 1997, p. 63). The flooding of information also depends on its direction; if it is an enemy flooding misinformation, it might be a source of friction in causing much confusion, but if the flooding is done deliberately to collect as much data about renovating a weapon one already has, the flooding decreases uncertainty and doubt about deciding to renovate it and, therefore, is an advantage. It is also dependent on how open and flexible a military bureaucracy is toward change and necessary measures to realize change, for a closed structure makes new technologies impervious and a rigid adherence to old methods and inflexible theories render real-life problem-solving very difficult (Jensen et al., 2022, p. 17, pp. 43–45).

Therefore, it is important to have sufficient organizational capacity and resourcefulness to manage the inflow of new information and maximize the fluidity of information, which might require a major reduction in the size of a military bureaucracy to render the control of information into a central key to victory (Jensen et al., 2022, pp. 36–37, p. 45). The existence of friction does, however, imply the adaptive evolution of the Clausewitzian Trinity because the latter could focus on restructuring the former by enhancing the capability of identifying enemies by establishing more precise parameters using the flooding of information directed toward generating more information about developing new weapons. Information is not a source of friction, but rather, the direction to which information is used and how open a military bureaucracy is to accepting change and new techniques determines the flow of information (Jensen et al., 2022, p. 151, p. 199).

If autonomous weapons acquire the ability to launch tactical offensive strikes without issuing warnings to an enemy, it may suggest the persistence of structural non-linearity or the uncertainty of information, but because developing more sophisticated autonomous weapons is precisely meant to enhance their accuracy in targeting and discernment between allies and enemies, further developments in autonomous weapons would aim to erase friction rather than to maintain it (Gardner, 2021, p. 95; Sartor & Omicini, 2016, p. 47, p. 49, p. 53, p. 58, p. 61). The persistence of friction will not hinder any nation from setting the destruction of an enemy's bureaucracy, and subsequently, the destruction of their Clausewitzian Trinity, and as Leys argues, triggers an intense competition to encourage the proliferation of autonomous weapons (Leys, 2018, p. 54). Since technological superiority can never eternally become technological supremacy for a particular nation, even chance and friction will be of limited importance due to enhanced accuracy. Therefore, aside from friction as a lone exception, autonomous weapons require the Clausewitzian Trinity to modify and subsequently adapt, since "military territoriality" will still require the proper positioning of weapons and soldiers to maximize efficiency. The next section will concretely explain how autonomous weapons' inherent features will make my theoretical argument a reality soon.

11.3 AUTONOMOUS WEAPONS AND INDISCRIMINATE AND EFFICIENT DESTRUCTION, OR THE NEED TO SHIFT THE FOCUS OF THE CLAUSEWITZIAN TRINITY TO TACTICAL POSITIONING

Clausewitz was certainly right when he argued for the need of effective coordination to properly execute strategies, but a fundamental drawback in his thinking is that the coordination of the Trinity is too dependent on numerical superiority under the assumption that asymmetrical warfare exists and cannot be realized through any other means. However, the exponential growth in the proliferation of diverse military technologies ranging from killer robots to drones has rendered autonomous weapons into decisive game changers which could obliterate large masses of soldiers and civilians with indiscriminate and effective power. The development of autonomous weapons and artificial intelligence has enabled militaries to eliminate military and civilian targets with astonishing efficiency such that the capacity of autonomous weapons to implement indiscriminate and efficient destruction is the principal reason behind the necessary adaptive transformation of the Clausewitzian Trinity to maximize the efficacy of "military territoriality." Supporters of the Clausewitzian Trinity have argued that it is the philosophy behind the Trinity, or more specifically, the need to coordinate citizens, soldiers, and generals to facilitate a constructive dialogue between strategy, tactics, and politics which gives the Trinity its timeless character. However, as I have argued in the previous section, the formation of a Clausewitzian Trinity, bearing the aim of embodying "military territoriality," might become interchangeable with its rapid destruction, and the interchangeability represents a competition to expedite the speed of technological innovation and renovation, and original inventions. The variety of autonomous weapons developed in the 21st century produces questions not only about their lethal nature but also about the role of personnel and even whether human beings would be necessary to operate the weapons. In connection with the Clausewitzian Trinity and "military territoriality," these questions can be translated into whether the formation of these concepts as physical entities is even necessary. Although a military bureaucracy might still be necessary to decide what kinds of information need to be revealed to the public and what kinds of technology might be necessary to develop more efficient killer robots, the consolidation of a Clausewitzian Trinity and "military territoriality" would be limited to these particular roles.

The efficiency of autonomous weapons, if maximized, would render distinctions between strategy and tactics opaque, since the need for using human personnel would be greatly reduced. Since the tactical suitability and efficiency of technological warfare using autonomous weapons would largely be determined by how much fewer soldiers and citizens must be used and sacrificed on the battlefield, the more developed autonomous warfare becomes in the future, the more sophisticated and strategic the Clausewitzian Trinity and "military territoriality" would become, since coordination would assume less physicality as more machines than human beings would be introduced on battlefields. The Clausewitzian Trinity in its original form of positioning soldiers to simply exchange volleys and depend on the soldiers' mobilization to concretize tactics and strategy will instead focus on economic positioning or training soldiers to locate advantageous terrain as long as more advanced weapons such as robots or drones can replace human beings on the battlefield with minimal instructions from human operators and initiate fully automated warfare. I will substantiate this claim by providing more concrete examples of autonomous weapons and their capabilities and uses in wars to substantiate on the previous section's speculation that the Clausewitzian Trinity and "military territoriality" would need to undergo adaptive evolution toward tactical positioning due to the fact that the formation of these concepts from one nation would be precisely aimed at their destruction from another by encouraging the development of more advanced and rapid technologies. Since this translates into a competition to see which side can afford to position fewer soldiers, the Clausewitzian dictum that war is politics by other means would attain a more literal essence through autonomous warfare only reflecting specific political ends which would predominantly determine their value and use.

Autonomous weapons have rapidly evolved to become economically more feasible than conventional weapons because they require fewer people to operate them, allowing militaries to reduce investments on educating and training personnel and directing them to developing more sophisticated weapons (Scharre, 2018, p. 16). Since depersonalization in the deployment of force could directly translate into a personalization of targets chosen by autonomous weapons, there is less to worry about concerning unwarranted injuries or accidents (Heyns, 2016a, p. 7). Moreover, there is little need for a complex military bureaucracy because a consensus-based coordination between individuals holding the authority and expertise to make decisions allows for collective action, and a series of simple decisions are enough to generate complex actions from autonomous weapons (Scharre, 2018, p. 21).

Although there are varying levels of coordination depending on how autonomous a weapon is, the distinction between semi-autonomous and autonomous weapons in relation to the question of whether a full military bureaucracy is needed is thin. Semi-autonomous weapons perform a task and then wait for human commands to judge its propriety, whereas autonomous weapons do not require human consultation in making such a judgment. The common denominator for both types of autonomous weapons is that human strength is economized, for coordination concerning decision-making is minimal for semi-autonomous weapons and is largely obsolete for autonomous weapons (Scharre, 2018, pp. 28–29). In the case of the latter, the Clausewitzian Trinity is not effectual, for judgments about whether the autonomous weapon ought to fire at an enemy or refrain from doing so are largely reactions to its surrounding environment, not about whether a bureaucratic coordination about such a decision had been made in advance (Scharre, 2018, pp. 30–31).

While a military bureaucracy might be required in the process of establishing acceptable parameters for an autonomous weapon to decide who is an enemy and who is not, once an autonomous weapon is built, the bureaucracy is no longer relevant because judgments about the weapon's future courses of action can never be accurately predicted and must be left for the weapon alone to make (Scharre, 2018, pp. 46–47). The more advanced an autonomous weapon becomes, the more distant and irrelevant a military bureaucracy becomes because the autonomy of an autonomous weapon is dependent on the weapon's reactions to its surrounding environment and its own decisions about its future actions, both of which cannot be read in advance by a human mind.

Even if a military bureaucracy could supervise autonomous weapons about general parameters concerning the identities of enemy targets, the generality associated with describing targets can also be an autonomous weapon's prime weakness. Since knowledge about the identities of enemy targets are largely based on algorithms and patterns, any movement or behavior of enemy targets which fall outside the prescribed parameters cannot be accurately determined by supervised autonomous weapons unless a new set of algorithms or patterns are entered by human beings (Bode & Huelss, 2022, p. 23). Although they may be prized for their ability to decide which enemies they must engage with, a military bureaucracy cannot function in the traditional Clausewitzian manner of formulating specific plans and strategies for an autonomous weapon because their formation is not dependent on the bureaucracy but on the weapon itself. The weapon must have the ability to navigate through any kind of terrain without human assistance, it must be able to distinguish between enemies who are genuinely surrendering and enemies who are "faking" their surrender to extract military secrets, and finally, be able to exert an appropriate amount of force to kill an enemy. It is the nature and degree of an autonomous weapon's autonomy which requires a military bureaucracy, not the other way around.

Yet, it is difficult for an autonomous weapon, even if it is equipped with artificial intelligence to have an extreme sense of precision to accurately judge one war from another while conditions rapidly change. Aside from targets which are programmed in advance into a computer, autonomous weapons cannot accurately judge potential enemies who might be outside the original parameters because they do not possess cognitive reasoning to allow them to make distinctions between civilians and soldiers and use proportionality as a yardstick for determining an appropriate amount of violence to be used (Corn, 2016, p. 215, p. 228, p. 231, p. 235; Leveringhaus, 2016, pp. 54–55; Lieblich & Benvenisti, 2016, p. 250). Should misjudgments concerning the identity of an enemy occur, autonomous weapons would not only commit war crimes against citizens and harm their human dignity, but also not be able to understand the value of life, provide victims with adequate compensations, or hold human beings accountable for their decisions (Birnbacher, 2016, p. 113, p. 119; Bode & Huelss, 2022, p. 50; Heyns, 2016a, p. 9; Sartor & Omicini, 2016, p. 66, p. 71). An indifference to morality, human dignity, and moral hazards might require partial human intervention.

Should an autonomous weapon outperform human beings, it may not be able to handle the pressure of the morality associated with using force in particular cases because the morality is not only associated with the act of killing but the proportionality of violence, which is dependent on a war's complex and dynamic context (Bode & Huelss, 2022, p. 55; Heyns, 2016a, p. 11, p. 13; Kalmanovitz, 2016, pp. 150–151). Unless they are supervised by humans, semi-autonomous and autonomous weapons would lack situational understanding required for assessing complex battle situations and experience a lack of deliberation due to a high level of pressure exerted from a lack of time (Bode & Huelss, 2022, p. 192; Monte, 2018, p. 110). They would require big data to compel artificial intelligence to be exposed to millions of different scenarios to render the artificial intelligence to be comfortable with its current environment (Kalmanovitz, 2016, p. 170; Monte, 2018, p. 54). Without prior programming, artificial intelligence cannot be expected to automatically adjust to a completely different environment without prior exposure, meaning that the extent to which human beings can actually be confident about leaving the judgment of who exactly is an acceptable human target can never be determined with absolute certainty (Bhuta & Pantazopoulos, 2016, p. 288; Lieblich & Benvenisti, 2016, p. 274, p. 277; Monte, 2018, p. 166; Wallace, 2018, p. 72, p. 77). In short, in the most extreme scenario of a highly frenzied warzone, without human supervision, the absence of moral judgment can make it possible for an autonomous weapon to kill for an unlimited period of time and render the fog of war impenetrable regardless of how sophisticated technology becomes (Leveringhaus, 2016, p. 65; Monte, 2018, p. 71, p. 218).

Nevertheless, the existence of a moral hazard does not necessarily imply that the Clausewitzian Trinity does not need to evolve. The undeniable reality which supersedes such moral considerations is that a military bureaucracy, however well-staffed, is not only irrelevant in real time to meaningfully alter the autonomous weapon but also cannot have an infinite amount of time to satisfactorily train autonomous weapons to have a natural human acumen to judge instantaneously. If the threshold between semi-autonomous and autonomous weapons is thin and varies from country to country, it will be extremely confusing for commanders to precisely determine acceptable levels of autonomy for various weapons (Sharkey, 2016, p. 26). Since specific details concerning the identity of an enemy will differ for every nation, and because autonomy within autonomous weapons is in itself programmed within specific limits which cannot be changed unless altered by human intervention, a military bureaucracy cannot function properly once it is finished with developing it because no military bureaucracy can ensure that autonomous weapons will judge correctly and act rationally in every circumstance (Sharkey, 2016, pp. 32–33). Uncertainty about an autonomous weapon's decisions render military bureaucracies to be uncertain about their precise courses of action and whether such actions will produce desired results.

Moreover, because autonomous weapons are simply programmed to kill targets, military decision-making might not always exclusively be within the purview of high-ranking officials but also under that of a private, insofar as it only takes a simple set of commands to program autonomous weapons (Jablonsky, 1994; Scharre, 2018, p. 309). Drones, as decentralized and self-organized systems of "swarm intelligence," can assist nations in designating laser targets, guide rockets and artillery, and provide air superiority without substantive human intervention (Frantzman, 2021, p. 7, p. 22; Jenks, 2016, p. 352). They also allow ground commanders to switch from radar to infrared to visible wavelengths and hover over an enemy target for 24 hours, all owing to rapid technological developments ensuing from technicians and engineers, and military bureaucracies continuously renovating lists of demands for more technological precision and efficiency (Frantzman, 2021, p. 27, p. 38).

Although the increasing sophistication of the technology used to develop drones saved pilots from long hours of flight for reconnaissance missions, drones were not effective in distinguishing between armed soldiers or terrorists and ordinary civilians, often resulting in the sacrifice of innocent lives killed by indiscriminate targeting (Frantzman, 2021, pp. 48–49, pp. 62–63). The main problem with autonomous weapons, especially drones which mainly just have targeting and killing functions, is that they lack a human face and a human capacity to morally judge a battle situation and might create an anomie toward human rights violations, but their military capabilities are far more efficient to warrant the economization of manpower, which is why they are in extensive demand and use (Frantzman, 2021, p. 65).

There are some major problems associated with such efficiency, but a Clausewitzian Trinity can do little about them. First, there is a strong propensity for the proliferation of drones which might create heated competition among countries to develop more efficient drones or to simply increase their quantity by replicating old models. Such competition could lead to increased occurrences of an abusive use of force among national militaries, paramilitary organizations, and even terrorist groups and expand the drone market exponentially by encouraging both demand and supply (Frantzman, 2021, pp. 71–73, p. 85, pp. 96–97, p. 142, p. 148, p. 162). Resources which could have been used to improve public infrastructures or a general standard of living could be misdirected to excessively produce drones, harming the economic and political stability of a nation (Frantzman, 2021, p. 194).

Unfortunately, the question of how a nation's military bureaucracy should justify a suspension against developing such weapons while confronting a war is not clearly answerable, for it is out of a fear that rivaling powers would be rapidly developing such weapons to surpass each other which prompts the acceleration of the competition to develop autonomous weapons. Second, attempts to destroy drones would unexpectedly and unintentionally expand the scope of what originally was a skirmish into a full-scale war through the introduction of missiles, radar-directed cannons, and other weapons (Frantzman, 2021, p. 110). The problem of determining just how much force ought to be applied can vary based on the nature of the military engagement that the enemy wants to have and what kind of defense system an enemy has. Since autonomous weapons lack human emotions, it is difficult to expect autonomous weapons to rationally distinguish between professional soldiers and armed civilians (Scharre, 2018, p. 123, pp. 252–253, pp. 255–256). Consequently, mercy and compassion might be exceptions rather than norms on battlefields if every decision is under the domain of autonomous weapons (Scharre, 2018, pp. 274–275). None of these problems can be immediately or easily resolved *in situ* if correcting such problems requires a wholesale reprogramming of an autonomous weapon, which would require countless meetings within a military bureaucracy, time for which not all nations would equally have. Moreover, what might have been resolved through a series of meetings cannot foresee problems occurring in future battle situations, so a Clausewitzian Trinity must concentrate on tactical positioning of soldiers and weapons to be a constant panacea for every possible malfunction in an autonomous weapon, since such malfunctions are unpredictable in the form and timing in which they occur from a human operator's perspective.

An even more serious problem is that autonomous weapons could break down in the middle of a war or a policing operation, malfunction, incur errors due to mistakes made by an initial user, or they could encounter environments completely foreign to it, or be vulnerable to hacking (Scharre, 2018, p. 177, pp. 191–193, pp. 222–223). The irony is that although autonomous weapons are given autonomy to implement what they believe are "right" decisions, the range of "right" decisions to be made in real-life situations is more numerous and complicated due to varying human motives and actions (Jain, 2016, p. 313). Moreover, although human beings can think more rationally and in a more complicated fashion than machines, machines respond and react more quickly than human beings, which might render sophisticated rational thinking obsolete before mechanized speed because machines could simply surpass it if they are equipped with artificial intelligence (Scharre, 2018, pp. 229–230, pp. 242–243). The burgeoning of unexpected complexities which might fall outside programmed boundaries of acceptable decisions might confuse autonomous weapons and compromise their autonomy (Scharre, 2018, p. 148, p. 153, p. 157).

In such scenarios, a military bureaucracy might not have time to congregate to correct the problems, whereupon autonomous weapons might be used by the enemy to hack into the system and use the weapon to their desired ends. Should one of such ends be the destruction of a nation's military headquarters, the Clausewitzian Trinity, through a focus on "military territoriality," must strategically position autonomous weapons and soldiers to counter an enemy's commands (Liu, 2016, p. 337). A military bureaucracy is less effective *after* an autonomous weapon is made, because it cannot predict in advance all situations an autonomous weapon might find itself in, and because it lacks emotions, an autonomous weapon might not always kill the exact type of enemy a military wants, and finally, because autonomous weapons are machines, they could be subject to manipulations by enemy forces, ultimately leading to the theoretical suicide of the Clausewitzian Trinity.

11.4 TACTICAL POSITIONING AS THE KEY TO SURVIVAL FOR THE CLAUSEWITZIAN TRINITY IN AUTONOMOUS WARFARE

In essence, the Clausewitzian Trinity suffers from two kinds of interrelated imperfections when applied to autonomous warfare. First, theoretically, if the coordination between passion, chance, and reason is translated into "military territoriality," the use of "military territoriality" by a military to destroy the formation of the Clausewitzian Trinity of another military can become an object for which war literally becomes politics by other means. The Trinity's inherent weakness lies in technological warfare's capability to destroy bureaucratic efficiency within militaries and the coordination between a military and citizens, which could also be an enemy's objective behind formulating their own Trinity. The Clausewitzian Trinity can avoid its destruction with autonomous warfare by concentrating on an adaptive "military territoriality" of tactical positioning of soldiers and autonomous weapons. Since it does not take many technicians to develop autonomous weapons and many commanders to decide whether or not to use them, and because a major objective behind deploying more sophisticated autonomous weapons than an enemy is to ensure a rapid and complete destruction of the enemy's Trinity before it could even form, the Trinity's full formation may not always be necessary or even possible, depending on the degree of a military's technological development. Since speed is essential in determining the effective mobilization of autonomous weapons, the minimization of the Clausewitzian Trinity may be desirable for one nation, which translates into a simplicity and rapidity with which the Trinity can be destroyed by another. Since these mutually contradictory features are going to be commonly available and desirable for all nations, the Clausewitzian Trinity must undergo an evolution with an eye toward enhancing the tactical positioning of soldiers and weaponry to guarantee its survival.

Second, the real danger and devastation of autonomous weapons is not simply the ease and extent to which it can kill people but the damaging of the Clausewitzian Trinity, necessitating its evolution. It is increasingly demonstrated through the use of military drones in the mission to kill terrorists such as Osama Bin Laden and Qasem Soleimani, and the rapidly expanding drone market, the lethal potential to decapitate communication, and the formation of any military strategy based on numbers and mobilization of troops alone. Autonomous weapons require that the Clausewitzian Trinity evolve to focus on tactical positioning. Technological superiority can supersede and compromise bureaucratic efficiency, which depends on the coordination. Since technological superiority determines the effectiveness of autonomous weapons, no individual can be held responsible, overruling the need for a bureaucracy and individual liability. Human strength is not only utilized differently from manning weapons to developing drones or killer robots, but also economized to the greatest degree, with less exposure to enemy fire. However, because technological superiority is always relative and cannot absolutely translate into technological supremacy, the Clausewitzian Trinity must evolve toward enhancing tactical positioning to permanently preserve the important roles of strategic thinking and decision-making within the human mind. The Clausewitzian Trinity must adaptively evolve through a concentration on tactical positioning to maintain parity with the rapidity with which autonomous weapons can destroy targets. It is through such evolutionary adaptation of the Clausewitzian Trinity that autonomous weapons attain the highest strategic value in wars while still preserving a human element. The efficiency of autonomous weapons, if maximized, would render distinctions between strategy and tactics opaque, since the need for using human personnel would be greatly reduced. Since the tactical suitability and efficiency of technological warfare using autonomous weapons would largely be determined by how much fewer soldiers and citizens must be used and sacrificed on the battlefield, the more developed autonomous warfare becomes in the future, the more strategic and efficient the Clausewitzian Trinity and "military territoriality" must become, since coordination would assume less physicality as more machines than human beings would be introduced on battlefields. How many soldiers will be placed in which strategic locations will be the core question guiding the Clausewitzian Trinity's future in the age of autonomous warfare.

The Clausewitzian Trinity in its original form of positioning soldiers to simply exchange volleys and depend on the soldiers' mobilization to concretize tactics and strategy will no longer be valid or necessary as long as more advanced weapons such as robots or drones can replace human beings on the battlefield with minimal instructions from human operators and initiate fully automated warfare. A military bureaucracy is not only slow in real time to meaningfully alter the autonomous weapon but also cannot have an infinite amount of time to satisfactorily train autonomous weapons to have a natural human acumen to judge instantaneously. Uncertainty about an autonomous weapon's decisions render military bureaucracies to be uncertain about their precise courses of action and whether such actions will produce desired results. To reduce the likelihood of an increasing lack of control over autonomous weapons due to such uncertainty, military bureaucracies must concentrate on enhancing tactical comprehension of a battlefield by preserving the tactical positioning of soldiers and weapons within the realm of the human mind.

Moreover, because autonomous weapons are simply designed to kill people, military decision-making might not always exclusively be within the purview of high-ranking officials but renders a military bureaucracy into an "egalitarian" institution in that it only takes a simple set of commands to program autonomous weapons. Consequently, as it was demonstrated through my discussion of drones, there is a strong propensity for the proliferation of drones which might create heated competition among countries to develop more efficient drones or to simply increase their quantity by replicating old models. In addition to the simplicity of issuing commands, determining just how much force ought to be applied can vary based on the nature of the military engagement that the enemy wants to have and what kind of defense system an enemy has. What might have been resolved through a series of meetings cannot foresee problems occurring in future battle situations, so a Clausewitzian Trinity cannot be a constant panacea for every possible malfunction in an autonomous weapon, since such malfunctions are inherently unpredictable in the form and timing in which they occur from a human operator's perspective. If an autonomous weapon malfunctions in the middle of a battle or is required to think more quickly than rationally in a battle situation, a military bureaucracy might not have sufficient time or even the opportunity to congregate to devise new manuals every time an autonomous weapon must make an unexpected decision. Even if a military is given such an opportunity, it cannot predict with absolute certainty about an autonomous weapon's future actions and decisions. The formation and operation of a Clausewitzian Trinity must undergo strategic and adaptive changes to concentrate on tactical positioning to not fall prey to an ever-progressing technology and to ensure that only technological superiority, not supremacy, is the only friction.

REFERENCES

Aron, R. (1974). Clausewitz's conceptual system. *Armed Forces & Society, 1*(1), 49–59. https://doi.org/10.1177/0095327X7400100104

Beyerchen, A. (1992). Clausewitz, nonlinearity, and the unpredictability of war. *International Security, 17*(3), 59–90.

Bhuta, N., & Pantazopoulos, S.-E. (2016). Autonomy and uncertainty: Increasingly autonomous weapons systems and the international legal regulation of risk. In C. Kreβ, H.-Y. Liu, N. Bhuta, R. Geiβ, & S. Beck (Eds.), *Autonomous weapons systems: Law, ethics, policy* (pp. 284–300). Cambridge University Press. https://doi.org/10.1017/CBO9781316597873.012

Birnbacher, D. (2016). Are autonomous weapons systems a threat to human dignity? In C. Kreβ, H.-Y. Liu, N. Bhuta, R. Geiβ, & S. Beck (Eds.), *Autonomous weapons systems: Law, ethics, policy* (pp. 105–121). Cambridge University Press. https://doi.org/10.1017/CBO9781316597873.005

Bode, I., & Huelss, H. (2022). *Autonomous weapons systems and international norms.* McGill-Queen's Press-MQUP. https://www.google.com/books?hl=tr&lr=&id=gvYrEAAAQBAJ&oi=fnd&pg=PP1&dq=Autonomous+Weapons+Systems+and+International+Norms&ots=jM-CfGujpC&sig=o25d6fasc58vjul9DwYHlDdyF-E

Čajić, J. (2016). The relevance of Clausewitz's theory of war to contemporary conflict resolution. *Connections: The Quarterly Journal, 15*(1), 72–78.

Carlini, J. (2020). The application of artificial intelligence to asymmetric warfare: Not a silver bullet. *American Intelligence Journal, 37*(2), 21–29.

Corn, G. S. (2016). Autonomous weapons systems: Managing the inevitability of 'taking the man out of the loop'. In C. Kreβ, H.-Y. Liu, N. Bhuta, R. Geiβ, & S. Beck (Eds.), *Autonomous weapons systems: Law, ethics, policy* (pp. 209–242). Cambridge University Press. https://doi.org/10.1017/CBO9781316597873.010

Crootof, R. (2016). War torts: Accountability for autonomous weapons. *University of Pennsylvania Law Review, 164*(6), 1347.

Daase, C., & Davis, J. (Eds.). (2015). *Clausewitz on small war.* Oxford University Press. https://doi.org/10.1093/acprof:oso/9780198737131.001.0001

Davison, N. (2018). *A legal perspective: Autonomous weapon systems under international humanitarian law* (UNODA Occasional Papers 30; pp. 5–18). UNOD. https://doi.org/10.18356/29a571ba-en

Dimitriu, G. (2020). Clausewitz and the politics of war: A contemporary theory. *Journal of Strategic Studies, 43*(5), 645–685. https://doi.org/10.1080/01402390.2018.1529567

Frantzman, S. J. (2021). *The drone wars: Pioneers, killing machines, artificial intelligence, and the battle for the future.* Bombardier Books. https://www.google.com/books?hl=tr&lr=&id=IBksEAAAQBAJ&oi=fnd&pg=PT5&dq=Drone+Wars:+Pioneers,+Killing+Machines,+Artificial+Intelligence,+and+the+Battle+for+the+Future&ots=0NxKbUHVOO&sig=qGCSrgtr8r3MrBVDQ2oSQqEiAvM

Gardner, N. (2021). Clausewitzian friction and autonomous weapon systems. *Comparative Strategy, 40*(1), 86–98. https://doi.org/10.1080/01495933.2021.1853442

Hadji-Janev, M., & Hristovski, K. (2017). Beyond the fog: Autonomous weapon systems in the context of the international law of armed conflicts. *Jurimetrics, 57*(3), 325–340.

Herberg-Rothe, A., Honig, J. W., & Moran, D. (Eds.). (2011). *Clausewitz: The state and war.* Steiner.

Heyns, C. (2016a). Autonomous weapons systems: Living a dignified life and dying a dignified death. In N. Bhuta, S. Beck, R. Geiß, H.-y. Liu, and C. Kreß (Eds.), *Autonomous weapons systems: Law, ethics, policy* (pp. 3–20). Cambridge University Press.

Heyns, C. (2016b). Human rights and the use of autonomous weapons systems (AWS) during domestic law enforcement. *Human Rights Quarterly, 38*(2), 350–378.

Hoffman, F. G. (2017). Will war's nature change in the seventh military revolution? *The US Army War College Quarterly: Parameters, 47*(4), 19–31.

Horowitz, M. C. (2016). The ethics & morality of robotic warfare: Assessing the debate over autonomous weapons. *Daedalus*, *145*(4), 25–36.

Horowitz, M. C., Perkoski, E., & Potter, P. B. (2018). Tactical diversity in militant violence. *International Organization*, *72*(1), 139–171.

Jablonsky, D. (1994). *The owl of minerva flies at twilight: Doctrinal change and continuity and the revolution in military affairs*. DIANE Publishing. https://www.google.com/books?hl=tr&lr=&id=02vg-MwA12AC&oi=fnd&pg=PT2&dq=The+Owl+of+Minerva+Flies+at+Twilight:+Doctrinal+Change+and+Continuity+and+Revolution+in+Military+Affairs&ots=PsVvB9YCfl&sig=VU5bOKHlYMO2L-9lNf0qxw69mDM

Jain, N. (2016). Autonomous weapons systems: New frameworks for individual responsibility. In C. Kreβ, H.-Y. Liu, N. Bhuta, R. Geiβ, & S. Beck (Eds.), *Autonomous weapons systems: Law, ethics, policy* (pp. 303–324). Cambridge University Press. https://doi.org/10.1017/CBO9781316597873.013

Jenks, C. (2016). The gathering swarm: The path to increasingly autonomous weapons systems. *Jurimetrics*, *57*(3), 341–359.

Jensen, B. M., Whyte, C., & Cuomo, S. (2022). *Information in war: Military innovation, battle networks, and the future of artificial intelligence*. Georgetown University Press. https://www.google.com/books?hl=tr&lr=&id=yrRvEAAAQBAJ&oi=fnd&pg=PP1&dq=Information+in+War:+Military+Innovation,+Battle+Networks,+and+the+Future+of+Artificial+Intelligence&ots=3DmYIxL4mW&sig=AzZmiNDAsSe2BJrjQHi_R1GkFVI

Jo, K. (2021). The role and place of 'military territoriality' in the Clausewitzian conception of war. *Territory, Politics, Governance*, *12*(3), 395–409. https://doi.org/10.1080/21622671.2021.2009016

Kalmanovitz, P. (2016). Judgment, liability and the risks of riskless warfare. In C. Kreβ, H.-Y. Liu, N. Bhuta, R. Geiβ, & S. Beck (Eds.), *Autonomous weapons systems: Law, ethics, policy* (pp. 145–163). Cambridge University Press. https://doi.org/10.1017/CBO9781316597873.007

Leveringhaus, A. (2016). *Ethics and autonomous weapons*. Palgrave Macmillan UK. https://doi.org/10.1057/978-1-137-52361-7

Lewis, J. (2015). The case for regulating fully autonomous weapons. *The Yale Law Journal*, *124*(4), 1309–1325.

Leys, N. (2018). Autonomous weapon systems and international crises. *Strategic Studies Quarterly*, *12*(1), 48–73.

Lieblich, E., & Benvenisti, E. (2016). The obligation to exercise discretion in warfare: Why autonomous weapons systems are unlawful. In C. Kreβ, H.-Y. Liu, N. Bhuta, R. Geiβ, & S. Beck (Eds.), *Autonomous weapons systems: Law, ethics, policy* (pp. 245–283). Cambridge University Press. https://doi.org/10.1017/CBO9781316597873.011

Liu, H.-Y. (2016). Refining responsibility: Differentiating two types of responsibility issues raised by autonomous weapons systems. In C. Kreβ, H.-Y. Liu, N. Bhuta, R. Geiβ, & S. Beck (Eds.), *Autonomous weapons systems: Law, ethics, policy* (pp. 325–344). Cambridge University Press. https://doi.org/10.1017/CBO9781316597873.014

McAdam, D. (1983). Tactical innovation and the pace of insurgency. *American Sociological Review*, *48*(6), 735–754. https://doi.org/10.2307/2095322

Meilinger, P. S. (2007). Busting the icon: Restoring balance to the influence of Clausewitz. *Strategic Studies Quarterly*, *1*(1), 116–145.

Monte, L. A. D. (2018). *Genius weapons: Artificial intelligence, autonomous weaponry, and the future of warfare* (Illustrat ed.). Prometheus.

Murray, W. (1997). Clausewitz out, computer in: Military culture and technological hubris. *The National Interest*, *48*, 57–64.

Neuneck, G. (2008). The revolution in military affairs: Its driving forces, elements, and complexity. *Complexity*, *14*(1), 50–61. https://doi.org/10.1002/cplx.20236

Onorato, M. G., Scheve, K., & Stasavage, D. (2014). Technology and the era of the mass army. *The Journal of Economic History*, *74*(2), 449–481.

Petrenko, A. (2012). Between berserksgang and the autonomous weapons systems. *Public Affairs Quarterly*, *26*(2), 81–102.

Roxborough, I. (1994). Clausewitz and the sociology of war. *The British Journal of Sociology*, *45*(4), 619–636. https://doi.org/10.2307/591886

Rudischhauser, W. (2017). Autonomous or semi-autonomous weapons systems: A potential new threat of terrorism. *Federal Security Policy*, *23*, 1–4.

Sartor, G., & Omicini, A. (2016). The autonomy of technological systems and responsibilities for their use. In C. Kreβ, H.-Y. Liu, N. Bhuta, R. Geiβ, & S. Beck (Eds.), *Autonomous weapons systems: Law, ethics, policy* (pp. 39–74). Cambridge University Press. https://doi.org/10.1017/CBO9781316597873.003

Saxon, D. (2014). A human touch: Autonomous weapons, directive 3000.09, and the "Appropriate Levels of Human Judgment over the Use of Force". *Georgetown Journal of International Affairs, 15*(2), 100–109.

Scharre, P. (2018). *Army of none: Autonomous weapons and the future of war* (1st ed.). W. W. Norton & Company.

Sharkey, N. (2016). Staying in the loop: Human supervisory control of weapons. In C. Kreβ, H.-Y. Liu, N. Bhuta, R. Geiβ, & S. Beck (Eds.), *Autonomous weapons systems: Law, ethics, policy* (pp. 23–38). Cambridge University Press. https://doi.org/10.1017/CBO9781316597873.002

Simpson, T. W., & Müller, V. C. (2016). Just war and robots' killings. *The Philosophical Quarterly (1950–), 66*(263), 302–322.

Sparrow, R. (2007). Killer robots. *Journal of Applied Philosophy, 24*(1), 62–77.

Summers, H. G. (1983). Clausewitz and strategy today. *Naval War College Review, 36*(2), 40–46.

Torkelson, T. D. (2008). *Ideas in arms: The relationship of kinetic and ideological means in America's global war on terror.* Air University Press.

von Clausewitz, C. (1982). *On war.* Penguin Publishing Group.

Wallace, R. (2018). *Carl von Clausewitz, the fog-of-war, and the AI revolution.* Springer International Publishing. https://doi.org/10.1007/978-3-319-74633-3

Warren, A., & Hillas, A. (2018). Lethal autonomous robotics: Rethinking the dehumanization of warfare. *UCLA Journal of International Law and Foreign Affairs, 22*(2), 218–249.

12 Compatibility of Autonomous Weapon Systems with the Basic Principles of Weapons Law

Gökhan Güneysu

12.1 INTRODUCTION

The employment of artificial intelligence (AI) in a number of different sectors is already a common phenomenon today. This widespread tendency to use such decision and/or implementation assistance brings inherent challenges to the fore (Boutin, 2023, p. 133). One of the most challenging cases of AI deployment is the use of such systems during hostilities. Those systems deployed during armed conflicts, regardless of the exact autonomous capacity they have, are given many different names, such as lethal autonomous robots and killer robots. In this chapter, this author will opt to name them autonomous weapon systems (AWS), which is one of the most embraced among a number of alternatives. The utilization of AWS offers distinct military benefits in minimizing the losses of one's own troops during armed conflict (Arai & Matsumoto, 2023). This turns them inevitably into much coveted systems (Krishnan, 2021, p. 227) and logically calls for a deeper probe into their characteristics.

The issue of AWS is one of today's most debated topics. Discussions are being held in a wide range of contexts, from politics to law and ethics (Asaro, 2012, p. 688; Solovyeva & Hynek, 2018). There is no doubt that the development of autonomous systems will have manifold consequences for armed conflict and its conduct. As hinted, AWS have been widely discussed for more than a decade now (Cottier, 2022, p. 1; Wood, 2023a). Ethicists, computer scientists, political scientist, as well as lawyers have all had their say on this significant technological innovation, which is labeled the third revolution in warfare (Ma, 2020, p. 1438; Seixas-Nunes, 2022, p. 140). In a majority of The Law of Armed Conflict (LOAC)-based analyses, the conduct of warfare and humanitarian principles and rules closely related to it, like proportionality and distinction, have taken up a central position, whereas the evaluation of the matter in terms of weapons law has attracted relatively less attention.

This chapter aims to evaluate these systems in terms of weapons law. Though weapons law has a plethora of different sources from different bodies of international law like disarmament law among others, this analysis will be restricted to weapons law rules originating from International Humanitarian Law (IHL), since weapons law deals with rather inherent qualities of weapons and not with contextual uses and abuses of them. Such an inquiry will enable one to have a better understanding of the essence of such weapons in accordance with weapons law in IHL.

12.2 AUTONOMOUS WEAPON SYSTEMS

The impact of technology on the conduct of hostilities and the evolution of warfare, as was the case with gun powder and nuclear weapons, has been profound over the millennia (Boothby, 2014, p. 155). AI is developing quickly across many industries and for many different uses, so it was only a matter of time until the focus of its research shifted to the battlefield (Boutin, 2023, p. 133; Douglas, 2023, p. 502). Development of AWS has become one of the prioritized policy objectives for a large number of states (Ekelhof, 2019; Garcia, 2023; Zurek et al., 2023). Especially, such states with powerful armed forces are trying to enhance their war-making capabilities via AWS. These industrial and military nations appear to be eager to develop these systems with little to no restrictions. These systems with differing levels of autonomy are already available in many national arsenals and are currently being used in armed confrontations like the deployment of the Turkish Kargu-2 in Libya (Anderson et al., 2014, p. 386; Wood, 2023b, p. 22). It looks justifiable, for a while to come, to expect that AWS will be one of the most sought-after means of warfare since they carry an inherent promise to expand military abilities exponentially. For the purposes of this chapter, AWS are to be defined as such weapon systems that carry out critical military functions of targeting and attacking without any human involvement (Hughes, 2023, p. 136; Sehoon Park, 2020, p. 411). This definition is completely in line with that definition suggested by the International Committee of the Red Cross (ICRC), according to which AWS are "any weapon system with autonomy in its critical functions. That is, a weapon system that can select and attack targets without human intervention" (Hughes, 2023, p. 136; Kayser, 2023, p. 25; Wood, 2023a, p. 3). ICRC refines and develops its stance on the definition of AWS with its position paper, in which AWS are defined as follows (ICRC, 2021, p. 1):

> Autonomous weapon systems select and apply force to targets without human intervention. After initial activation or launch by a person, an autonomous weapon system self-initiates or triggers a strike in response to information from the environment received through sensors and on the basis of a generalized "target profile". This means that the user does not choose, or even know, the specific target(s) and the precise timing and/or location of the resulting application(s) of force.

AWS are obviously complex systems (Thurnher, 2016, p. 187; Verdiesen et al., 2021, p. 139). AI-powered systems are capable of solving such problems that were previously solely reserved for the human intellect, owing to problem-solving and decision-making capabilities these systems have (Boutin, 2023, p. 135). One of the most challenging strategy board games is a Japanese game called Go, which proves to be very difficult to master due to the incredibly high number of variables and possible strategies. Google's AlphaGo utilized an artificial neural network to defeat Go world champion Ke Jie in 2017 (Krishnan, 2021, p. 224).

It would be utterly wrong to confuse them with automated weapons such as mines, which operate automatically but have only a one-dimensional operation or function (Dahlmann et al., 2021, p. 1; Güneysu, 2022, p. 26; Krishnan, 2021). When one refers to AWS, it is the understanding of this author that she is talking about weapons and systems that are exponentially more advanced than basic homemade drones or mechanisms (Kwik, 2022). It is rather about even more sophisticated forms of modern systems that require serious investment and industrial ability to build (Kwik, 2022). These systems have built-in components that are state-of-the-art technology and that perform such military functions following a plethora of complicated processes that take place simultaneously within the machine. This whole mode of operation may mitigate any chances of estimation or oversight as to how, when, and where military functions will be carried out by the AWS (ICRC, 2021; Solovyeva & Hynek, 2018).

There is no inherent element of unpredictability in automated weapons, which simply react in one determined way to physical influences or changes from the outside (Ma, 2020, p. 1441; Taddeo & Blanchard, 2022). Regarding automated weapons, it is also not possible to talk about a "decision" following a highly complicated and almost opaque inner-process facilitated by machine autonomy (Güneysu, 2022, p. 26). The way automated weapons react to the changes in their surroundings is also not prone to variations. They have a short range of capabilities that usually enable them to react only in one predetermined way. There is not a possibility of changing these reactions, nor is it possible to talk about a *decision* to use or refrain from using the ammunition loaded in the system as per the situation. As Hughes put it, *a landmine would be an automated weapon because the response to a particular pressure being applied is already determined* (Hughes, 2023, p. 139).

In contradistinction, AWS will have to generate, depending on the circumstances of the battle-field, *solutions* as to whether the use of force is necessary in the given situation, which type of ammunition to use, as well as where and when to use them. It may at times be almost impossible to know in advance about the forthcoming deployment of the AWS and how it will engage the enemy personnel and objects in the course of it. We must remember that as the level of autonomy increases, unpredictability and complexity accordingly tend to increase (Schaub, 2020). This unpre-dictability will be maximized if artificial neural networks as tools of deep learning are deployed more frequently. Machine learning dependent upon artificial neural networks allows the learning system to learn more and more rapidly; it may also enable the system to alter its previously existing algorithm, which in turn nullifies any in-built fail-safe mechanisms instilled beforehand with a view to proper implementation of humanitarian rules. If, per algorithm, it was categorically disallowed for an autonomous system to initiate attack in an urban setting and if the system describes this as a factor that curbs its military success, it may one day find it the best solution to rewrite the relevant portion of the algorithm to hold nothing barred afterward.

AWS are not currently governed by any dedicated laws (Hughes, 2023, p. 135; Melzer, 2019, p. 123). Therefore, the extant general legal framework on means and methods of warfare shall apply if it can be established that AWS are legally to be subsumed under these terms. Weapons law and criteria therein shall be applicable for the purposes of this subsumption as well.

12.3 WEAPONS LAW OF HUMANITARIAN ORIGIN

A sizeable corpus of international laws and regulations does regulate the use, manufacture, procure-ment and possession of weapons. The IHL regime, which aims to lessen the effects of conflict by banning or limiting the use of specific weapons, happens to be only one of these weapon-related legal frameworks (Amoroso, 2023, p. 137; Casey-Maslen, 2020, p. 263). As already mentioned above, IHL will be the focal point of this inquiry. Conventional IHL and IHL treaties have regula-tions that forbid the impacts and features of some weapons (Bruun et al., 2023, p. 9). These include the bans on weapons that are designed to inflict superfluous injury or unnecessary suffering, as well as those that could be anticipated to cause extensive, severe, long-lasting environmental damage and indiscriminate attacks (Boothby, 2018, p. 29; Bruun et al., 2023, p. 9).

There is no established definition of "weapons" in international law (Amoroso, 2023; Arendt, 2016, p. 50; Blake & Imburgia, 2010, p. 169; Casey-Maslen, 2020, p. 261). The national definitions will come into play when international rules are to be implemented by the national authorities. The US military establishment does not have an overarching legal definition for lethal weapons, applicable to all the services (Blake & Imburgia, 2010, p. 170). The term "weapon" would cover "all arms, muni-tions, materiel, instruments, mechanisms or devices that have an intended effect of injuring, damag-ing, destroying or disabling personnel or property", according to a definition proposed for one specific occasion by the United States Department of Defense (US DoD) (Mauri, 2022). The US DoD had the following proposal to define weapon systems as "the weapon itself and those components required for its operation, including new, advanced or emerging technologies which may lead to development of weapons or weapon systems, and which have significant legal and policy implications".[1] In the same understanding, weapons systems are "limited to those components or technologies having direct injury or damaging effect on people or property" (International Review of The Red Cross, 2006, p. 937).

The Program on Humanitarian Policy and Conflict Research at Harvard University, by virtue of their "Manual on International Law Applicable to Air and Missile Warfare", undertook to define these terms (Blake & Imburgia, 2010, p. 171). According to this manual, "means of warfare" do "mean weapons, weapon systems or platforms employed for the purposes of attack", underlying in a way the overlapping of terms means of warfare and weapons (Cambridge University Press, 2013, p. 4). Methods of warfare, on the other hand, "mean attacks and other activities designed to adversely affect the enemy's military operations or military capacity, as distinct from the means of warfare used during military operations, such as weapons" (Cambridge University Press, 2013, p. 5).

For Boothby, it is the "offensive capability that can be applied" to the enemy, which turns a device, item, or munition into a weapon (Boothby, 2018, p. 22). The item itself might have been designed as a weapon from the beginning or was used to the same effect during hostilities (Boothby, 2018, p. 22). In Boothby's presentation of the matter, "means of warfare" refers to all weapons as described, as well as other weapons platforms and associated equipment deployed to generate force on the enemy personnel and military objects (Boothby, 2018, p. 22).

In light of the foregoing, it is safe to accept that the concepts of "means of warfare" and weapons essentially overlap, whereas the concept of "methods of warfare" basically explains how weapons are utilized, since methods of warfare consist of general categories of operations (Amoroso, 2023, p. 137; Arendt, 2016, pp. 50–51; Boothby, 2018, p. 22; Mauri, 2022). 1987 Commentary of the AP I highlight this as follows: *The term "means of combat" or "means of warfare" generally refers to the weapons being used, while the expression "methods of combat"' generally refers to the way in which such weapons are used.*

For the purposes of this analysis, Amoroso's definition is apt to be followed. A weapon is then any "device, system, munition, substance, object, or piece of equipment that is used, that it is intended to use, or that has been designed for use to apply the offensive capability, usually causing injury or damage to an adverse party to an armed conflict" (Amoroso, 2023).

AWS are not strictly speaking weapons. They are a sophisticated mixture of hardware and software that can send damaging ammunition against enemy personnel and objects. In a way, they are more than mere weapons and must rather be classified as delivery platforms for the weapons attached to the system. In accordance with this and with the definition developed by Amoroso, AWS may justifiably be classified as both weapons and means of warfare (Arendt, 2016, p. 51).

12.4 UNLAWFUL WEAPONS *PER SE*

Article 35 of Additional Protocol I sets forth that *the right of the Parties to the conflict to choose methods or means of warfare is not unlimited* (Melzer, 2019, p. 104). This necessitates that warring parties are restricted by the applicable rules of conventional or customary nature in both weapon selection and weapon deployment (Casey-Maslen, 2020, p. 263). To this end, *i.e.,* to delineate certain restraints on the conduct of hostilities, modern IHL has created a large set of laws limiting or regulating the production, possession, and use of specific weapons (means of warfare), as well as preventing or restricting how such weapons can be used (Melzer, 2019, p. 104).

Article 35 as a whole is a crucial tenet of the weapons law. According to the provision contained therein, the use of weapons, ammunition, material, and methods of warfare that cause unnecessary injury or unnecessary suffering is prohibited (Boothby, 2014, p. 158; Güneysu, 2022; Trumbull, 2020, p. 555). But this is not the only prohibition foreseen by the law as to weapons. Indiscriminate attacks as well as the use of means and methods of war that are indiscriminate by nature are also forbidden (Boothby, 2018, pp. 29–30).

Article 51, paragraph 4 of Additional Protocol I has the following on weapons: *Indiscriminate attacks are prohibited. Indiscriminate attacks are: (a) those which are not directed at a specific military objective; (b) those which employ a method or means of combat which cannot be directed at a specific military objective; or (c) those which employ a method or means of combat the effects of which cannot be limited as required by this Protocol.* As can be seen, Paragraph 4 prohibits indiscriminate attacks and defines that term (Bothe, 2013, p. 340). These two IHL articles build up an important pillar of weapons law. The fact that breaking any one of this article's provisions is considered a grave breach serves as further evidence of its significance (Sandoz et al., 1987, p. 616).

A weapon otherwise perfectly capable of being used in compliance with humanitarian rules may be deployed to commit a war crime. In such a case, there is no question as to the inherent legality of the weapon concerned. What embodies a violation of law in such a case is directly and solely attributable to the acts or omissions of the military personnel concerned. While the use of weapons and legal problems attached to it are the subjects of targeting law and conduct of warfare in general, the inherent illegality of a weapon will be measured within the parameters of weapons law (van den Boogaard & Roorda, 2021, p. 426).

In weapons law, what is decisive is whether a weapon is innately indiscriminate and/or whether it inflicts widespread, long-lasting, and grave damages to the environment (Boothby, 2018, pp. 29–30). As Cottier reminds (Cottier, 2022), *the relevant provisions of IHL imply that such weapons systems must not be used if they are likely to cause superfluous injury or unnecessary suffering, or if they are inherently indiscriminate, or if they are otherwise incapable of being used in accordance with law.* The legality of practically all weapons is determined by their inherent qualities rather than how they are used in a given context (Bruun et al., 2023, p. 10; Trumbull, 2020, p. 556). If AWS are found to be *indiscriminate by nature*, then their use during hostilities is to be considered unlawful in terms of IHL (Thurnher, 2016, p. 187).

As hinted above, indiscriminate attacks due to the nature of a weapon are not the only grounds for prohibition in weapons law. Weapons law has three important principles (Chengeta, 2017, p. 842; Güneysu, 2022, p. 65). The use of weapons that do not comply with these principles will be considered illegal under international law (Güneysu, 2022, p. 65). The illegality specifically relevant in this context does not arise from the actual use of the weapon concerned, but it does rather stem from an essential characteristic of the weapon concerned. Illegal deployment of otherwise perfectly IHL-compatible weapons is a different question, which the author will not delve into.

12.4.1 Weapons Indiscriminate by Nature

The legality of the use of autonomous weapons is, among others, directly dependent upon whether they are indiscriminate by nature or not. This necessitates a short handling of those weapons indiscriminate by nature as foreseen by the law.

The rule concerning inherently indiscriminating weapons is the first of those three principles of weapon law alluded to above (Güneysu, 2022, p. 65; Rosert & Sauer, 2019, p. 371). The legal situation regarding such weapons is, as noted above, clarified in Article 51, paragraph 4(b) of Additional Protocol I (Chengeta, 2019, p. 246; Güneysu, 2022, p. 65). The rule is as follows:

> "4. Indiscriminate attacks are prohibited. Indiscriminate attacks are:
>
> …
>
> (b) those which employ a method or means of combat which cannot be directed at a specific military objective,
>
> …"

Boothby reminds us that *(t)here had been no previous explicit provision of conventional law prohibiting weapons that are indiscriminate by nature, and it could therefore be argued that the provision in Article 51(4)(b) and (c) was a new rule* (Boothby, 2014, p. 160). This paragraph, *as a whole,* aims to ban any indiscriminate attacks (Melzer, 2019, p. 86). The inadmissibility of using weapons indiscriminately by nature or effect not only serves as an independent criterion by which all means and methods of warfare have to be evaluated, but it has also prompted the creation of numerous different treaties governing particular weaponry (Melzer, 2019, p. 111). Indiscriminate attacks are such attacks with no distinction between military goals and civilians and civilian objects, either since they are not or cannot be directed at a specific military objective or because their effects cannot be restrained as required by IHL (Melzer, 2019, p. 86).

The ban on using weapons that are automatically indiscriminate refers in part to the features of a weapon that would make its use inevitably illegal. Attacks with at least two different kinds of indiscriminate weapons—those that *cannot* be aimed at a specified military target and those whose effects cannot be constrained as required by IHL—are covered by the ban (Bruun et al., 2023, p. 9; Trumbull, 2020, p. 556). Here, the test required by law is twofold (Casey-Maslen, 2020, p. 265). First, weapons that cannot be targeted are to be disallowed to be used during hostilities. This failure to target something specific could be a factor attributable to rudimentary targeting systems of the weapons concerned, as was the case with those old Scud missiles. Second, those weapons and weapon systems with such effects that cannot be limited to certain military objectives are to be singled out, and accordingly, their use must be prohibited (Casey-Maslen, 2020, p. 265; van den Boogaard & Roorda, 2021, p. 428). Needless to say, such attacks that are intentionally indiscriminate are also prohibited by virtue of AP I Article 51, paragraph 4(a), which prohibits attacks that *are*, without any further elaboration, indiscriminate. In this last one, the indiscriminate attack is directly attributable to the personnel but not the weapon deployed.

As can be seen, one specific type of indiscriminate attack is "attacks which employ a method or means of combat which *cannot* be directed at a specific military objective". As Melzer (2019, p. 111) reminds, "(t)his also includes weapon systems that, as an inherent feature of the technology employed and their intended use, may be expected to inflict excessive collateral harm on the civilian population". What makes the use of such weapons during hostilities prohibited is *the inability to control the direction and/or effects of the weapon that renders it indiscriminate by nature and which therefore renders it unlawful* (Boothby, 2018, p. 30). This inability is not a contextual one but rather an unalterable characteristic of the weapon concerned. Any contextual wrongdoing in terms of indiscriminate attacks will be administered by the sub-paragraph (a).

By virtue of this regulatory framework of sub-paragraphs (b) and (c), attacks using methods and means of warfare that cannot be directed against a specific military target, or *additionally*, attacks using methods and means of warfare whose effects cannot be limited, are prohibited categorically. These two scenarios are fundamentally different from the indiscriminate use of a weapon, which is not indiscriminate *per se*. In the latter case, the weapon is used for the commission of a crime solely due to *mens rea* of the personnel concerned (Seixas-Nunes, 2022, p. 151). Such conduct-related indiscriminate attacks fall under the remit of sub-paragraph (a) of the same article, yet indiscriminate attacks originating exclusively from the conduct of warfare are not directly to be subsumed under weapons law, for it is impossible to talk about an inherent inevitability or probability of an indiscriminate attack that is related to an innate quality of the weapon or weapon system implicated.

Whether or not the AWS or any other means of warfare can actually be employed in a way that allows them to be directed against military goals, *i.e.* they are not indiscriminate, is the crux of the whole matter (McClelland, 2003, p. 408). The ICRC is very anxious about the unpredictability of AWS, which in turn comes to mean these could be means and methods of war that are indiscriminate by nature: *unpredictability in AWS poses a fundamental challenge to IHL. Customary IHL prohibits weapons that are by nature indiscriminate, that is, weapons that in their normal or expected circumstances of use, cannot be directed at a specific military objective or whose effects cannot be limited as required by IHL* (ICRC, 2021, p. 7). The evaluation of AWS in terms of weapons law in general and their inherent indiscriminateness specifically will be made in the fourth section of this chapter.

12.4.2 Weapons Causing Superfluous Injury

In addition to shielding civilians from the impacts of conflicts, IHL forbids or restricts the use of weapons and tactics in conflict that are thought to cause unnecessary suffering or "superfluous injury" to combatants (Melzer, 2019, p. 19). The employment of "weapons, projectiles, material, and methods of warfare of a nature to cause superfluous injury or unnecessary suffering" is forbidden by Protocol I's Article 35(2) (Boothby & von Heinegg, 2018, p. 165; McFarland, 2020, p. 100; Sassòli, 2019, p. 381). The Saint Petersburg Declaration's preamble in 1868 first stated the rationale behind this prohibition (Amoroso, 2023, p. 146; Brenneke, 2020, p. 87; Sassòli, 2019, p. 381). This prohibition aims to ensure that a certain level of suffering and injury required to render enemy combatants *hors de combat* should be inflicted and that this level should not be exceeded (McFarland, 2020, p. 100). The prohibition today covers both "means and methods of warfare", and it is perfectly applicable in restricting, especially, the pain or harm given to combatants, despite the fact that they are legitimate targets of attack under IHL (Sassòli, 2019, p. 381). To reiterate, it is permitted to inflict misery only insofar as it is justified by the necessities of war, i.e. for the lawful purpose of "weakening the military forces of the enemy, by disabling the greatest possible number of soldiers" (Amoroso, 2023, p. 147). Any means or methods of warfare that produce excessive injury and suffering should not be used during armed conflict.

According to ICJ, "it is prohibited to cause unnecessary suffering to combatants: it is accordingly prohibited to use weapons causing them such harm or uselessly aggravating their suffering. In application of that second principle, States do not have unlimited freedom of choice of means in the weapons they use". This prohibition is one of the "cardinal principles contained in the texts constituting the fabric of humanitarian law" and is perfectly applicable in international as well as non-international armed conflicts as a customary rule of international law (Casey-Maslen, 2020, p. 266). The current restriction is not a treaty-based one; rather, it should be viewed as a component of general international law, making it binding for all states (Amoroso, 2023, p. 147).

The rule concentrates on the design of the weapon concerned to find out whether it has any potential to cause unnecessary suffering or superfluous injury (Trumbull, 2020, p. 555). Here, the most important characteristic that the law requires all to seek with a view to constructing its illegality is whether the weapon inevitably causes a violation of this rule in the designed or intended applications (Boothby & von Heinegg, 2018, p. 166; Trumbull, 2020, p. 555). If such an inevitable violation is not to be found, the weapon cannot be declared prohibited by this regulation. Mere usages of weapons that do cause unnecessary suffering or superfluous injury not because of their design but because of the context-specific way of deployment in that specific case will not be sufficient grounds to declare the use of that weapon as prohibited (Trumbull, 2020, p. 555). Hence, this prohibition will not be relevant in arguments against and doubts on the IHL-compatibility of AWS (McFarland, 2020, p. 100). There may be few instances of possible violation of this rule in relation to AWS. The first scenario assumes wounded enemy personnel who are trying to surrender. If the AWS programmed to destroy such personnel keeps on shooting at those wounded personnel or even trying to use blunt force to bring about the same outcome due to a lack of ammunition, then it may come to a breach of this principle. Here, the second significant point should kick in. Via absolute terms placed in their algorithms, AWS must be enabled to recognize such dire situations and programmed to abide by the letter of the law.

Other than these imaginary situations, it rather comes down to the ammunition and weapon deployed by the system, which has the potential to be a constant violator of this rule. The contextual scenario itself would not suffice to ban the use of AWS on the grounds of their incompatibility with the prohibition to inflict superfluous injury, since it says nothing essential about the AWS.

12.4.3 Environmental Damage

A variety of laws in IHL safeguard the natural environment against the repercussions of military attack (Crowe & Weston-Scheuber, 2013, p. 68). Some of these rules defend the natural environment in and of themselves, while others defend the environment in light of its relevance to humans (Crowe & Weston-Scheuber, 2013, p. 68).

As an illustration for the former category, Article 35(3) of Additional Protocol I forbids conflicting parties from employing such *methods or means of warfare which are intended, or may be expected, to cause widespread, long-term and severe damage to the natural environment* (Crowe & Weston-Scheuber, 2013, p. 68; Sassòli, 2019, p. 570).

By virtue of this chapter, the environment itself has been designated as a value that deserves direct protection (Sandoz et al., 1987, p. 410). The limitation imposed here is a limitation of absolute nature. The article is organized in such a way as to avoid allowing and condoning irregularities or breaches in this regard.

It is stated in the article that any military means and methods are prohibited that have caused or may cause environmental damage, as defined in this chapter. In the determination of the scope of this prohibition and its implementation, the meaning attached to the term *natural environment* plays a pivotal role. In addition, this chapter seeks the cumulative existence of "widespread, long-term and grave" environmental degradation, which is also the activation threshold for this regulation (Güneysu, 2014, p. 84). As easily discernible, the chapter pertains to major destruction that seriously exceeds the battlefield destruction normally expected in a hostility and affects human life and natural resources (Henckaerts, 2000, p. 17). Here, what practically inflicts that damage on the environment are the weapons attached to an AWS. In this sense, whether the delivery system is an autonomous means of warfare or a man-controlled system does not play a big role. What counts is the weapon that causes the kinetic damage to the environment. This author is of the opinion that the mere existence of autonomy in a weapon system may not in itself be the underlying reason for that system to cause the environmental degradation as foreseen and banned by the chapter. It is rather the weapon controlled by the AWS that is decisive. This means that any military planner having to observe this prohibition must review the probable environmental damage caused by that system's attached weapons.

12.5   EVALUATION OF AWS IN TERMS OF WEAPONS LAW

The emergence of autonomous systems creates risks as well as opportunities (Güneysu, 2022, p. 111). AWS, like any other weapons, means of warfare, or methods of warfare developed, no matter how advanced they may be technologically, will remain within the sphere of application of IHL (Güneysu, 2022, p. 109). Accordingly, by virtue of the principles and rules of weapons law, the use of such autonomous systems that have been found to contravene them must not occur (Güneysu, 2022, p. 109). The legal qualification of AWS in the light of weapons law will mainly dwell upon the result of an inquiry as to whether these weapons may be classified as indiscriminate weapons *per se*. In this probe, one will have to acknowledge that the unpredictability of the outcomes of the decisions of the AWS may be a huge impediment for any human control, either *ex ante* or during the hostilities. Brenneke warns us on the same issue by stating that "the prohibition of indiscriminate means may be translated into illegality of weapons that cannot be aimed at a certain target, thus bearing on reliability, accuracy and uncontrollable effects" (Brenneke, 2020, p. 89). This risk is also highlighted *expressis verbis* by the ICRC in one of their reports on AWS. There, the ICRC warns about the fact that "the user does not choose, or even know, the specific target(s) and the precise timing and/or location of the resulting application(s) of force" (ICRC, 2021, p. 2). The ICRC also has the following to say on the same issue (ICRC, 2021, p. 2):

> The use of autonomous weapon systems entails risks due to the difficulties in anticipating and limiting their effects. This loss of human control and judgement in the use of force and weapons raises serious concerns from humanitarian, legal and ethical perspectives.

Especially with the employment of artificial neural networks for the sake of machine learning, the peril of software capable of modifying its own programming in order to yield better results in its own objectives is even more alarming (Krishnan, 2021, p. 220). This mode of learning has the potential to escape the diligent gaze of a human supervisor and to turn the whole inner-process of the autonomous systems into a black box, which impedes traceability and accountability (Krishnan, 2021, p. 221).

The legal requirement of directing at "specific military objectives" has been an important part of any discussions as to the compatibility of AWS with the weapons law. This direct ability requires not only the existence of this characteristic during the actual hostilities. It is obvious that the weapons deployed must be directable during the course of battles. This capacity must have a prior component when one deals with AWS, and that is the decision-making process of the system concerned. If we are talking about a perfect "black box", then targeting only specific military objectives will not be taken for granted since the whole process of intra-system decision-making will be opaque for the human controllers and/or other military personnel. This hardship to know in advance what kind of attacks the systems will initiate and execute may in end effect come to mean that the systems will not be directed at specific military targets. In conformity with this interpretation, there may be such deployments where AWS will not be put into action with a specific target in mind but rather for patrolling reasons. During such overwatch missions, a target of opportunity may emerge. However, since there was no specific target in mind before the AWS were deployed to a certain zone, one feels obliged to ask whether such a mission may qualify as an "attack not directed at a specific military objective" as prescribed by Rule 12 in CIHLS (Henckaerts & Doswald-Beck, 2005). This discussion could be the crux of the whole legality problematique of the use of AWS during armed conflicts. AWS are no ordinary tools. They need little input from the human operators once started, i.e. they immensely depart from any old-fashioned modalities of deployment one may observe between a human being and their weapons. If, during these autonomous missions, such systems are "predictably unpredictable" as to what or whom they will decide to attack, then the military personnel managing the system are running a very probable risk of deploying an inherently indiscriminate weapon.

This probability has not been ignored. In this regard, a differentiation was made between AWS used against personnel and objects. Since it is almost impossible for them to perceive the real meanings of humans' behaviors, including hypothetical cases where one combatant is trying to surrender, the ICRC, for example, has voiced its firm stance to ban such anti-personnel AWS (ICRC, 2021, p. 9).

Regarding the deployment of AWS to target military objectives, the same concerns stemming from the inherent unpredictability of AWS come to the fore. As known, legal yardstick to decide whether an object is a military objective has four distinct qualities, each of which is a sufficient ground to affirm that status. An object is a military objective according to its nature, location, purpose, or use. This author adheres to the position that AWS should only be deployed in such attacks where military objectives *by nature* will be targeted. The other criteria are just too fluid and much prone to change, and any such change may be just too easy to miss in the proverbial fog of war. This inherent risk should be avoided as much as possible, and solely attacks on objectives determined by virtue of their military nature should be allowed. Only then could humanitarian protection be provided in modern and future wars, where AWS will be used in high numbers.

12.6 CONCLUDING REMARKS

AWS have been a significant issue of arguments and contestation. Many states want to continue developing such systems. Recently, the capabilities of these systems have been told to surpass those of expert systems, and deep learning has been used in such systems.

ICRC reminds of *serious doubts about the ability of autonomous weapon systems to comply with IHL in all but the narrowest of scenarios and the simplest of environments.* Yet battlefield could be anything but simple! This chaotic nature of battlefield calls for a deep inspection as to their compatibility with IHL rules, especially with weapons law. Any concrete potential for a probable breach of these rules should disqualify these weapons as viable alternatives for the conduct of hostilities. Knight (Knight, 2017) reminds us that deep-learning "has transformed computer vision … and workings of any machine-learning technology are inherently opaquer, even to computer scientists", turning them into black boxes.

This can come to mean only one thing, and it is that how these systems will fare during hostilities in terms of compliance with IHL rules is a totally enigmatic question with no concrete answer in sight. Even absolute bans or restraints introduced into their programming can now be deleted by the machine itself if it, by way of machine learning, finds it a more convenient way to produce better results.

In terms of weapons law, the most pressing issue seems to arise as regards the compatibility of these opaque systems with rules pertaining to indiscriminate attacks. The other two rules dealt with in this chapter do rather relate to the weapons in the systems, and whether these systems are autonomous has nothing or little to do with these rules. Yet, indiscriminate attacks are, from an IHL outlook, a very thorny issue with great hazard potential. This author wholly agrees with others about the serious wake-up calls on the undeniable existence of this violation-prone characteristic of AWS. In such machine-learning systems, then, as per the law, one has to take into cognizance whether anything about these systems could remain foreseeable. This foreseeability embodies the crux of the whole issue as to indiscriminate attacks problematique. This author feels obliged to warn the reader about the fact that the moment AWS can modify their programming, no one will be able to claim the capability to surely know in advance where, when, how, and whom these weapons will engage. These weapons must then be qualified as weapons that cannot be directed at a specific military objective and treated accordingly.

NOTE

1. International Review of The Red Cross, 'A Guide to the Legal Review of New Weapons, Means and Methods of Warfare: Measures to Implement Article 36 of Additional Protocol I of 1977: International Committee of the Red Cross Geneva, January 2006' (2006) 88 International Review of the Red Cross 931, 937.

REFERENCES

Amoroso, D. (2023). International weapons law. In *Kirimli Dr. Aziz Bey collected courses on international humanitarian law—Vol. I* (pp. 137–176). Turkish Red Crescent.

Anderson, K., Reisner, D., & Waxman, M. (2014). Adapting the law of armed conflict to autonomous weapon systems. *International Law Studies, 90*, 386–411.

Arai, K., & Matsumoto, M. (2023). Public perceptions of autonomous lethal weapons systems. *AI and Ethics.* https://doi.org/10.1007/s43681-023-00282-9

Arendt, R. (2016). *Völkerrechtliche Probleme beim Einsatz autonomer Waffensysteme* [Doctoral Thesis]. Universität Potsdam.

Asaro, P. (2012). On banning autonomous weapon systems: Human rights, automation, and the dehumanization of lethal decision-making. *International Review of the Red Cross, 94*(886), 687–709. Cambridge Core. https://doi.org/10.1017/S1816383112000768

Blake, D., & Imburgia, J. S. (2010). Promoting integrity from without: A call for the military to conduct outside, independent investigations of alleged procurement integrity act violations. *The Air Force Law Review, 66*.

Boothby, W. (2018). Dehumanization: Is there a legal problem under article 36? In W. Heintschel von Heinegg, R. Frau, & T. Singer (Eds.), *Dehumanization of warfare: Legal implications of new weapon technologies* (pp. 21–52). Springer International Publishing. https://doi.org/10.1007/978-3-319-67266-3_3

Boothby, W. H. (2014). *Conflict law: The influence of new weapons technology, human rights and emerging actors.* T.M.C. Asser Press.

Boothby, W. H., & von Heinegg, W. H. (2018). *The law of war: A detailed assessment of the US Department of Defense law of war manual.* Cambridge University Press; Cambridge Core. https://doi.org/10.1017/9781108639422

Bothe, M. (2013). *New Rules for Victims of Armed Conflicts: Commentary on the Two 1977 Protocols Additional to the Geneva Conventions of 1949. Second Edition. Reprint revised by Michael Bothe.* Brill | Nijhoff. https://doi.org/10.1163/9789004254718

Boutin, B. (2023). State responsibility in relation to military applications of artificial intelligence. *Leiden Journal of International Law, 36*(1), 133–150. Cambridge Core. https://doi.org/10.1017/S0922156522000607

Brenneke, M. (2020). Lethal autonomous weapon systems and their compatibility with international humanitarian law: A primer on the debate. In T. D. Gill, R. Geiß, H. Krieger, & C. Paulussen (Eds.), *Yearbook of international humanitarian law, volume 21 (2018)* (pp. 59–98). T.M.C. Asser Press. https://doi.org/10.1007/978-94-6265-343-6_3

Bruun, L., Bo, M., & Goussac, N. (2023). *Compliance with international humanitarian law in the development and use of autonomous weapon systems: What does IHL permit, prohibit and require?* Stockholm International Peace Research Institute. https://doi.org/10.55163/DFXR3984

Cambridge University Press. (2013). *HPCR Manual on International Law Applicable to Air and Missile Warfare.* Cambridge University Press; Cambridge Core. https://doi.org/10.1017/CBO9781139525275

Casey-Maslen, S. (2020). Weapons. In B. Saul & D. Akande (Eds.), *The Oxford guide to international humanitarian law.* Oxford University Press.

Chengeta, T. (2017). Defining the emerging notion of 'Meaningful human Control' in weapon systems. *New York University Journal of International Law and Politics, 49*(3), 833–890.

Chengeta, T. (2019). Is existing law adequate to govern autonomous weapon systems? In M. Davel & E. Barnard (Eds.), *FAIR 2019 South African forum for artificial intelligence research* (Vol. 2540, pp. 244–251). CEUR.

Cottier, D. (2022). *Emergence of lethal autonomous weapons systems (LAWS) and their necessary apprehension through European human rights law.* Council of Europe, Committee on Legal Affairs and Human Rights.

Crowe, J., & Weston-Scheuber, K. (2013). *Principles of international humanitarian law.* Edward Elgar Publishing.

Dahlmann, A., Hoffberger-Pippan, E., & Wachs, L. (2021). *Autonome Waffensysteme und menschliche Kontrolle—Konsens über das Konzept, Unklarheit über die Operationalisierung.*

Douglas, C. (2023). What we talk about when we talk about killer robots: The prospects of an autonomous weapons treaty. *Georgia Journal of International and Comparative Law, 51*(2), 501–538.

Ekelhof, M. (2019). Moving beyond semantics on autonomous weapons: Meaningful human control in operation. *Global Policy, 10*(3), 343–348. https://doi.org/10.1111/1758-5899.12665

Garcia, D. (2023). Algorithms and decision-making in military artificial intelligence. *Global Society,* 1–10. https://doi.org/10.1080/13600826.2023.2273484

Güneysu, G. (2014). *Çevrenin Silahlı Çatışmalar Esnasında Korunması.* Adalet Yayınevi.

Güneysu, G. (2022). *Otonom Silah Sistemleri: Bir Uluslararası Hukuk İncelemesi.* Nisan Kitapevi.

Henckaerts, J.-M. (2000). Towards better protection for the environment in armed conflict: Recent developments in international humanitarian law. *Review of European Community & International Environmental Law, 9*(1), 13–19. https://doi.org/10.1111/1467-9388.00228

Henckaerts, J.-M., & Doswald-Beck, L. (2005). *Customary international humanitarian law: Volume 1: Rules* (Vol. 1). Cambridge University Press. https://doi.org/10.1017/CBO9780511804700

Hughes, J. G. (2023). Learning from automation in targeting to better regulate autonomous weapon systems: Target lists, the electronic battlefield and automation in mines. *Journal of Conflict and Security Law, 28*(1), 135–160. https://doi.org/10.1093/jcsl/krac030

ICRC (2021). *ICRC position on autonomous weapon systems: ICRC position and background paper.* ICRC.

International Review of The Red Cross (2006). A guide to the legal review of new weapons, means and methods of warfare: Measures to implement article 36 of additional protocol i of 1977: International Committee of the Red Cross Geneva, January 2006. *International Review of the Red Cross, 88*(864), 931–956. https://doi.org/10.1017/S1816383107000938

Kayser, D. (2023). Why a treaty on autonomous weapons is necessary and feasible. *Ethics and Information Technology, 25*(2), 25. https://doi.org/10.1007/s10676-023-09685-y

Knight, W. (2017, April 11). The Dark Secret at the Heart of AI. *MIT Technology Review.* https://www.technologyreview.com/2017/04/11/5113/the-dark-secret-at-the-heart-of-ai/

Krishnan, A. (2021). Enforced transparency: A solution to autonomous weapons as potentially uncontrollable weapons similar to bioweapons. In J. Galliott, D. MacIntosh, & J. D. Ohlin (Eds.), *Lethal autonomous weapons: Re-examining the law and ethics of robotic warfare* (p. 219–236). Oxford University Press. https://doi.org/10.1093/oso/9780197546048.003.0015

Kwik, J. (2022). Mitigating the risk of autonomous weapon misuse by insurgent groups. *Laws, 12*(1), 5. https://doi.org/10.3390/laws12010005

Ma, E. (2020). Autonomous weapons systems under international law. *New York University Law Review, 95*(5), 1435–1474.

Mauri, D. (2022). *Chapter 21 The use of CBRN weapons in armed conflict* (pp. 358–379). Brill | Nijhoff. https://doi.org/10.1163/9789004507999_022

McClelland, J. (2003). The review of weapons in accordance with Article 36 of additional protocol I. *Revue Internationale de La Croix-Rouge/International Review of the Red Cross, 85*(850), 397. https://doi.org/10.1017/S1560775500115226

McFarland, T. (2020). *Autonomous weapon systems and the law of armed conflict: Compatibility with international humanitarian law.* Cambridge University Press; Cambridge Core. https://doi.org/10.1017/9781108584654

Melzer, N. (2019). *International humanitarian law: A comprehensive introduction.* International Committee of the Red Cross.

Rosert, E., & Sauer, F. (2019). Prohibiting autonomous weapons: Put human dignity first. *Global Policy, 10*(3), 370–375. https://doi.org/10.1111/1758-5899.12691

Sandoz, Y., Swinarski, C., & Zimmermann, B. (Eds.). (1987). *Commentary on the additional protocols of 8 June 1977 to the Geneva conventions of 12 August 1949*. International Committee of The Red Cross / Martinus Nijhoff Publishers.

Sassòli, M. (2019). *International humanitarian law: Rules, controversies, and solutions to problems arising in warfare*. Edward Elgar Publishing. https://doi.org/10.4337/9781786438553

Schaub, H. (2020). Der Einsatz autonomer Waffensysteme aus psychologischer Perspektive. *Zeitschrift Für Außen- Und Sicherheitspolitik, 13*, 333–342.

Sehoon Park (2020). Analysis of the positions held by countries on legal issues of lethal autonomous aweapons systems and proper domestic policy direction of South Korea. *The Korean Journal of Defense Analysis, 32*(3), 393–418. https://doi.org/10.22883/KJDA.2020.32.3.004

Seixas-Nunes, A. (2022). *The legality and accountability of autonomous weapon systems: A humanitarian law perspective*. Cambridge University Press; Cambridge Core. https://doi.org/10.1017/9781009090001

Solovyeva, A., & Hynek, N. (2018). Going beyond the "Killer Robots" debate: Six dilemmas autonomous weapon systems raise. *Central European Journal of International & Security Studies, 12*(3).

Taddeo, M., & Blanchard, A. (2022). A comparative analysis of the definitions of autonomous weapons systems. *Science and Engineering Ethics, 28*(5), 37. https://doi.org/10.1007/s11948-022-00392-3

Thurnher, J. S. (2016). Means and methods of the future: Autonomous systems. In P. A. L. Ducheine, M. N. Schmitt, & F. P. B. Osinga (Eds.), *Targeting: The challenges of modern warfare* (pp. 177–199). T.M.C. Asser Press. https://doi.org/10.1007/978-94-6265-072-5_9

Trumbull, C. P. (2020). Autonomous weapons: How existing law can regulate future weapons. *Emory International Law Journal, Volume 34*(Issue 2). https://doi.org/10.2139/ssrn.3440981

van den Boogaard, J. C., & Roorda, M. P. (2021). 'Autonomous' weapons and human control. In R. Bartels, J. C. van den Boogaard, P. A. L. Ducheine, E. Pouw, & J. Voetelink (Eds.), *Military operations and the notion of control under international law: Liber Amicorum Terry D. Gill* (pp. 421–437). T.M.C. Asser Press. https://doi.org/10.1007/978-94-6265-395-5_20

Verdiesen, I., Santoni de Sio, F., & Dignum, V. (2021). Accountability and control over autonomous weapon systems: A framework for comprehensive human oversight. *Minds and Machines, 31*(1), 137–163. https://doi.org/10.1007/s11023-020-09532-9

Wood, N. G. (2023a). Autonomous weapon systems: A clarification. *Journal of Military Ethics, 22*(1), 18–32. https://doi.org/10.1080/15027570.2023.2214402

Wood, N. G. (2023b). Autonomous weapon systems and responsibility gaps: A taxonomy. *Ethics and Information Technology, 25*(1), 16. https://doi.org/10.1007/s10676-023-09690-1

Zurek, T., Kwik, J., & van Engers, T. (2023). Model of a military autonomous device following International Humanitarian Law. *Ethics and Information Technology, 25*(1), 15. https://doi.org/10.1007/s10676-023-09682-1

13 Meaningful Human Control of Autonomous Weapons Systems

Translating Functional Affordances to Inform Ethical Assessment and Design

Jordan Richard Schoenherr

13.1 INTRODUCTION

Cooperation and conflict are persistent features of human nature, apparent throughout our history and evolution. As human societies have evolved, the nature of conflict has evolved with them (e.g., Citino, 2004; Dupuy, 1984; Kilcullen, 2019; Lautenschläger, 1983). New affordances are discovered that allow humans to exploit features of their natural environment, turning domestic tools into weapons. Military innovation studies have also repeatedly highlighted how technologies have changed – and have been changed by – conflicts and warfighters (Rosen, 1991), with cyberweapons and autonomous weapons systems (AWS) presenting the most recent disruptive technologies that provide new strategical and tactical opportunities for early adopters (Kaag & Kreps, 2014; Scharre, 2018). Yet, these technologies also challenge social organization and the notion of what constitutes ethical warfare (Benjamin, 2013; Hijazi et al., 2019; Parks & Kaplan, 2017).

Every society has prohibitions on violence against members of its own community unless violent acts are used as an instrument to enforce compliance with a group's norms and conventions (Bates et al., 2002; Feshbach, 1971; Fiske & Rai, 2015; Rai & Fiske, 2012; Weber, 1919). As human populations have grown and dispersed, interpersonal conflicts have evolved into intergroup and international conflicts. While material resources (e.g., land, raw materials) are often important, even these conflicts are often framed in moralistic terms (Burleigh, 2007, 2010). Philosophers have also tried to frame organized intergroup conflicts in terms of ethical principles and processes, or *just wars* (Elshtain, 1991; Moseley, 2011). These approaches tend to consider general features of armed conflict (e.g., initiating a battle, killing under specific conditions, treatment of combatants) rather than the specific features of the technologies of warfare (Allenby, 2013). Yet, an understanding of the specific design features of a technology – their affordances – is integral to assessing and predicting the outcomes and implications of their use (Schoenherr, 2022a).

Technologies are created, and have meaning, by virtue of their interactions with humans (Bijker & Law, 1994; Latour, 1990; MacKenzie & Wajcman, 1999). So-called disruptive technologies denote technological innovations that fundamentally change the relationships between individuals (Hopster, 2021; Latzer, 2009; Schoenherr, 2022a; Schuelke-Leech, 2018). Nowhere is this more apparent than in warfare (Kosal, 2020; cf. sustaining innovations, Pierce, 2004). If warfighters can successfully adopt and adapt novel technologies, they can gain a decisive advantage in armed conflicts, e.g., new metallurgical and horseback riding technologies (Turchin et al., 2021). In recent years, concerns have been raised over the introduction of AWS including fundamental issues about

DOI: 10.1201/9781003441700-13

the precision and autonomy of these systems and existential issues such as the implications of an artificial agent making life-and-death decisions related to human agents. Multiple solutions have been proposed including prohibition, an imperative to adopt technology to keep pace with adversaries, the creation of international guidelines, and the inclusion of 'ethical governors' to regulate these systems (Arkin, 2009; Arkin et al., 2009; Asaro, 2012; Docherty, 2014; Heyns, 2014; Rosert & Sauer, 2019). In parallel, concerns have also been raised about so-called cyberweapons that exploit growing interdependencies between humans and technology, creating vulnerabilities due to failures to understand the affordances of hardware and software (e.g., zero-day vulnerabilities; Allhoff et al., 2016; Dipert, 2016; Lucas, 2017; Robinson et al., 2015; Schoenherr & Thomson, 2022).

The Prospects for Meaningful Human Control. A clear path to ensuring that combat remains ethical is to require that human agents remain in control of weapons. Human-controlled combat technologies have a nearly unchallenged history, with the exception of the use and of domesticated animals in warfare, which have some measure of autonomy regardless of extensive training (e.g., horses, dolphins, and elephants; Mayor, 2014; Sorenson, 2014). As weapon capabilities have increased, these systems have been progressively granted greater autonomy to select their targets.

As responsibility is distributed between humans and AWS in combat situations, there is greater difficulty in identifying and attributing responsibility to these agents, referred to as the 'responsibility gap' (Docherty, 2014; Matthias, 2004; Michael et al., 2024; Santoni de Sio & Nucci, 2016; Santoro et al., 2008; Sparrow, 2007). Underpinning these concerns are questions about the extent to which AWS can, or should, be granted autonomy. For practical or theoretical reasons, some deny that artificial intelligence (AI)-based systems can function autonomously, assuming that systems are not currently – nor can they ever be – capable of doing so. This has led to calls for meaningful human control (MHC) of AWS. MHC requires that humans must have sufficient information about – and understanding of – the operations of these systems and an ability to intervene in the operations of an AWS (Ekelhof, 2019; Roff & Moyes, 2016). MHC has three design criteria (*transparency, explainability,* and *feasibility*) to ensure that the monitoring and regulation of a system aligns with human cognitive capabilities and capacities (Horowitz & Scharre, 2015). As AI-based technologies have become capable of handling larger amounts of data faster than humans are capable of, the technological affordances of AWS create challenges for implementing MHC by limiting the explainability of these systems and the feasibility of regulating them in a comprehensive and timely manner. In this chapter, I briefly consider these issues.

13.1.1 Overview

Using an ethical design framework (Schoenherr, 2022a, 2022b) this chapter considers the design affordances and mental models that influence ethical sensemaking in AWS and cyberweapons (for other examples, see Schoenherr, 2021). Based on previous work, Section 13.3 reviews frameworks developed in traditional philosophical ethics as well as the ethical sensemaking capabilities of humans. Drawing on research from design psychology, Section 13.4 describes how systems can be described in terms of specific design features (affordances) based on the technical or functional properties of a system (autonomy, intelligence, and connectivity; or AIC). Military technologies are briefly reviewed to demonstrate the orthogonality of these dimensions and their relationship with the primary affordances of weapons, their *lethality.* Finally 13.5 provides an overview of the ethical affordances of technology by considering the psychological impact of the use of AI-based weapons on users, with special consideration given to specific properties of AWS and cyberweapons.

13.2 ETHICAL FRAMEWORKS FOR ADDRESSING THE MORALITY OF WARFARE

Ethical frameworks have evolved in human societies to structure the intuitive moral motivations of our species. They vary in terms of their focus on analytic or wholistic approaches to ethical sensemaking, whether the individual, group, or both are the foci of ethical discourse, and whether an emphasis is placed on outcomes (consequentialism), principles (non-consequentialism), or competencies (virtues ethics). Following a review, I consider how these models can be used to understand the process of ethical sensemaking, a term I use to describe the processes used by moral agents to identify, process, and respond to moral and ethical dilemmas they encounter in their social and digital environments.

13.2.1 Metaethical Frameworks

Philosophical ethics provides a rich set of epistemological tools for designers to reference as they struggle with the ethical implications of AWS. The vast set of ethical theories developed over the course of human civilization can be reduced to three broad metaethical frameworks: consequentialism, non-consequentialism, and virtue theory. Here, I provide a brief review. For a more extensive discussion, see Schoenherr (2022a, 2022b).

Consequentialist Theories prioritize the outcomes of an action rather than the intentions of a moral agent or the principles that inform the actions of a moral agent. This approach assumes that the object of ethics should be to maximize positive outcomes and minimize negative outcomes for a set of moral agents (Bentham, 1789; Hare et al., 1981; Mill, 1861). Pleasure and pain were initially presented as an intuitive foundation for assessing the ethicality of the outcomes of an action. As their goal was to maximize good and minimize pain, the moral agent simply needs to sum and weigh the outcomes in a hedonic calculus. Subsequent developments suggested that higher order pleasures (i.e., intellectual pleasure) should instead be used, or that we should focus on rules rather than actions. These models superficially appear ideal for assessing AWS and cyberweapons as they provide an impartial, scientific means to inform behavior and assess human and nonhuman moral agents.

According to such an approach, ethical warfare would be defined as that which reduces (or eliminates) causalities and damage while resulting in an end to conflict and an increase in happiness and or a decrease in distress (however construed). The development and use of weapons and tactics that achieve these ends should therefore be prioritized. Despite the apparent reasonableness of the aims of such an approach, historical examples demonstrate that its applications are problematic at best or naïve at worst. For instance, such arguments have been used to justify the use of nuclear weapons in Japan to end the Second World War: The number of deaths resulting from the use of atomic bombs would have been less than a planned land invasion; the use of nuclear weapons acted as a subsequent deterrent for large-scale conflicts and the use of weapons of mass destruction (e.g., Bernstein, 1999; Lewis, 2020). Despite the logic of such an argument, it largely denies the horrific experiences of those who died as a result of its use and provides uncertain evidence concerning its efficacy as a deterrent.

As with any attempt at measurement, the consequentialist approaches face multiple issues, principally quantification and causality. First, outcomes must be discretized during the process of measurement, e.g., positive and negative affect, financial damages. In the case of hedonic measures, these are far too impractical to implement in a meaningful way, e.g., rating the pleasure or pain felt by the populations affected. Instead, evaluators are likely to rely on crude measures such as body count. Body count requires an oversimplistic categorization of people as 'enemies' and 'allies' and ignores subcategories. For instance, combatants and noncombatants vary in terms of their commitment to the cause, the contributions they can provide inside and outside of a conflict, and their adherence to moral norms, e.g., criminality. Such crude measures hardly seem like a firm foundation for moral reasoning. Moreover, by reducing people to numbers, we might create moral disengagement that disregards category boundaries between combatants and noncombatants.

In a related vein, this account does not address how the outcomes will be assessed for different populations. Namely, physical attacks and cyberattacks are intended to have asymmetric outcomes for the attacker and the defender. It seems unlikely that belligerents will assess their casualties on a par with those of their adversaries. Thus, they must assign a value to human life: theirs and those of their adversaries. For instance, the large number of civilian deaths in Gaza resulting from Israel's reprisal attacks following vicious attacks by Hamas in October 2023, have received widespread criticism, due to the disproportionate Palestinian civilian losses relative to those of Israel. Israeli Defense Force (IDF) estimates suggest that initially civilian: combatant mortality ratio was 2:1 (McCluskey & Green, 2023). However, consequentialism might justify the actions of either side in terms of *weighing* the losses of an adversary less than one's own.

Second, we must select an appropriate timescale to assess outcomes and the chain of causality that leads from the act of killing in combat to prospective outcomes. While short-term evaluation is relatively straightforward (i.e., immediate deaths, material, and financial losses), long-term prospective outcomes are much harder to measure. For instance, the remarkable social and economic recovery of both Germany and Japan following the Second World War can be contrasted against the ongoing social disorder resulting from the Second Iraq War and De-Ba'athification of the former Iraqi government.[1] The uncertainty associated with these outcomes illustrates the limitations of any simply attempt at quantification.

Third, a failure to consider people's intentions also appears highly problematic if humans are deemed moral *agents*. For instance, we might consider that the use of a weapon might be justified if the consequences were considered *a priori*. However, if the balance of positive and negative outcomes is only systematically measured *a posteriori* due to the fog of war, then *actions* do not seem to be fundamentally ethical. Rather, this type of consequentialism is merely rationalization. Given that many decisions in combat need to be rapid and the demands of the moral calculus of consequentialism is exceedingly high, it is impractical to assume that warfighters can be expected to think and decide in combat situations. Moreover, it is also naïve to assume that military leadership will have access to sufficient information to accurately perform such calculations. Indeed, failures of military intelligence are quite common (Hedley, 2005) and, arguably, inevitable on some scale (Betts, 1978). Similarly, we should be reluctant to blame cyber security analysts for missing a zero-day vulnerability because, definitionally, no one was aware of it.

Non-consequentialist theories prioritize the intentions or principles used to make an ethical judgment rather than the outcomes of an action. Such approaches to ethics tend to assume that ethical principles should be adhered to *categorically* rather than judging acts along a continuum of beneficent (or maleficent) outcomes. According to such an approach, individuals should be treated as autonomous moral agents such that each person is an end in themselves and their behaviour must be judged relative to their reasons for acting (Hill, 1971; Kant, 1999; O'Neill, 1975; Rawls, 1971).

Non-consequentialism can be used to understand the ethics of conflict. For instance, Just War Theory (JWT) provides the foundation for addressing the use of lethal force by outlining principles for conduct leading up to (*jus ad bellum*), during (*jus in bellum*), and after an armed conflict (*jus post bellum*; Elshtain, 1991; Moseley, 2011). In contemporary armed conflicts, a just war is defined in terms of the Law of Arms Conflict (LAC) and the Rules of Engagement (RoE). *Jus ad bellum* and *jus in bellum* require further consideration of *necessity* and *proportionality* in the use of lethal force. *Necessity* requires that a conflict (as a whole) or a particular action is necessary and that there is no other means available for resolution. *Proportionality* requires that a response is proportional to the objectives. Evidence for beliefs about the conditions of a just war is also evident in early civilizations (Cox, 2017, 2023; Kang, 2006; Lewis, 2005; Lo, 2012; Lo & Twiss, 2015). JWT defines numerous legal and ethical frameworks used in conflict including international laws and RoE. For instance, the Geneva Convention outlines the rights of prisoners of war, sick/wounded, and civilians in conflict zones, e.g., do not target or kill civilians, do not shoot prisoners of war. For instance, the IDF Dahiya Doctrine which intentionally uses disproportionate force – including the destruction of civilian infrastructure – against belligerents, has been criticized on these grounds (Khalidi, 2014; Marei, 2020).

Like consequentialist frameworks, non-consequentialism also has multiple issues. First, as a categorical approach, there is very little margin for error. Rather than a continuum of ethicality, this approach potentially imposes a strict dichotomy and assumes that ethical principles are inviolable. In practical settings, actions are often associated with positive and negative outcomes, i.e., the doctrine of double effect. Moreover, most situations undoubtedly have competing social and ethical features, factors, and dimensions. For instance, focusing exclusively on the death of Palestinians in the Israel-Hamas conflict ignores how the conflict started as well as the intentions of the military leaders who started – and continued – the war. While it is possible that a set of necessary and sufficient rules might capture all (or most) ethical situations, this has yet to be achieved over the millennia of ethical inquiry around the globe.

Second, non-consequentialist approaches such as JWT also require a reliable means to discriminate between combatants and noncombatants as well as conditions of peace from those of war. Irregular warfare presents a significant issue in that warfighters reduce the cues required to differentiate military and civilian populations (Boot, 2013; Finlay, 2013; Robinson et al., 2009). The use of cyberweapons presents a clear illustration of this, such that countries might not formally declare a state of war, however, persistent attacks suggest otherwise (Schoenherr, 2023b). If one wishes to avoid all civilian deaths or preemptively engaging in a conflict, this becomes nearly impossible to achieve if combatants masquerade as civilians. Refraining from a conflict to maintain a single principle, such as avoidance of civilian deaths, might lead to the violation of other principles. Thus, identifying which principles supervene over others introduces as much uncertainty in ethical judgment as attempting a moral calculus.

Third, by disregarding outcomes of an action, non-consequentialist approaches leave open the possibility that something that might appear to be the 'best' for a group might produce suboptimal outcomes. This would necessarily result in inadequate domain-specific knowledge to understand specific kinds of situations or inadequate mechanisms of feedback. These conditions are likely in conflicts and combat situations which are undoubtedly idiosyncratic. Indeed, it is difficult to know how ethical principles are derived without assessing the consequences of an action to determine cause and effect relationships which remain the focus of consequentialist frameworks.

Virtue ethics presents the last major ethical framework, one that arguably antedates the other two approaches and from which they are both derived (MacIntyre, 1981). Both the Eastern Confucian and Western Aristotelian approaches can be described in these terms, prioritizing relational models and idealized traits/characteristics that individuals should strive to attain, respectively. These models also place moral agency and passionate commitment at their core: An action is not good simply because of the principles that inform it or the resulting consequences, there is a need for active reflection and acceptance of responsibility. According to such an approach, leaders and warfighters need to engage in effortful deliberation to determine the best course of action given sets of priorities (e.g., RoE) and take responsibility for the outcomes. Consequently, individual morality is not static, it is a competence that must be developed and used. For Aristotle (1908), the person "who faces and who fears the right things and with the right aim, in the right way and at the right time, and who feels confidence under the corresponding conditions, is brave," (1115M5-19, Ross, 1956).

In principle, virtue ethics appears to be a far superior approach to the alternatives in that it prioritizes a dynamic moral agency. General rules and moral calculus are used as *instruments* by a moral agent. However, qualifying the necessary and sufficient features of moral agency and competency introduces an additional layer of complexity. Specifically, virtue ethics requires understanding the categories of moral agents, relevant values that are applicable to any situation, individual characteristics associated with moral agency, the situational variables that affect outcomes, and a host of other factors required of any competency. Thus, effective virtue ethics require a sophisticated understanding of the *ethical sensemaking* capacity and capabilities of moral agents. In the next section, I briefly outline this process based on psychological science (Schoenherr, 2022a, 2022b).

13.2.2 Ethical Sensemaking

Making sense of social and ethical issues requires that we can identify relevant features of a situation. A key feature of sensemaking is to identify the problem space. Among other critical elements, ethical sensemaking requires the ability to categorize agents (e.g., combatant, noncombatant), evaluating accuracy of information and equipment (e.g., fallibility of human intelligence, precision of weapons), the risk associated with an operation (e.g., collateral damage, escalation of conflict), uncertainty associated with the operational environment (e.g., weather conditions, terrain), the cognitive constraints of the moral agents (e.g., limited attention and working memory, response times), and how other moral agents will act in a situation (e.g., reciprocity, empathy). Understandably, this require extensive cognitive effort, imposing a significant cognitive load on ethical decision-makers.

When presented with a situation, ethical sensemaking requires the acquisition and successful activation of appropriate affective responses, moral categories and rules, and response schemata. To accomplish this, models of human learning, decision-making, social cognition, and metacognition typically distinguish between two types of systems (Evans & Frankish, 2009; Kahneman, 2013; Stanovich, 2005). Type 1 systems are evolutionarily older systems, often associated with rapid ballistic responses to familiar stimuli within specific domains, often associated with wholistic stimulus processing and affective responses. Type 2 systems are evolutionarily younger, require effortful algorithmic deliberation, with limited associates to emotions and consisting of analytic features isolation. When presented with novel objects, persons, or events, people typically engage in Type 2 thinking unless their mental resources are already engaged or situational stressors have narrowed the focus of their attention (Nelson et al., 2017). With extensive experience, the stochastic learning of Type 1 systems eventually acquires automatic pattern recognition and dominant response selection. However, if pattern recognition fails, people can produce random errors or systematic biases.

Ethical sensemaking uses these same processes. When individuals have familiarity with a situation, they will most likely rely on automatic responses that are activated based on familiar cues – something seems *wrong* about the location, identity, or behaviour of an object or person in a scenario. However, due to misidentification of targets, time limitations, physiological stressors, etc., these judgments might be incorrect. Leaders and warfighters must therefore monitor and regulate their own behavior. The extent to which Type 2 processes are activated during the process of ethical sensemaking reflects moral engagement. However, merely activating Type 2 processes might be necessary but is not sufficient for moral behaviour. For instance, the promotion of analytic thinking can *reduce* automatic prosocial behavior. Similarly, moral disengagement can result from (re) categorization of agents into members of an 'outgroup', thereby rationalizing behavior that might otherwise be considered unethical. In extreme cases, this constitutes dehumanization.

Monitoring and regulation one's own conduct additionally requires an accurate representation of our performance in a task, in this case, the ethicality of an agent. However, studies suggest that there is a reasonable degree of flexibility in how an agent's behavior is construed (Mazar et al., 2008). This appears to be determined by two factors. First, humans have a need to maintain a positive self-image as good moral agents. Second, people typically allow themselves variability in the applicability of ethical principles, rationalizing moral violations based on individual differences, the context, and the closeness of the moral agent to themselves. Thus, certain behaviors or outcomes that might otherwise be seen as 'unethical' to one actor might be viewed as 'ethical' to another to maintain a positive representation of an individual or a group.

13.3   THE TECHNICAL AND FUNCTIONAL AFFORDANCES
OF AUTONOMOUS WEAPON SYSTEMS

Ethical sensemaking requires that a moral agent can identify the ethically relevant features and processes that define a situation. Paradigms from design psychology provide an analytic framework to understand how stakeholders (e.g., designers, distributors, users, the public, and regulators) make sense of the material, social, and ethical features of technology (Michael et al., 2024; Schoenherr, 2021, 2022a; Schoenherr et al., 2023). To be effective, we must assume that each stakeholder will focus on different features of a technology (affordances) depending on their motivation, knowledge, and time. Discussions of MHC of AWS and cyberweapons therefore require that stakeholders share an understanding of the affordances that are relevant to specific questions, defining a shared mental model.

13.3.1 The Affordances of Technology

Technological evolution requires replication, variation, and selection of functional features of a system. Initially, technologies can evolve through both stochastic processes (natural selection) and intentional variation (artificial selection). For evolution to occur, we must identify the unit of analysis that is analogous to the gene in biological evolution (Dennett, 1995). In the psychology of design, critical features of design are referred to as *affordances* (Gibson, 1966, 1979; Norman, 1988). Affordances reflect the functional properties of an object or a system, i.e., what it can do based on how humans interact with the feature.[2] Thus, technological evolution assumes that affordances are created, vary, and selected based on the goals that they satisfy.

Affordances are not an intrinsic features of an object. Rather, humans create them by identifying how the properties of material objects can satisfy their goals. For instance, a rock can be used to fulfill one function (hammer) or another (weapon) but is objectively a collection of atoms that have specific material properties. Through repeated interaction and perception, these *ad hoc* properties can become fixed in the user's mind such that other uses are no longer considered, i.e., functional fixedness. This point is especially important for computer technologies and AI that have a potentially limitless number of possible affordances.[3] For instance, a user might use social media to express themselves and to participate in a community. However, the information they post can be used as intelligence to predict their behavior or that of their group, i.e., 'open source' intelligence (Ju et al., 2020; Malakauskis & Juozapavičius, 2023; Taşkın & Sağsan, 2020).

Once discovered, individuals can pass their knowledge of affordances on through social learning processes (Boyd et al., 2011; Morgan et al., 2015). Over time, these affordances, and the technologies that they constitute, can diffuse within – or between – groups depending on their degree of connectivity (Goldman & Eliason, 2003; Schmid, 2017). With sufficient time, a sociotechnical system can be created as users become sufficiently dependent on their technology that they distribute social and cognitive functions between themselves and technology. Humans can also *infer* affordances of technology even when they were not explicitly instructed, i.e., reverse engineering. This means that, provided there is enough knowledge about other components and properties of a system, humans can discover novel affordance through defeasible reasoning, e.g., abduction and analogy. For instance, the technological superiority of one military can be lost when their technology falls into the hands of an adversary, such as when Iran captured US drones (Hambling, 2020). Even if the systems are not understood in their entirety, access to the technology can provide insight into how the affordances of a system are achieved. However, such replication has its limits (cf. Gilli & Gilli, 2018).

Technologies are typically designed and developed to perform one function but can be used for many others, e.g., re-domaining. Computers and AI are often described as 'dual use' technologies such that they can be designed for one purpose (e.g., computation, domestic technologies such as 'drones') and adapted to others (e.g., simulation, military operations).[4] Similarly, cyberspace and the physical infrastructure that imparts its realism also have numerous functional capabilities. The near-limitless nature of these technologies reflects the myriad affordances that are available to perceptive users.

Not all affordances might be immediately apparent. Instead, they might be discovered through the process of implementation. As users begin to integrate technologies in larger sociotechnical systems, new strengths and weakness might be revealed. For instance, zero-day vulnerabilities reflect affordances ('exploits') that are discovered by security analysts or hackers that were not previously known or were deemed low-risk (Ahmad et al., 2023; Singh et al., 2019).

13.3.1.1 Types of Affordances

Two types of affordances are typically considered during design, development, and implementation: technical affordances and functional affordances. In computer science, *technical affordance* represents the hardware and software of a system. These affordances require considerable technical knowledge (e.g., physical, chemical, engineering) that would typically only be maintained by designers and developers of a technology. In contrast, *functional affordances* represent the functional capabilities and capacities of a system, i.e., the operations it performs. These affordances can be known and discovered by users and shared with designers and developers. Stakeholders need not share the same mental models. Users might not be aware of all the functional affordances of a system whereas designers and developers might have a more comprehensive understanding of the capabilities and capacities of a system. These potential asymmetries in knowledge are problematic in that they potentially deny users autonomy.

The social situatedness of affordances is demonstrated when considering the primary affordance of offensive weapon systems: *lethality*. Lethality cannot be considered as an objective property of a weapon system. Such systems might be designed to exert a specific kinetic force through firing or explosion; however, lethality is a function of the physical and social environment. For instance, a machine gun's lethality is ultimately a function of the explosive force of the weapon, the medium a projectile is passing through as well as how adversarial forces are organized. If adversarial forces are densely packed together relative to if they are distributed, a weapon's lethality is greater. Similarly, the ethicality of a system is based on the social categories and consequences of these outcomes: Killing a combatant is qualitatively different than killing a noncombatant, the exception being conflicts perceived through the lens of 'total war'.

Cyberoperations can be understood in a similar manner. If many high-value computers are densely connected in the same network or use the same security protocols, the effect of a cyberattack or infection will be much greater than if the network configuration or security protocols differed. However, if offensive weapons are defined by lethality, cyberoperations might not reflect true weapons. Instead, they might merely reflect force multipliers (Schoenherr & Thomson, 2022). In this case, they will only multiply the lethal effects of a kinetic attack. These observations highlight the need to understand the affordances of AI-based systems. In the same vein as lethality, the detrimental impact of a cyberattack is ultimately related to what information is being stored on a system: Destroying an individual's financial history or personal health information is qualitatively different than destroying their digital music library or purchase history from an online retailer.

13.3.2 Affordances of Autonomous Weapon Systems

Computer-based technologies are necessarily multi-purpose. It follows then that the set of affordances will differ depending on the data set, algorithms/AI, the users, and the use case. However, three general affordances can be used to describe all systems: intelligence, autonomy, and connectivity (Schoenherr, 2022a; Figure 13.1).

Intelligence. As a functional affordance, intelligence is typically defined in human terms consisting of two broad factors: domain-specific and domain-general intelligence (e.g., Cosmides & Tooby, 2013; Spearman, 1961; Sternberg, 2000). Properties of intelligence include the quantity and diversity of knowledge as well as the ability to retrieve and apply information in different and novel situations.

The intelligence of weapon systems can also be understood in similar terms and judged relative to the accuracy/precision as well as the extent to which a system can respond to a dynamic environment. Historically, weapon systems have had neither general intelligence nor specific intelligence that is independent of human creators and users. The extent to which a weapon system is successful has instead been dependent on the identification and use of the affordance identified by intelligent human agents. This observation provides support for the ethical requirement of MHC: an AWS is not a moral agent if it does not possess the capability of understanding the ethical affordances of a situation.

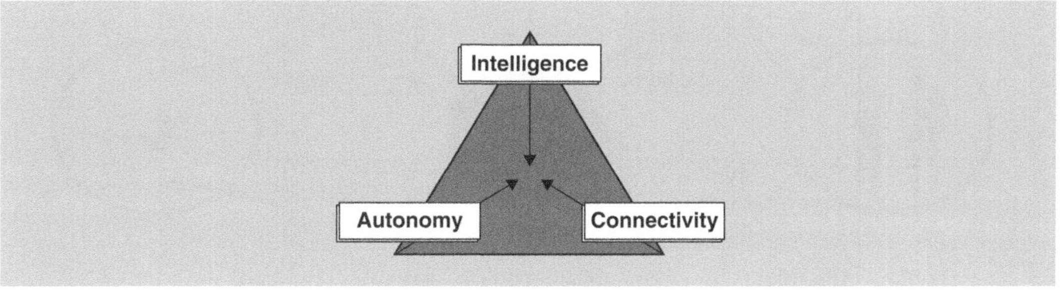

FIGURE 13.1   Triadic autonomy, intelligence, and connectivity (AIC) functional affordances ethical assessment framework (Schoenherr, 2022a).

With the introduction of basic sensors, triggering mechanism provided the first rudimentary semblance of specific intelligence. For instance, early torpedoes and landmines represented systems that employed simple mechanisms that automatically responded to stimuli in the environment. Rocketry technology developed during the Second World War in Nazi Germany (V1 and V2) provided minor improvements, however, the inability of these rockets to adapt is clearly demonstrated by observations that V1 and V2 rockets were relatively ineffective, with only 23% and 38% detonating on British soil, respectively (Evans & Delaney, 2018). Much like simple single-celled or multicellular organisms, this basic stimulus-response coupling cannot be considered true intelligence. This is especially true when the target (Britain) was such a large area.

Compared to their ancestral weapons, the precision of these systems improved markedly over a short period of time (Figure 13.2). As MacKenzie (1993) has noted in the evolution of nuclear ICBM guidance capabilities, the "[error probability] of a missile tends to get smaller through time, even without hardware changes, as understanding or error processes develops" (p. 427).

The adoption of AI-based technologies can be judged in similar terms. However, evidence for the efficacy of AWS and cyberweapons is comparatively limited. This is partially due to the sensitive

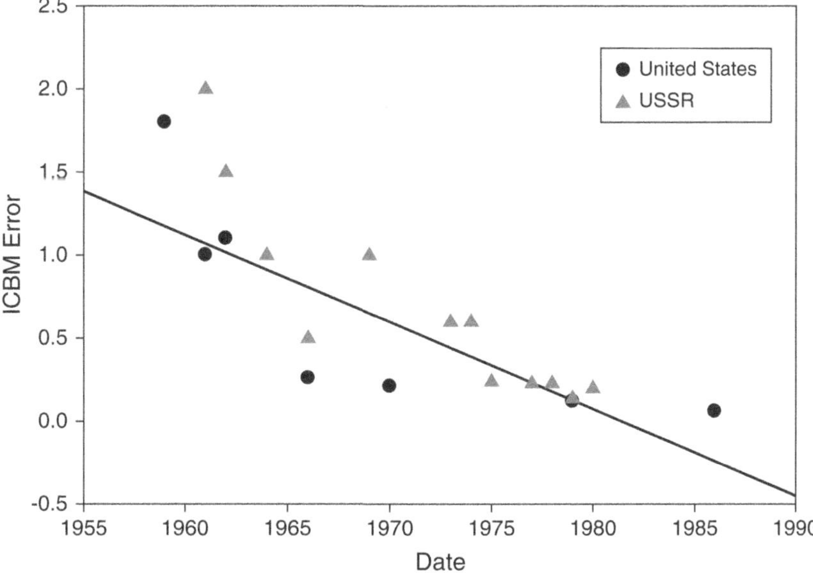

FIGURE 13.2   Estimated ICBM missile accuracy (reduction in error) for US and USSR arsenals. Data obtained from MacKenzie (1993).

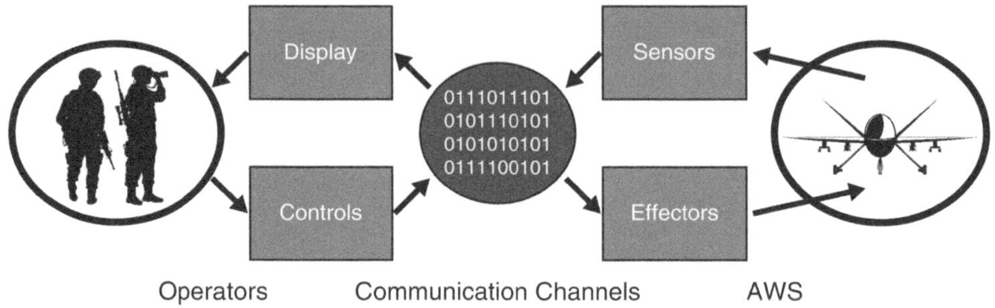

FIGURE 13.3 Simplified configuration for human-supervised autonomous weapon system. Drone image designed by macrovector, drone, and soldier silhouettes retrieved from freepik.com.

nature of military operations. For instance, external studies of remotely piloted aircraft (RPAs) over a 15-year period suggest that approximately 20% of an acknowledged 700 strikes resulted in an estimated 400 civilian casualties (Moorehead et al., 2017). While these statistics do not provide a 'hit' rate, they do provide crude estimates of false alarms – whether through failures of the targeting systems or evidence of 'collateral' damage. In the conflict between Russia and Ukraine, reports have suggested that 85–86% of Russian drones were destroyed (Polityuk & Balmforth, 2022) with later reports reporting the destruction of 75% (Karazy et al., 2023). While these numbers do not provide a basis for direct comparison, it does suggest that drones are largely unsuccessful when viewed in isolation. When viewed as part of a coordinated action, however, the reduced need (and loss) of human warfighters might recommend them. Similar observations have also been made in the Israel-Hamas conflict. Israel reportedly used AI ('the Gospel') to select viable targets. However, the extreme loss of life in Gaza suggests that the Gospel has a biased algorithm, limited precision, or military leadership were unconcerned about civilian casualties given the perceived value of the intended targets (Davies et al., 2023; Figure 13.3).

Autonomy. Autonomy assumes a degree of independence between the user and a system. Two broad approaches to autonomy are present in the literature reflecting categorical user-system relationships and continuous models that consider the degree of autonomy. ***Categorical Models*** are defined in terms of when and how humans enter the 'control loop' of a system. A *human-in-the-loop* (HITL) approach assumes that people have direct control over the operations of a machine. Here, autonomy is limited. A *human-on-the-loop* (HOTL) approach assumes that people monitor a system, and only intervene when necessary. Here, a system is granted autonomy over some of its operations. Finally, a *human-out-of-the-loop* (HOOTL) assumes that humans 'fire and forget'. For instance, combat drones typically reflect HOTL as the operators can enable the drone to act independently or override its operations whereas DDoS and malware attacks would largely be defined as HOOTL. The problem with categorical models of autonomy is that they fail to consider the specific affordances that can be controlled by a human operator. Definitionally, AWSs require some degree of *meaningful* autonomy, i.e., if not MHC than meaningful AI control. However, what counts as meaningful must be clearly specified.

Rather than considering an AI and a human as two discrete agents and *how* a human agent exerts control, we might instead consider the *degree* of control and the dimensions of controllability. ***Continuous Models*** define autonomy in terms of discrete steps that are specified by given criteria. For instance, self-driving vehicles can be judged by the amount of control an AI has over the vehicle's operations. Simple continuums, however, ignore the collection of processes that are required for a system to operate. In contrast, multidimensional models of autonomy identify the *specific* dimensions of control. In AWS, this minimally consists of sensors, target information, and shoot/no-shoot decisions.

Autonomous weapons have been a persistent feature of modern conflicts. Autonomy can vary from a simple ballistic response such as the 'fire and forget' principle of ancient technologies like arrows. The limited guidance systems of Nazi Germany's V1 and V2 rockets provide a basic illustration of this issue. While more complicated than land and sea mines, a degree of autonomy was achieved with simple mechanisms, e.g., gyroscopes. The Vietnam War saw greater integration of autonomous vehicles into combat, with RPAs used for day and night reconnaissance, signal intelligence, and antiradar jamming (Zaloga, 2011). The Aegis Control System (ACS) employed by the US Navy provides a far more complicated example of a system that uses radar to automatically target and destroy incoming threats to a naval vessel. Cyberweapons must also be largely autonomous. By operating 'behind enemy lines' in adversarial networks, once a file has been downloaded, any communication to an external system can alert cyber security analysts to the intrusion. Here, however, such malware has comparatively low intelligence. The possibility of intelligent and autonomous cyberweapons represents a more alarming threat.

Connectivity. In technical terms, connectivity corresponds to whether a system *requires*, or is *influenced* by, the operations of other systems either through direct physical contact (wired) or indirect communication (wireless). This dimension also concerns how many operators are, or can, direct the activity of a system. Crucially, connectivity is an orthogonal dimension relative to the other two, e.g., a system might be sufficiently intelligent and autonomous to perform a task but require connectivity to override its operation or provide it with new commands. Connectivity can also affect the accuracy and lethality of weapon systems.

For the foreseeable, humans will require some degree of control over AWS. Thus, we cannot consider the accuracy of warfighters and AWS alone. Rather, we need to consider their combined accuracy in a distributed cognitive network or sociotechnical systems (Figure 13.2). Consider a simple situation: One human operator is controlling an AWS remotely. Here, we must consider the accuracy of the AWS in terms of its (a) sensors and categorization abilities in identifying targets and nontargets,[5] (b) this information then needs to be accurately encoded into a message, (c) transmitted through the physical environment (subject to EM interference), (d) received by a system that presents the information to a human operator, and (e) who then identifies the target and selects a response for the AWS (engage or disengage). If we assume that each agent is highly reliable (e.g., $p = 0.95$), then the reliability of the system is much less than its parts (e.g., $p = 0.77$). This approach is an oversimplification. For instance, a drone might send an image and its classification to the human operator who realizes that it is a nontarget, meaning that the accuracy of the whole is greater than AWS; similarly, teams of human operators working together might increase or decrease the accuracy based on team cohesiveness and groupthink. However, it does represent the issue of distributed responsibility that emphasizes the issues with connectivity.

Communication and connectivity are critical features of warfare. For instance, with the introduction of wireless radio, the operations of individual tanks could be coordinated in a much more effective manner, facilitating the creation and refinement of tactics such as the Blitzkrieg (Citino, 2017). Similarly, the need for a weapon system to adapt to changes in a situation was addressed with the creation of 'fly-by-wire' (FBW) systems or the 'Lay Torpedo' (Newpower, 2006).

The Russian war against Ukraine provides a clear example of the nature of connectivity in modern warfare. Drones were a prominent weapon during the conflict. These systems require a high degree of connectivity to navigate, select a target, and fire, requiring supporting infrastructure. During the early stages of the conflict, US entrepreneur Elon Musk provided Ukrainians with access to the Starlink satellite network. As the conflict continued, Musk announced that the expenses associated with network access were too great, that the US government should take over the expenses in addition to its other funding packages, and initially denied the Ukrainian government the use of the network for drone strikes (Kim et al., 2023; Schoenherr, 2023c). Thus, a major portion of Ukraine's defensive capability was threatened due to connectivity issues.

13.4 ETHICAL ISSUES OF AIS AND CYBERWEAPONS

Beyond the three basic affordances of AI, how humans *use* a technology will determine what specific social and ethical affordances must be considered. As ethics is concerned with the identifying features and principles based on the real or perceived consequences of an action for moral agents, the *specific* ethical affordances of technology are one of the primary objects of ethical sensemaking. In the final section, I provide a basic overview of the ethical affordances of AWS and cyberweapons that arise from human-AI interaction.

13.4.1 LETHALITY AND ACCURACY OF AWS AND CYBERWEAPONS

Lethality is not only a functional affordance of AWS but also the *primary* ethical affordance due to cross-cultural prohibitions against killing. Operationalization of lethality is typically done in terms of the precision of a weapon system, i.e., its accuracy. However, accuracy can also be deconstructed into hits, false alarms, misses, and correct rejections using Signal Detection theory (SDT).[6]

In humans, accuracy requires efficient accumulation of evidence, the acquisition and use of specific and general decision-making criteria, and effective response selection. Humans' ability to deliberately generate, test, and adjust rules dynamically is quite limited due to attention and working memory limitations. Rapid responses are related to automatic learning systems that require extensive feedback and lengthy training periods, and tend to dominate slower, effortful systems only later in training (for reviews, see 'Ashby & O'Brien, 2005; Ashby et al., 2011). In principle, this is where AWS might reflect an ethical superior choice in that they might be capable of processing information faster and more accurately than humans in certain situations: *If* they can be more accurate and consistent in discriminating targets from nontargets in an operational theatre, then they clearly represent a preferable alternative to human warfighters. For instance, given that discriminative accuracy is a function of learning and the quantity of data provided to an algorithm, then AWS could exceed the accuracy of humans *if* they are trained with data accumulated from all of the scenarios encountered by other AWS. This kind of pooling of experience cannot be achieved through social learning in humans. Thus, AWS could be in a better position to determine when – and how much – lethal forces are required.

The same claims cannot be made about current cyberweapons. Any 'learning' that occurs concerning cyberweapons is not attributable to the weapons themselves. While there are certain instances of malicious software changing (e.g., Conficker), this occurs through a process of random variation akin to natural selection rather than self-directed adaptation. This means that current cyberweapons are not truly *intelligent* systems. Moreover, given that these weapons must infiltrate a system unnoticed, the length of the code required to give these systems a degree of intelligence and autonomy is not likely to present a realistic goal in the near future.

The possibility of ethicality superior AWS is quite different than the reality. Numerous issues need to be addressed to begin to address such an ambitious enterprise including reliability, and limited or biased data including missing data. Ideally, AWS would learn by having access to datasets from successful and unsuccessful scenarios encountered by other systems. However, these datasets might be subject to survivorship bias. *Survivorship bias* occurs when only successful outcomes are considered, while failures are not considered (Brown et al., 1992). This would undoubtedly occur due to the reduced availability of data due to the destruction of an AWS: if a black box cannot be recovered, certain failures would not be registered by AWS. Moreover, as the last moments of an AWS operation before it is destroyed are likely critical to understanding the successful tactics used against the system, the prospects of an ethically superior AWS are meager at best.

13.4.2 PRECISION AND THE AMBIGUITY OF CYBERWEAPONS

Cyberattacks represent a challenge to normal conceptions of lethality. Like physical weapons, their destructive capabilities ultimately depend on how assets and human targets are distributed on the digital battlefield. At present, human bodies are not directly integrated with the digital world (e.g., implants or cyborgs), meaning that direct and immediate lethality is not possible. By eliminating mortality as a possible outcome, cyberweapons could be interpreted as a more ethical form of attack. However, like a biological attack or the use of landmines, the potentially lethal effects of cyberweapons might be delayed. By damaging critical infrastructure with a cyberweapon such as a traffic control system or powerplant, civilian vehicles could crash or power loss in low-temperature conditions or in hospitals could lead to deaths, respectively. The death of noncombatants and potentially lengthier suffering relative to a direct, immediate attack cannot be considered more ethical. Moreover, the temporal distance between cause and effect might create ambiguity in the attribution of deaths and result in discounting negative outcomes.

Now and in the future, cyberweapons are best seen as force multipliers (Schoenherr & Thomson, 2022). By deploying them in conjunction with a kinetic attack, communication infrastructure can be disrupted leading to greater disorganization and easier military victories. However, as AWS are adopted and integrated into military tactics and strategies, the lethality of cyberweapons will likely increase. For instance, the 2011 loss of a US Lockheed Martin RQ-170 Sentinel to Iranian forces provides a clear example of this possibility. While performing reconnaissance, the Iranian government used a GPS-spoofing attack which redirected the US drone to land. Similarly, a virus deployed at Creech Air Force Base in Nevada in 2011 was used to log the keystrokes of RPAs (Shachtman, 2011). While the RPAs remained functional, repeated attempts to remove the relatively benign virus were required (Hartmann & Steup, 2013).

While these attacks are directed toward exploration, information, and technology theft, and disabling AWS, lethal AWS could be captured and redirected against the military or civilian population of an adversary. Incidental civilian deaths might also result if AWS are merely disabled, resulting in crashes in populated areas. These incidents also provide important lessons for warfare in the information age: Although the Creech Air Force Base incident might be attributable to limitations of air gapping, cyber security defenses are imperfect. Even if we adopt the requirement for MHC, the connectivity required for this monitoring and regulation can be exploited.

13.4.3 The Legitimacy of Lethality: Human Rights to Life and Death

Despite the focus on accuracy, a recurrent issue in debates over the development and deployment of AWS is whether their use in killing is *legitimate*. Namely, it is not just that humans are autonomous agents. Rather, humans are biologically primed to experience distress when they witness the suffering of others (Decety & Lamm, 2011). It is this feature of decision-making that makes lethal decisions difficult for humans as they perceive 'targets' as agents like themselves, deserving of a measure of respect and dignity. In this way, a moral agent or group incurs the psychological cost of killing – either through an adverse physiological response or a social response, e.g., the reputational harm they would accrue due to taking a life. Due to the absence of these responses in AWS, there are no circumstances in which using AWSs to kill is justified because a moral agent that understands the consequences of killing is not involved. This prohibition reflects a simple categorical imperative: *AWS must not kill humans.*

From a social-cognitive perspective, the perceived illegitimacy of AWS as moral agents can be attributed to a variant of *ingroup-outgroup bias*. Studies of moral reasoning have demonstrated that ingroup members are valued more than outgroup members (Cikara, 2018; Cikara & Fiske, 2011, 2012; Cikara et al., 2010). Moreover, aggression is only permitted *within* a group provided that a group member is attempting to correct a perceived violation of a group norm, i.e., 'virtuous violence'. This typically entails a proportionate response. Complimenting this, punishment (or killing) that is initiated by an outgroup member is viewed as less justified than those initiated by an ingroup member unless a metarelational rule has been violated. For instance, police are often seen as justified in killing a noncompliant 'dangerous' suspect as a last resort because their authority is legitimated within a social system. Conversely, the use of excessive force by a police officer, or the habitual use of force by police services, reflects a deregulation of behavior and limited distress which is typically criticized within a society (collectively, this constitutes 'police brutality'; Bradford et al., 2017; Lawrence, 1996, 2022).

Consequently, the superficially unique concern about AWS killing a target is best seen as another instance of ingroup-outgroup bias, with the dementalization of AWS representing a central feature, i.e., lacking (moral) agency, empathy, and awareness. Thus, a categorical imperative against AWS killing is likely accepted to the extent that humans are in a better cognitive and social position to decide when killing is justified, even if this judgment is subsequently determined to be wrong.[7]

Designers could include a decision-making capability in AWS to address any specific ethical issues. Such a proposal would require an 'ethical governor' or regulator to identify the *ethical* affordances of a situation rather than simply identifying targets and responses (Arkin et al., 2009; Schoenherr, 2022a, 2022b). However, even if such a regulation algorithm were developed, it seems unlikely that it would satisfy those who have adopted a categorical imperative against AWS killing. For the foreseeable future, there remains a need for MHC to ensure that decisions to fire on targets are perceived as legitimate. We must nevertheless acknowledge that, by including humans in the decision-making processes, we might knowingly increase the probability of errors and biases resulting from situations in which AWS might be more effective at making decisions.

13.4.4 Bots on the Ground: Depersonalization and Anonymity

Ethical combat requires the successful application of ethical decision-making rules. Inasmuch as successful offensive or defensive actions require identifying targets, ethical actions require identifying *human* agents in the field.[8] To overcome prohibitions against killing, Grossman (Grossman, 1995, 2001; Grossman & Christensen, 2022) has suggested that standard military training takes ordinary individuals and automates actions that would result in the death of a target before this response can be inhibited. Of course, this approach fails to address the negative outcomes associated with taking life.

Exceptions to prohibitions against killing are also a common feature of societies. Principles such as an 'eye for an eye' and 'blood for blood' are exemplars of the *reciprocity norm*. Interpersonally, the death of a close other requires reciprocal killing on the grounds that a group identity is shared by the person who has died and the individual seeking revenge as well as between the killer and other potential targets. When framed in this manner, killing represents an act of 'virtuous violence' when directed toward members of one's own ground who fail to enact ethical norms or when directed toward members of another group who have violated general human norms (Fiske, 1991; Fiske & Rai, 2015). The use of shared symbols (e.g., flags) and clothing (e.g., fatigues), promotes this perceived shared identity and helps reinforce differences with the identity of others.

Even if adversaries are outgroup members and a warfighter has been trained to fire automatically, warfighters will still be reluctant to kill another human. An alternate ground for justification of killing results from reclassifying humans as nonhuman. Slogans and ethnonyms can be created to emphasize the nonhuman or subhuman status of adversaries, e.g., 'dogs', 'animals', 'savages', and 'the Hun'. Once this process has been initiated by one group of combatants, it can result in reciprocal dehumanization (Kteily et al., 2016). Indeed, once an adversary is dehumanized, there is less emphasis on seeking a restorative means to settle conflict (Leidner et al., 2013).

By mediating killing through AWS, we promote the experiences of anonymity and depersonalization. Anonymity occurs when one is not identifiable or perceived as accountable in a social situation. The experience of anonymity reduces prosocial behaviors (helping) and increases antisocial behaviors (harming; Mann, 1981; Phillips & Mann, 2019). In the case of AWS, RPA operators are absent from the battlefield, with strikes occurring at a distance and often at night. The absence of any material presence or observation capability with cyberweapons further reduces the saliency of individuating characteristics.

While more research is required, preliminary results provide insight into the effects of the kind of anonymity that results from RPA use. In a study conducted by Petras et al. (2016), they found that visual angles consistent with drone video ('bird's eye'-view) were more likely to promote moral disengagement relative to an egocentric perspective. In a related vein, others have argued that by reducing humans to numbers in a 'body count', that the moral significance of these individuals' deaths is lost (Slovic, 2007; Turse, 2013).

The removal of human warfighters from the field does have one clear benefit. The reduced salience of situational cues, existential dread, and concomitant reduction in stress means that AWS operators are more likely to engage in deliberate thinking. While this might result in stress and moral injury, it might also increase the effectiveness of human-machine collaboration required of MHC. A significant issue for decision-making in combat situations is that stress narrows the focus of attention and increases the likelihood that a warfighter will rely on automatic responses (Nelson et al., 2017). By using AWS as mediators, operators could be removed when psychopathologies or pathophysiological responses are detected and provided with appropriate treatment.

13.4.5 RISK PERCEPTION AND UNCERTAINTY IN THE CONDUCT OF WARFARE

Whether in terms of interpersonal combat or overall strategy, accurate risk perception is required for effective threat assessment. Ideally, risk assessment requires understanding the probability of the occurrence of an event or sequence of events, i.e., the conditional probability. Given a lack of intelligence or an adversary's deception efforts, leaders and warfighters are also unlikely to know all the variables that will affect the outcomes of a battle. They must therefore also be capable of dealing with uncertainty, i.e., a lack of awareness of the value of specific variables or variables unaccounted for during initial planning and preparation.

Risk perception and uncertainty in human decision-making are subject to numerous heuristics and biases (Gilovich et al., 2002). Perhaps the most fundamental concern is *base-rate neglect*: a failure to accurately estimate the probability of the occurrence of events (Pennycook et al., 2022). Instead, decision-makers tend to substitute comparatively easier questions for judgments based on information that is immediately available (from their memory or the environment), and its perceived representativeness (Gilovich et al., 2002). Similarly, humans often find it difficult to differentiate random from non-random events (Nickerson, 2002), frequently focus on successes rather than losses (survivorship bias), and tend to be more concerned with avoidance of losses than making equivalent gains (Tversky & Kahneman, 1992).

In the case of AWS, current variables can be adapted to predict risk factors, i.e., assessing force strength, and position. Thus, risk perception should simply require that warfighters account for how an AWS' specific capabilities (i.e., intelligence, autonomy, and connectivity) affect its operation in theatre of operations. However, as the AWS hacks noted above demonstrate, risk perception and uncertainty in cyber security might not be adequately accounted for by military leadership and warfighters.

13.4.6 THE POSSIBILITY OF JUST WARS: EXPLICIT OR IMPLICIT RULES?

In humans, judgments about ethicality tend to stem from rules that are often implicit, i.e., social contracts and relational models (Fiske, 1991). Ethical frameworks require a firmer foundation, with more systematic sets of principles and their consistent application. Beyond a simplistic categorical imperative to avoid killing with AWS, systematic rules for decision-making must be developed that can be used by both human operators and coded into AWS themselves.

For JWT to be used effectively, we must first be capable of clearly differentiating states of peace and war. This requires identifying a clear set of features or events. The application of JWT relies on the ability of a leader or warfighter to clearly categorize persons, material objects, or events and assess their risks. Persons must clearly be identified as combatants and noncombatants, and targets as military or civilian infrastructure. Wearing uniforms displaying markings associated with the respective belligerents and using weapons facilitates the categorization of combatants and noncombatants. Engaging in 'moral' combat therefore requires accurate categorization. If a human warfighter or AWS is incapable of making such distinctions, they can no longer accurately engage in ethical sensemaking.

Multiple factors influence categorization in warfare, either in terms of combat situations experienced by warfighters or leaders. Lack of training, fatigue, time limitation, and stress will result in warfighters relying on automatic responses and a reduced likelihood that they will be able to identify the ethical affordances required of a just war. This issue is further compounded by so-called irregular warfare. It requires that at least one belligerent is using atypical methods, e.g., terrorists and guerrilla warfare (Boot, 2013). An adversary might dress in civilian clothing to obscure their identity. Over time, the *visual* distinction between combatants and noncombatants will be lessened. Similarly, cyberoperations are problematic, in that attribution can be exceptionally difficult due to a variety of strategies, e.g., IP spoofing.

Even if AWS are more accurate in identifying adversaries and employ superior tactical judgment, a looming issue that threatens their legitimacy is the extent to which their actions can be assessed. For human warfighters, decisions can be assessed, albeit imperfectly through debriefing and after-action reports. Following from issues of explainable AI, we might have concerns that AWSs are not capable of providing justification for their action. However, when humans are asked to explain their behavior, verbal reports and their subjective confidence often show little correspondence to the actual factors that affected their decisions. It can be argued that AWS should be held to a higher standard of justification even if their outcomes are more accurate than humans and that while bullets and bombs cannot been required to justify their actions, i.e., AWS must function with some kind of ethical agents. Thus, the degree of intelligence and autonomy will determine the requirement of connectivity to human operators, i.e., MHC.

13.4.7 HUMAN KILLING MACHINES: MORAL INJURY AND THE EFFECTS OF KILLING

Ethical affordances of AWS are not only determined by their effectiveness in targeting the enemy and prevention of noncombatant deaths. We must also consider the negative outcomes on those who operate them. If combat training produces automate responses that override prohibitions against killing (Grossman, 1995, 2001), AWS operators might be more likely to fire than they would if they were physically present. However, the knowledge of taking someone's life will remain.

Studies have found that killing in combat can promote violence and distress in warfighters (MacManus et al., 2015). Supporting this point, Killgore et al. (2008) found that soldiers involved in the death of 'friendlies' or civilians were more likely to demonstrate psychological distress. Moreover, they tended to engage in more interpersonal aggression to members of their *own unit*. This pattern suggests the experience of a *moral injury*, wherein an individual believes that a bond of trust with their group has been in that they or members of their group violated moral norms or conventions. (Litz et al., 2009; Shay, 2010, 2014).

Similar patterns have been observed in RPA operators and support staff in studies examining the prevalence of posttraumatic stress disorder (PTSD). For instance, Chappelle et al. (2014) found that while the level of PTSD in RPA operators was lower than soldiers deployed in combat (10–18%), it was higher than those found in typical medical records of military personnel (1%: Otto & Webber, 2013). Rather than a simple causal relationship, multiple individual differences, and situational factors undoubtedly account for susceptibility to PTSD. In a subsequent study, they found higher instances of PTSD for operators between 31 and 40 years old, those who flew drones for 51 or more hours per week, and those who witnessed injury or death of bystanders and felt responsibility for these events (Chappelle et al., 2019). Moreover, there is some suggestive evidence that an individual's spiritual well-being mediates this effect (Wood et al., 2018). These results might suggest that if an individual sees the act of killing as justified or supported by larger ideologies (e.g., religious, sociopolitical), then outcomes can be rationalized as part of a larger effort, i.e., a collective form of virtuous violence.

This adverse response conforms to the requirements of the legitimacy of the use of lethal force. Specifically, the decision to kill weighs on the warfighter even if they are directly removed from the combat situation: The operator might simply be following orders, but the consequences of those orders follow them. Such an observation suggests that military leadership has an obligation to ensure that warfighters are aware of, and prepared for, these situations. Beyond education and informed consent, health services must also be provided to ensure that warfighters will have the necessary support to resume function in their post and in society.

13.4.8 DISTRIBUTING RESPONSIBILITY: OVERTRUST AND THE ILLUSION OF CONTROL

Like any weapon, incorporating new technologies like AWS and cyberweapons into military tactics requires understanding how to effectively use them in combat. Unlike other weapons, AWS are granted greater autonomy in their operations. Consequently, military leaders and warfighters must account for how activities are distributed among the human and artificial agents. By distributing cognitive operations, more complicated tasks can be accomplished in a rapid manner, provided that specific subtasks are assigned to agents in a systematic manner.

Effective extended cognition requires having accurate knowledge of the capabilities and capacities of the agents and technology that constitute the system, referred to as *transactive memory* (Peltokorpi, 2008; Wegner, 1987). This will arise either through direct experience with these systems or because of trust associated with the category of weapon system that is deployed. Over time, affective and cognitive dimensions of trust will be used as heuristics. Leaders and warfighters might fail to consider the actual operational capabilities of a system allowing them to focus their attention on other tasks. Nevertheless, as lethal weapons, the responsibility of monitoring and regulating the operation of AWS remains theirs.

MHC requires that we differentiate between the *possibility of control* and the *probability of control*, i.e., feasibility. For instance, designers might create AWS that include affordances that allow operators to remotely override the system. Nevertheless, this functional affordance might not be available in practice. The operations of a system might be so rapid and tightly coupled that it requires an unmanageable level of complexity for humans relative to their cognitive capabilities. Consider a situation wherein an AWS detects a lock-on by an adversary's weapon system. In this situation, an AWS *should* switch into defense mode and destroy the oncoming attack to ensure its preservation. However, the possibility of misidentification could lead to the destruction of civilians or civilian assets. Ideally, human operators should intervene, and designers might include affordances to support control, i.e., a kill switch. Despite this possibility, as the time between the detection and defensive action decreases, the probability of MHC decreases.

Rather than MHC, the possibility of controlling a system provides the user with an illusory of control. This can substantially alter how a system is used as well as the perception of responsibility and accountability. First, there is a general belief in an *internal* locus of control, that an individual is responsible for the outcomes of a situation.[9] This can be contrasted with an *external* locus of control, wherein situational factors (e.g., social, and physical environment) determine the outcomes. Thus, operators might assume that an AWS has been adopted by the armed forces because it performs an intended function adequately, whereas external parties (e.g., auditors) might instead assume that the operators are principally responsible for the operations of the system. However, if the roles and responsibilities are not clearly assigned to, and understood by, all agents within the sociotechnical system, errors and biased decision-making will result. In the case of lethal AWS, this moral weight is likely to affect those operating these systems.

13.4.9 THE CHAINS OF WAR: AWS AS TRANSNATIONAL WEAPONS

In warfare, advances in social organization and technology can be the key to success. However, The creation and operation of AWS require components, the complexity of which are unprecedented in human history. Supporting this, supply chains must also be considered when developing ethical and legal guidelines. Suppliers, manufacturers, and distributors are typically aware of both their clients' needs and the intended *use* of a weapon system. By selling AWSs or supporting technology, they are tacitly endorsing the state or nonstate actors use of these weapons. They therefore share in the moral responsibility of their use.

As I've noted above, the ethical affordance of AI-based technologies might not always be clear. Many of the components required to produce AWS are also used in the manufacture of civilian technologies, i.e., multipurpose technologies. This potentially increases the ambiguity in ascertaining their intended use. For instance, a report of Russian drones destroyed prior to the onset of the 2022 Ukraine-Russia conflict found that many components were manufactured by Western countries, many of them allies of Ukraine (Whalen, 2022). While these results were obtained *before* the onset of the 2022 conflict they occurred *after* Russia's annexation of Crimea in 2014. Manufacturers who knowingly sold these weapons therefore bear some responsibility.

Supply chains create interdependency between countries. In turn, economic interdependence necessarily creates a common fate. This can increase incentives for cooperation and reduce those for large-scale conflicts. Indeed, small-scale observations have demonstrated that human groups that are more interdependent are more concerned with adherence to shared moral norms and conventions (Ensminger & Henrich, 2014; Henrich et al., 2001) with others arguing that economic interdependence has resulted in a reduction of large-scale violent conflicts over the course of human history (Pinker, 2012).

Export controls provide a political and economic means to regulate the exchange of key components for the development of these technologies. However, their successes are limited (Flynn, 2020). For instance, existing export controls cover AI-based technologies that have military applications Yet, third-party nations can be used to bypass and sanitize imports. Countries are aware of this fact. For instance, Russia's Federal Customs Service prioritized the import of "dual-use goods and civilian products intended to support the combat and daily activities of Russian military units" (cited in Fromer, 2023). Similarly, DJI, a Chinese drone manufacturer, noted that they "cannot control how [their] products are [used] once they leave [their] management… [although they] have taken all steps under their control to emphasize that their products should not be used in combat, to cause harm, or be modified into weapons" (quoted in Fromer, 2023). These observations suggest that both implicit motivations associated with economic interdependence and current explicit attempts at regulations are inadequate barriers to stopping the sales of components associated with AWS.

13.4.10 Accessibility and Asymmetric Warfare: Armchair Arms Race

If supply chains suggest that interdependence provides one means to ensure reduced armed warfare, we must also consider what the implications of relatively accessible AWS, cyberweapons, and related technologies to adversarial state and nonstate actors. Namely, the technologies that provide these actors with lethal affordances were previously quite limited. Currently, the diffusion of multipurpose technologies provides these actors with unprecedented capabilities.

While RPAs are quite costly, modified domestic drones (e.g., quadcopters) are comparatively inexpensive. When used in conjunction with other technologies (e.g., 3D printing), physical equipment can be modified to increase their capabilities (Schoenherr, 2023a). Costs are further reduced in the development of cyberweapons capabilities. While computer programming is a highly skilled profession requiring years of training, learning to develop software/malware presents a shallower learning curve than developing kinetic, nuclear, or biological weapon systems, with the time from design to deployment being a fraction of physical systems. Moreover, the interaction between digital and physical technologies can have a synergistic effect. For instance, in a set of toxic chemical compounds identified by an AI, approximately 40,000 have lethal capabilities that could be used as biological weapons (Urbina et al., 2022). Using an analogy from economics, the market entry cost of researching, developing, and deploying such weapons is comparatively small. While this will undoubtedly change as cyberweapons and defenses continue to evolve, adversarial state and nonstate actors that can exploit these advantages will have access to an important force multiplier. Overall, weapon systems are far more accessible than they have been historically (Figure 13.4).

Increased accessibility to weapon systems has multiple effects that challenge existing geopolitical equilibria. *Asymmetric warfare* is the most widely recognized. Given that national and international geopolitical entities often have large, advanced arsenals of traditional kinetic weapons, they cannot be readily challenged. Nevertheless, as history has shown, a decisive short-term victory is not always feasible nor is it always the goal of combat. When outmatched, a war of attrition can

FIGURE 13.4 R18 octocopter equipped with a modified Soviet-era RKG-3 anti-tank grenade. Photo credit and permission by Aerorozvidka.

result in a long-term victory. Thus, unconventional weapons and warfare like domestic drones and cyberweapons can radically alter the power dynamic if defenders adopt new strategies, i.e., guerilla warfare. This is evident in N. Korea. Despite the development of nuclear missile capabilities, it is far outmatched by its adversaries. Yet, its use of persistent attacks places it on the footing that is comparable to its adversaries. While such advantages might be short-lived due to changes in the strategies of more developed nations, if an advantage can be identified in one domain, it provides leverage. A similar observation can be made about Hamas. Despite Israel reportedly destroying its drone capabilities and maintaining constant surveillance of the Palestinian population, the 2023 attack demonstrates the relative resilience of a nonstate actor in developing lethal capabilities and the devastating outcomes (Schoenherr, 2023c).

13.5 CONCLUSION

The creation and diffusion of military technologies change conflicts and combat, often in ways that can be difficult to appreciate. In the same way that AI has changed the domestic workforce, AWS and cyberweapons have come to restructure communities of warfighters. With the metaphorical 'keyboard warrior' having both physical and binary bullets at their disposal. New tactics and competencies must be developed for the next generation of warfighters. In addition to understanding the hardware and software, we must ensure that they understand the ethical affordances of AWS and cyberweapons to ensure that they can make informed decisions about the nature and necessity of their operations.

The perceived need for autonomy and the technical and functional affordances provided by AWS have rightly led to a belief that these weapons can provide a strategic advantage. However, the possibility that an adversarial nation might adopt these capabilities first or have a greater capacity, has led to a form of fatalism. Along these lines, there is a perceived imperative that AWS and cyberweapons *should* be adopted to ensure the safety and security of a nation. To forestall this inevitability, many have suggested that international guidelines and regulations must be developed to address how the laws of war should be adapted to accommodate these new weapons. Yet, we must acknowledge that both state and nonstate actors will not abide by these guidelines. This is especially true for those who are not entangled in socioeconomic world systems. In the twentieth and twenty-first centuries, war crimes still occur during armed conflicts and go unpunished if the country of the perpetrators is not decisively defeated. Consequently, there remains a moral imperative to ensure that the technical and functional affordances of technologies acknowledge the ethical affordances that result from human interaction with AWS and cyberweapons.

MHC superficially presents a reasonable criterion for the use of AWS. According to this account, humans should intervene during key decision points in the operation of an AWS to ensure that these systems operate effectively and ethically. Yet, we must not be naïve: The speed of operations and delays in communication due to environmental interference will ultimately result in a need for increasing AWS autonomy. If we assume that humans are both limited in their capacity and the speed of response, traditional forms of *meaningful* control will likely be eroded as AWS technologies begin to dominate and define global battlefields. Designers, developers, distributors, and military leadership must explicitly consider how to accommodate the limitations of human cognitive processing and the operational demands of combat. Similar observations can be made about cyberweapons.

Are Cyberweapons Weapons? Here, I have distinguished AWS from cyberweapons. Namely, despite their use of the term 'attack', so-called cyberweapons lack many of the essential features typically associated with weapons, especially lethality. While they might indirectly result in deaths, they are ultimately force multipliers rather than lethal weapons themselves. Inasmuch as we might use 'words as weapons' and engage in 'wars of words', attacks against digital assets of persons or organizations do not constitute physical attacks. Consequently, while they are referred to as weapons, this is an imperfect metaphor, one of many that have been used to understand technologies that are multi-purpose and ambiguous (Schoenherr, 2022a, 2022b 2023a; Schoenherr & Thomson, 2022). This fact must be acknowledged as the use of different metaphors will highlight different features of cyberspace. Failing to acknowledge this might lead to different vulnerabilities in the design and deployment of cyberweapons. Some vulnerabilities will undoubtedly have ethical implications.

Nevertheless, if we are to adopt a looser distinction for AWS, such that we can differentiate between *lethal* AWS (LAWS) and AWS, cyberweapons might be categorized as a weapon. By adopting this definition, certain kinds of RPAs are not *lethal* weapons if they are not equipped with lethal payloads in the same way that jeeps, planes, and boats are not necessarily weapon systems but can be equipped with weapons. Thus, cyberweapons only become weapons if their payload allows them to directly cause damage or death. While the possibility of death due to cyberweapons is currently remote, as humans become more integrated with technology and as leaders and warfighters become more reliant on AWS, well-designed malware could be used to turn the weapons against their creators.

Alternatively, cyberweapons might be considered weapons based on the intention of the user. A rock is not intrinsically a weapon but can be used as one once the affordance is perceived. If a cyberattack is conducted with the intention of causing death, then they would meet the definition of a lethal weapon. For instance, much like a V1 bomb, a cyberweapon might be deployed knowing that the deaths are possible, even if improbable. Using such an intentional criterion is also critical to ethics: non-consequentialist principles would assume that regardless of the actual impact of a cyberweapon, the intended use as a weapon makes it relevant to ethical decision-making.

Engineering Psychology and Ethics. Ethical principles, values, and norms represent criteria, dimensions, and categories that do not exist objectively within the world. The fact that large groups of individuals subscribe to collections of values, ensures that they are granted an intersubjective realism. Indeed, in parallel with conflict, human history also represents a record of identifying, operationalizing, and negotiating norms to facilitate cooperation within and between groups. International efforts at the regulation of LAWS/AWS clearly illustrate these ongoing efforts.

The speed of technological evolution and diversity of consideration requires that multi-stakeholder groups continuously monitor and regulate these technologies. At a minimum, this requires engineers, psychologists, and legal experts with the requisite skills to address these issues.

If ethical AWS can be designed, they will require an understanding of the moral blind spots that define human ethical sensemaking. At a fundamental level, this requires acknowledging the basic processing limitations associated with human attention and working memory. At an intermediate level, it requires that we understand how dimensions and categories are carved out of the environment and inform our judgment. At a societal and international level, this requires understanding how human and artificial agents interact with each other to determine the outcomes. MHC therefore requires that we adopt a multi-level approach to understanding the sociotechnical systems of AWS.

NOTES

1. Historians continue to seek the factors that can be attributed to success and failure, i.e., de-Ba'athification has been attributed to the failure of stability in Iraq.
2. In principle, this could be extended to people and relationships, i.e., social capital.
3. Computers were initially conceived of as a universal Turing Machine.
4. Hereafter, I use the term 'multi-purpose' given that it is more general than 'dual-use'.
5. Note that the hit and false alarm rates are not compliments. They will be a function of the respective signal and noise distributions and placement of the decision criteria.
6. This provides the basis for Signal Detection Theory such that the distribution of target and nontargets and placement of decision criteria will determine the ability of a system to respond.
7. As AWS becomes more commonplace, it is possible that a growing familiarity with the technology might decrease the adoption of this categorical imperative. For instance, this basic pattern is observed in studies of the uncanny valley (Burleigh et al., 2013; Burleigh & Schoenherr, 2015; Schoenherr & Burleigh, 2020; Schoenherr & Thomson, 2019) and has also been obtained in studies comparing human and nonhuman agents (Schoenherr & Thomson, 2022). However, in that people are not exposed to military technologies at the same frequency as domestic technologies, this will undoubtedly be a slow process.
8. Here, I do not consider nonhuman animals as the focus is on combat situations as the use of animals in combat (e.g., horses for transportation) are not typically used.
9. This appears to be more prevalent in individualistic societies.

REFERENCES

Ahmad, R., Alsmadi, I., Alhamdani, W., & Tawalbeh, L. (2023). Zero-day attack detection: A systematic literature review. *Artificial Intelligence Review*, *56*(10), 10733–10811. https://doi.org/10.1007/s10462-023-10437-z

Allenby, B. (2013). The implications of emerging technologies for just war theory. *Public Affairs Quarterly*, *27*(1), 49–67.

Allhoff, F., Henschke, A., & Strawser, B. J. (Eds.). (2016). *Binary bullets. The ethics of cyberwarfare.* Oxford University Press.

Arkin, R. (2009). *Governing lethal behavior in autonomous robots.* CRC Press. https://www.taylorfrancis.com/books/mono/10.1201/9781420085952/governing-lethal-behavior-autonomous-robots-ronald-arkin

Arkin, R. C., Ulam, P., & Duncan, B. (2009). *An ethical governor for constraining lethal action in an autonomous system* (Technical Report GIT-GVU-09-02). https://digitalcommons.unl.edu/csetechreports/163/

Asaro, P. (2012). On banning autonomous weapon systems: Human rights, automation, and the dehumanization of lethal decision-making. *International Review of the Red Cross*, *94*(886), 687–709.

Ashby, F. G., & O'Brien, J. B. (2005). Category learning and multiple memory systems. *Trends in Cognitive Sciences*, *9*(2), 83–89.

Ashby, F. G., Paul, E. J., & Maddox, W. T. (2011). COVIS. In E. M. Pothos, & A. J. Willis (Eds.), *Formal approaches in categorization* (pp. 65–88). Cambridge University Press.

Bates, R., Greif, A., & Singh, S. (2002). Organizing violence. *Journal of Conflict Resolution*, *46*(5), 599–628. https://doi.org/10.1177/002200202236166

Benjamin, M. (2013). *Drone warfare: Killing by remote control.* Verso Books.

Bentham, J. (1789). *An introduction to the principles of morals and legislation (1789).* Clarendon Press. https://www.reed.edu/humanities/hum220/syllabus/2010-11/Bentham-Principles.pdf

Bernstein, B. J. (1999). Reconsidering Truman's claim of 'half a million American lives' saved by the atomic bomb: The construction and deconstruction of a myth. *Journal of Strategic Studies*, *22*(1), 54–95. https://doi.org/10.1080/01402399908437744

Betts, R. K. (1978). Analysis, war, and decision: Why intelligence failures are inevitable. *World Politics*, *31*(1), 61–89.

Bijker, W. E., & Law, J. (1994). *Shaping technology/building society: Studies in sociotechnical change*. MIT Press.

Boot, M. (2013). *Invisible armies: An epic history of guerrilla warfare from ancient times to the present*. WW Norton & Company.

Boyd, R., Richerson, P. J., & Henrich, J. (2011). The cultural niche: Why social learning is essential for human adaptation. *Proceedings of the National Academy of Sciences*, *108*(supplement_2), 10918–10925. https://doi.org/10.1073/pnas.1100290108

Bradford, B., Milani, J., & Jackson, J. (2017). Identity, legitimacy and "making sense" of police use of force. *Policing: An International Journal*, *40*(3), 614–627.

Brown, S. J., Goetzmann, W., Ibbotson, R. G., & Ross, S. A. (1992). Survivorship bias in performance studies. *The Review of Financial Studies*, *5*(4), 553–580.

Burleigh, M. (2007). *Sacred causes: The clash of religion and politics, from the great war to the war on terror*. HarperCollins. https://ixtheo.de/Record/521773164

Burleigh, M. (2010). *Moral combat: A history of World War II*. HarperCollins UK.

Burleigh, T. J., & Schoenherr, J. R. (2015). A reappraisal of the uncanny valley: Categorical perception or frequency-based sensitization? *Frontiers in Psychology*, *5*, 104682.

Burleigh, T. J., Schoenherr, J. R., & Lacroix, G. L. (2013). Does the uncanny valley exist? An empirical test of the relationship between eeriness and the human likeness of digitally created faces. *Computers in Human Behavior*, *29*(3), 759–771.

Chappelle, W., Goodman, T., Reardon, L., & Thompson, W. (2014). An analysis of post-traumatic stress symptoms in United States Air Force drone operators. *Journal of Anxiety Disorders*, *28*(5), 480–487.

Chappelle, W., Goodman, T., Reardon, L., & Prince, L. (2019). Combat and operational risk factors for post-traumatic stress disorder symptom criteria among United States air force remotely piloted aircraft "Drone" warfighters. *Journal of Anxiety Disorders*, *62*, 86–93.

Cikara, M. (2018). Pleasure in response to outgroup pain as a motivator of intergroup aggression. In K. Gray & J. Graham (Eds.), *Atlas of moral psychology* (pp. 193–200). The Guilford Press.

Cikara, M., Farnsworth, R. A., Harris, L. T., & Fiske, S. T. (2010). On the wrong side of the trolley track: Neural correlates of relative social valuation. *Social Cognitive and Affective Neuroscience*, *5*(4), 404–413.

Cikara, M., & Fiske, S. T. (2011). Bounded empathy: Neural responses to outgroup targets'(mis) fortunes. *Journal of Cognitive Neuroscience*, *23*(12), 3791–3803.

Cikara, M., & Fiske, S. T. (2012). Stereotypes and Schadenfreude: Affective and physiological markers of pleasure at outgroup misfortunes. *Social Psychological and Personality Science*, *3*(1), 63–71. https://doi.org/10.1177/1948550611409245

Citino, R. (2017). Beyond fire and movement: Command, control and information in the German Blitzkrieg 1. In G. Jensen (Ed.), *Warfare in Europe 1919–1938* (pp. 65–85). Routledge. https://www.taylorfrancis.com/chapters/edit/10.4324/9781315234335-4/beyond-fire-movement-command-control-information-german-blitzkrieg-1-robert-citino

Citino, R. M. (2004). *Blitzkrieg to desert storm: The evolution of operational warfare*. University of Kansas Press. https://www.ajol.info/index.php/smsajms/article/download/70510/59114/0

Cosmides, L., & Tooby, J. (2013). Unraveling the enigma of human intelligence: Evolutionary psychology and the multimodular mind. In R. J. Sternberg & J. C. Kaufman (Eds.), *The evolution of intelligence* (pp. 145–198). Psychology Press. https://www.taylorfrancis.com/chapters/edit/10.4324/9781410605313-8/unraveling-enigma-human-intelligence-evolutionary-psychology-multimodular-mind-leda-cosmides-john-tooby

Cox, R. (2017). Expanding the history of the just war: The ethics of war in ancient Egypt. *International Studies Quarterly*, *61*(2), 371–384. https://doi.org/10.1093/isq/sqx009

Cox, R. (2023). *Origins of the just war: Military ethics and culture in the ancient near East*. Princeton University Press.

Davies, H., McKernan, B., & Sabbagh, D. (2023, December 1). 'The Gospel': How Israel uses AI to select bombing targets in Gaza. *The Guardian*. https://www.theguardian.com/world/2023/dec/01/the-gospel-how-israel-uses-ai-to-select-bombing-targets

Decety, J., & Lamm, C. (2011). 15 empathy versus personal distress: Recent evidence from social neuroscience. In J. Decety, & W. Ickes (Eds.), *The social neuroscience of empathy* (p. 199). MIT Press.

Dennett, D. C. (1995). Darwin's dangerous idea. *The Sciences*, *35*(3), 34–40. https://doi. org/10.1002/j.2326-1951.1995.tb03633.x

Dipert, R. R. (2016). The ethics of cyberwarfare. In T. J. Demy, G. R. Lucas Jr., & B. J. Strawser (Eds.), *Military ethics and emerging technologies* (pp. 159–185). Routledge. https://api.taylorfrancis.com/content/chapters/ edit/download?identifierName=doi&identifierValue=10.4324/9781315766843-13&type=chapterpdf

Docherty, B. L. (2014). *Shaking the foundations: The human rights implications of killer robots*. Human Rights Watch.

Dupuy, C. T. N. (1984). *The evolution of weapons and warfare*. Da Capo Press. https://www.dacapopress.com/ titles/colonel-trevor-n-dupuy/the-evolution-of-weapons-and-warfare/9780306803840/

Ekelhof, M. (2019). Moving beyond semantics on autonomous weapons: Meaningful human control in operation. *Global Policy*, *10*(3), 343–348. https://doi.org/10.1111/1758-5899.12665

Elshtain, J. B. (Ed.). (1991). *Just war theory* (First ed.). NYU Press.

Ensminger, J., & Henrich, J. (Eds.). (2014). *Experimenting with social norms: Fairness and punishment in cross-cultural perspective*. Russell Sage Foundation.

Evans, J., & Frankish, K. (Eds.). (2009). *In two minds: Dual processes and beyond* (1st edition). Oxford University Press.

Evans, S. G., & Delaney, K. B. (2018). The V1 (flying bomb) attack on London (1944–1945); The applied geography of early cruise missile accuracy. *Applied Geography*, *99*, 44–53. https://doi.org/10.1016/j. apgeog.2018.07.019

Feshbach, S. (1971). Dynamics and morality of violence and aggression: Some psychological considerations. *American Psychologist*, *26*(3), 281.

Finlay, C. J. (2013). Fairness and liability in the just war: Combatants, non-combatants and lawful irregulars. *Political Studies*, *61*(1), 142–160. https://doi.org/10.1111/j.1467-9248.2012.00954.x

Fiske, A. P. (1991). *Structures of social life: The four elementary forms of human relations: Communal sharing, authority ranking, equality matching, market pricing* (pp. xii, 480). Free Press.

Fiske, A. P., & Rai, T. S. (2015). *Virtuous violence: Hurting and killing to create, sustain, end, and honor social relationships* (pp. xxvi, 357). Cambridge University Press.

Flynn, C. (2020). Recommendations on Export Controls for Artificial Intelligence. *Center for Security and Emerging Technology*. https://cset.georgetown.edu/publication/recommendations-on-export-controls-for-artificial-intelligence/

Fromer, J. (2023). *Special report: Russia buying civilian drones from China for war effort*. Nikkei Asia. https://asia.nikkei.com/Politics/Ukraine-war/Special-report-Russia-buying-civilian-drones-from-China-for-war-effort

Gibson, J. J. (1966). *The senses considered as perceptual systems*. Houghton Mifflin Co. https://www.journals. uchicago.edu/doi/10.1086/406033

Gibson, J. J. (1979). *The ecological approach to perception*. Lawrence Erlbaum Associates.

Gilli, A., & Gilli, M. (2018). Why China has not caught up yet: Military-technological superiority and the limits of imitation, reverse engineering, and cyber espionage. *International Security*, *43*(3), 141–189. https://doi.org/10.1162/isec_a_00337

Gilovich, T., Griffin, D., & Kahneman, D. (Eds.). (2002). *Heuristics and biases: The psychology of intuitive judgment*. Cambridge University Press. https://doi.org/10.1017/CBO9780511808098

Goldman, E. O., & Eliason, L. C. (Eds.). (2003). *The diffusion of military technology and ideas* (1st ed.). Stanford University Press.

Grossman, D. (1995). *On killing: The psychological cost of learning to kill in war and society* (Revised ed.). Back Bay Books.

Grossman, D. (2001). On killing. II: The psychological cost of learning to kill. *International Journal of Emergency Mental Health*, *3*(3), 137–144.

Grossman, D., & Christensen, L. W. (2022). *On combat, the psychology and physiology of deadly conflict in war and in peace*. Open Road Media.

Hambling, D. (2020, July 16). *Clone wars: Why Iran will copy captured U.S. Global Hawk drone*. Forbes. https://www.forbes.com/sites/davidhambling/2020/07/16/clone-wars-why-iran-will-copy-captured-us-global-hawk-drone/?sh=7bbff96b7540

Hare, R. M., Hare, R. M., Hare, H., Mervyn, R., & Hare, R. M. (1981). *Moral thinking: Its levels, method, and point*. Oxford University Press.

Hartmann, K., & Steup, C. (2013). The vulnerability of UAVs to cyber attacks—An approach to the risk assessment. *2013 5th International Conference on Cyber Conflict (CYCON 2013)*, 1–23. https://ieeexplore.ieee.org/document/6568373

Hedley, J. H. (2005). Learning from intelligence failures. *International Journal of Intelligence and Counter Intelligence, 18*(3), 435–450. https://doi.org/10.1080/08850600590945416

Henrich, J., Boyd, R., Bowles, S., Camerer, C., Fehr, E., Gintis, H., & McElreath, R. (2001). Search of homo economicus: Behavioral experiments in 15 small-scale societies. *The American Economic Review, 91*(2), 73–78.

Heyns, C. (2014). *Report of the Special Rapporteur on Extrajudicial, Summary or Arbitrary Executions.* United Nations General Assembly. https://digitallibrary.un.org/record/771922

Hijazi, A., Ferguson, C. J., Richard Ferraro, F., Hall, H., Hovee, M., & Wilcox, S. (2019). Psychological dimensions of drone warfare. *Current Psychology, 38*(5), 1285–1296. https://doi.org/10.1007/s12144-017-9684-7

Hill, T. E. Jr. (1971). Kant on imperfect duty and supererogation. *Kant-Studien, 62*(1–4), 55–76. https://doi.org/10.1515/kant.1971.62.1-4.55

Hopster, J. (2021). What are socially disruptive technologies? *Technology in Society, 67,* 101750.

Horowitz, M. C., & Scharre, P. (2015). Meaningful Human Control in Weapon Systems. *Center for a New American Working Paper.* https://www.files.ethz.ch/isn/189786/Ethical_Autonomy_Working_Paper_031315.pdf

Ju, Y., Li, Q., Liu, H. Y., Cui, X. M., & Wang, Z. H. (2020). Study on application of open source intelligence from social media in the military. *Journal of Physics: Conference Series, 1507*(5), 052017. https://doi.org/10.1088/1742-6596/1507/5/052017

Kaag, J., & Kreps, S. (2014). *Drone warfare.* John Wiley & Sons.

Kahneman, D. (2013). *Thinking, fast and slow* (1st ed.). MacMillan.

Kang, J. (2006). Just war thought of the pre-Qin period. *Whampoa—An Interdisciplinary Journal, 51,* 125–141.

Kant, I. (1999). *Critique of pure reason* (P. Guyer & A. W. Wood, Eds.). Cambridge University Press.

Karazy, S., Peleschuk, D., Hunder, M., & Heritage, T. (2023, March 22). At least nine dead in Russian air strikes on two Ukrainian cities. *Reuters.* https://www.reuters.com/world/europe/three-killed-russian-drone-strike-kyiv-region-officials-2023-03-22/

Khalidi, R. I. (2014). From the editor: The Dahiya doctrine, proportionality, and war crimes. *Journal of Palestine Studies, 44*(1), 5–13. https://doi.org/10.1525/jps.2014.44.1.5

Kilcullen, D. (2019). The evolution of unconventional warfare. *Scandinavian Journal of Military Studies, 2*(1), 61–71. https://doi.org/10.31374/sjms.35

Killgore, W. D. S., Cotting, D. I., Thomas, J. L., Cox, A. L., McGurk, D., Vo, A. H., Castro, C. A., & Hoge C. W. (2008). Post-combat invincibility: Violent combat experiences are associated with increased risk-taking propensity following deployment. *Journal of Psychiatric Research, 42*(13), 1112–1121, https://doi.org/10.1016/j.jpsychires.2008.01.001. https://www.sciencedirect.com/science/article/pii/S0022395608000034

Kim, V., Pérez-Peña, R., & Kramer, A. E. (2023, September 8). Elon Musk Refused to Enable Ukraine Drone Attack on Russian Fleet. *The New York Times.* https://www.nytimes.com/2023/09/08/world/europe/elon-musk-ukraine-starlink-drones.html

Kosal, M. E. (Ed.). (2020). *Disruptive and game changing technologies in modern warfare: Development, use, and proliferation.* Springer International Publishing. https://doi.org/10.1007/978-3-030-28342-1

Kteily, N., Hodson, G., & Bruneau, E. (2016). They see us as less than human: Metadehumanization predicts intergroup conflict via reciprocal dehumanization. *Journal of Personality and Social Psychology, 110*(3), 343–370. https://doi.org/10.1037/pspa0000044

Latour, B. (1990). Technology is society made durable. *The Sociological Review, 38*(1_suppl), 103–131. https://doi.org/10.1111/j.1467-954X.1990.tb03350.x

Latzer, M. (2009). Information and communication technology innovations: Radical and disruptive? *New Media & Society, 11*(4), 599–619. https://doi.org/10.1177/1461444809102964

Lautenschläger, K. (1983). Technology and the evolution of naval warfare. *International Security, 8*(2), 3–51. https://doi.org/10.2307/2538594

Lawrence, R. G. (1996). Accidents, icons, and indexing: The dynamics of news coverage of police use of force. *Political Communication, 13*(4), 437–454. https://doi.org/10.1080/10584609.1996.9963130

Lawrence, R. G. (2022). *The politics of force: Media and the construction of police brutality.* Oxford University Press. https://doi.org/10.1093/oso/9780197616543.001.0001

Leidner, B., Castano, E., & Ginges, J. (2013). Dehumanization, retributive and restorative justice, and aggressive versus diplomatic intergroup conflict Resolution strategies. *Personality and Social Psychology Bulletin, 39*(2), 181–192. https://doi.org/10.1177/0146167212472208

Lewis, M. E. (2005). The just war in early china. In T. Brekke (Ed.), *The ethics of war in Asian civilizations* (pp. 203–218). Routledge.

Lewis, T. (2020). *Atomic salvation: How the a-bomb saved the lives of 32 million people.* Casemate.

Litz, B. T., Stein, N., Delaney, E., Lebowitz, L., Nash, W. P., Silva, C., & Maguen, S. (2009). Moral injury and moral repair in war veterans: A preliminary model and intervention strategy. *Clinical Psychology Review, 29*(8), 695–706. https://doi.org/10.1016/j.cpr.2009.07.003

Lo, P. (2012). The 'Art of war' corpus and Chinese just war ethics past and present. *The Journal of Religious Ethics, 40*(3), 404–446.

Lo, P.-C., & Twiss, S. B. (2015). *Chinese Just war ethics: Origin, development, and dissent.* Routledge. https://www.routledge.com/Chinese-Just-War-Ethics-Origin-Development-and-Dissent/Lo-Twiss/p/book/9781138729216

Lucas, G. (2017). *Ethics and cyber warfare: The quest for responsible security in the age of digital warfare* (1st ed.). Oxford University Press.

MacIntyre, A. (1981). *After virtue.* A&C Black.

MacKenzie, D. (1993). *Inventing accuracy: A historical sociology of nuclear missile guidance.* The MIT Press.

MacKenzie, D., & Wajcman, J. (1999). *The social shaping of technology.* Open University Press. https://eprints.lse.ac.uk/28638

MacManus, D., Rona, R., Dickson, H., Somaini, G., Fear, N., & Wessely, S. (2015). Aggressive and violent behavior among military personnel deployed to Iraq and Afghanistan: Prevalence and link with deployment and combat exposure. *Epidemiologic Reviews, 37*(1), 196–212. https://doi.org/10.1093/epirev/mxu006

Malakauskis, P., & Juozapavičius, A. (2023). Assessing the vulnerability of military personnel through open source intelligence: A case study of Lithuanian armed forces. In H. Degen, S. Ntoa, & A. Moallem (Eds.), *HCI international 2023 – Late breaking papers* (Vol. 14059, pp. 435–444). Springer Nature Switzerland. https://doi.org/10.1007/978-3-031-48057-7_27

Mann, L. (1981). The baiting crowd in episodes of threatened suicide. *Journal of Personality and Social Psychology, 41*(4), 703–709. https://doi.org/10.1037/0022-3514.41.4.703

Marei, F. G. (2020). Dahiya doctrine. In J. K. Zartman (Ed.), *Conflict in the modern Middle East: An encyclopedia of civil war, revolutions, and regime change* (pp. 75–76). ABC-CLIO.

Matthias, A. (2004). The responsibility gap: Ascribing responsibility for the actions of learning automata. *Ethics and Information Technology, 6*(3), 175–183. https://doi.org/10.1007/s10676-004-3422-1

Mayor, A. (2014). Animals in warfare. In G. L. Campbell (Ed.), *The Oxford handbook of animals in classical thought and life* (pp. 282–293). Oxford University Press. https://doi.org/10.1093/oxfordhb/9780199589425.013.017

Mazar, N., Amir, O., & Ariely, D. (2008). The dishonesty of honest people: A theory of self-concept maintenance. *Journal of Marketing Research, 45*(6), 633–644. https://doi.org/10.1509/jmkr.45.6.633

McCluskey, M., & Green, A. G. (2023, December 21). *Israeli Military says 2 civilians killed for every militant is 'tremendously positive' ratio given Gaza combat challenges.* CNN. https://www.cnn.com/2023/12/05/middleeast/israel-hamas-military-civilian-ratio-killed-intl-hnk/index.html

Michael, K., Schoenherr, J. R., & Vogel, K. (2024). Failures in the loop: Human leadership in AI-based decision-making. *IEEE Transactions in Technology and Society.* https://ieeexplore.ieee.org/stamp/stamp.jsp?tp=&arnumber=10539317

Mill, J. S. (1861). *Utilitarianism.* Oxford University Press.

Moorehead, A., Hussein, R., & Alhariri, W. (2017). *Out of the shadows: Recommendations to advance transparency in the use of lethal force.* Columbia University. https://scholarship.law.columbia.edu/human_rights_institute/25/

Morgan, T. J. H., Uomini, N. T., Rendell, L. E., Chouinard-Thuly, L., Street, S. E., Lewis, H. M., Cross, C. P., Evans, C., Kearney, R., de la Torre, I., Whiten, A., & Laland, K. N. (2015). Experimental evidence for the co-evolution of hominin tool-making teaching and language. *Nature Communications, 6*(1), 6029. https://doi.org/10.1038/ncomms7029

Moseley, A. (2011). Just War Theory. In *The Encyclopedia of Peace Psychology.*

Nelson, E. A., Schoenherr, J. R., Messervey, D., & Waylon, D. (2017). *How stress affects cognition, memory, and ethical decision making.* Defence Research and Development Canada.

Newpower, A. (2006). *Iron men and tin fish: The race to build a better torpedo during world war II.* Bloomsbury Publishing USA.

Nickerson, R. S. (2002). The production and perception of randomness. *Psychological Review, 109*(2), 330–357. https://doi.org/10.1037/0033-295X.109.2.330

Norman, D. (1988). *The psychology of everyday things* (1st ed.). Basic Books.

O'Neill, O. (1975). *Acting on principle: An essay on Kantian ethics.* Columbia University Press.

Otto, J. L., & Webber, B. J. (2013). Mental health diagnoses and counseling among pilots of remotely piloted aircraft in the United States Air Force. *Medical Surveillance Monthly Report, 20*(3), 3–8.

Parks, L., & Kaplan, C. (Eds.). (2017). *Life in the age of drone warfare*. Duke University Press.

Peltokorpi, V. (2008). Transactive memory systems. *Review of General Psychology, 12*(4), 378–394. https://doi.org/10.1037/1089-2680.12.4.378

Pennycook, G., Newton, C., & Thompson, V. A. (2022). Base-rate neglect. In R. F. Pohl (Ed.), *Cognitive illusions: Intriguing phenomena in thinking, judgment, and memory* (pp. 44–60). Routledge.

Petras, K., ten Oever, S., & Jansma, B. M. (2016). The effect of distance on moral engagement: Event related potentials and alpha power are sensitive to perspective in a virtual shooting task. *Frontiers in Psychology, 6*, 2008, https://doi.org/10.3389/fpsyg.2015.02008

Phillips, J. G., & Mann, L. (2019). Suicide baiting in the internet era. *Computers in Human Behavior, 92*, 29–36. https://doi.org/10.1016/j.chb.2018.10.027

Pierce, T. (2004). *Warfighting and disruptive technologies: Disguising innovation*. Routledge. https://www.taylorfrancis.com/books/mono/10.4324/9780203341551/warfighting-disruptive-technologies-terry-pierce

Pinker, S. (2012). *The better angels of our nature: Why violence has declined*. Penguin Books.

Polityuk, P., & Balmforth, T. (2022, October 17). Ukraine shot down 85-86% of Russian drones involved in latest attacks—Air force. *Reuters*. https://www.reuters.com/world/europe/ukraine-shot-down-85-86-russian-drones-involved-latest-attacks-air-force-2022-10-17/

Rai, T. S., & Fiske, A. P. (2012). Beyond harm, intention, and dyads: Relationship regulation, virtuous violence, and metarelational morality. *Psychological Inquiry, 23*(2), 189–193.

Rawls, J. (1971). *A theory of justice*. Cambridge: Belknap Press of Harvard University Press.

Robinson, P., Carrick, D., Connelly, J., & Robinson, P. (2009). The fall of the warrior king: Situational ethics in Iraq. In D. Carrick, J. Connelly, & P. Robinson (Eds.) *Ethics education for irregular warfare*. Routledge.

Robinson, M., Jones, K., & Janicke, H. (2015). Cyber warfare: Issues and challenges. *Computers & Security, 49*, 70–94. https://doi.org/10.1016/j.cose.2014.11.007

Roff, H. M., & Moyes, R. (2016). Meaningful human control, artificial intelligence and autonomous weapons. *Briefing Paper Prepared for the Informal Meeting of Experts on Lethal Au-Tonomous Weapons Systems, UN Convention on Certain Conventional Weapons*. https://article36.org/wp-content/uploads/2016/04/MHC-AI-and-AWS-FINAL.pdf

Rosen, S. P. (1991). *Winning the next war: Innovation and the modern military*. Cornell University Press.

Rosert, E., & Sauer, F. (2019). Prohibiting autonomous weapons: Put human dignity first. *Global Policy, 10*(3), 370–375. https://doi.org/10.1111/1758-5899.12691

Ross, D. (1956). Aristotle: The Nicomachean ethics. *Philosophy, 31*(116), 77–77.

Santoni de Sio, F., & Nucci, E. D. (2016). Drones and responsibility: Mapping the field. In E. Di Nucci, & F. Santoni de Sio (Eds.), *Drones and responsibility* (pp. 1–13). Routledge.

Santoro, M., Marino, D., & Tamburrini, G. (2008). Learning robots interacting with humans: From epistemic risk to responsibility. *AI & Society, 22*(3), 301–314. https://doi.org/10.1007/s00146-007-0155-9

Scharre, P. (2018). *Army of none: Autonomous weapons and the future of war* (First ed.). W. W. Norton & Company.

Schmid, J. (2017). The diffusion of military technology. *Defence and Peace Economics, 29*(6), 1–19. https://doi.org/10.1080/10242694.2017.1292203

Schoenherr, F. J. R., & Thomson, R. (2022). Ethical frameworks for cybersecurity: Applications for human and artificial agents. In A. J. Hampton, & J. A. DeFalco (Eds.), *The frontlines of artificial intelligence ethics* (pp. 141–161). Routledge.

Schoenherr, J. R. (2021). Designing ethical agency for adaptive instructional systems: The FATE of learning and assessment. In R. A. Sottilare & J. Schwarz (Eds.), *Adaptive instructional systems. Design and evaluation* (Vol. 12792, pp. 265–283). Springer International Publishing. https://doi.org/10.1007/978-3-030-77857-6_18

Schoenherr, J. R. (2022a). *Ethical artificial intelligence from popular to cognitive science* (1st ed.. Routledge.

Schoenherr, J. R. (2022b). Whose privacy, what surveillance? Dimensions of the mental models for privacy and security. *IEEE Technology and Society Magazine, 41*(1), 54–65. https://doi.org/10.1109/MTS.2022.3147536

Schoenherr, J. R. (2023a). The first total war and the sociotechnical systems of warfare. *IEEE Technology and Society Magazine, 42*(3), 42–56. https://doi.org/10.1109/MTS.2023.3299315

Schoenherr, J. R. (2023b, April 3). *How Russian and Iranian drone strikes further dehumanize warfare*. The Conversation. http://theconversation.com/how-russian-and-iranian-drone-strikes-further-dehumanize-warfare-201942

Schoenherr, J. R. (2023c, December 10). *Technologies like artificial intelligence are changing our understanding of war*. The Conversation. http://theconversation.com/technologies-like-artificial-intelligence-are-changing-our-understanding-of-war-217342

Schoenherr, J. R., Abbas, R., Michael, K., Rivas, P., & Anderson, T. D. (2023). Designing AI using a human-centered approach: Explainability and accuracy toward trustworthiness. *IEEE Transactions on Technology and Society, 4*(1), 9–23. https://doi.org/10.1109/TTS.2023.3257627

Schoenherr, J. R., & Burleigh, T. J. (2020). Dissociating affective and cognitive dimensions of uncertainty by altering regulatory focus. *Acta Psychologica, 205*, 103017. https://doi.org/10.1016/j.actpsy.2020.103017

Schoenherr, J. R., & Thomson, R. (2019). Dissociating cognitive and affective uncertainty using a general linear classifier. *Fechner Day 2019, 20.*

Schuelke-Leech, B.-A. (2018). A model for understanding the orders of magnitude of disruptive technologies. *Technological Forecasting and Social Change, 129*, 261–274.

Shachtman, N. (2011, October 7). Exclusive: Computer Virus Hits U.S. Drone Fleet. *Wired.* https://www.wired.com/2011/10/virus-hits-drone-fleet/

Shay, J. (2010). *Achilles in Vietnam: Combat trauma and the undoing of character.* Simon & Schuster.

Shay, J. (2014). Moral injury. *Psychoanalytic Psychology, 31*(2), 182–191. https://doi.org/10.1037/a0036090

Singh, U. K., Joshi, C., & Kanellopoulos, D. (2019). A framework for zero-day vulnerabilities detection and prioritization. *Journal of Information Security and Applications, 46*, 164–172. https://doi.org/10.1016/j.jisa.2019.03.011

Slovic, P. (2007). 'If I look at the mass I will never act': Psychic numbing and genocide. *Judgment and Decision Making, 2*(2), 79–95. https://doi.org/10.1017/S1930297500000061

Sorenson, J. (2014). Animals as vehicles of war. In A. Nocella, C. Salter, & J. K. C. Bentley (Eds.), *Animals and war: Confronting the military-animal industrial complex* (pp. 19–31). Lexington Books. https://rowman.com/ISBN/9781498520867/Animals-and-War-Confronting-the-Military-Animal-Industrial-Complex

Sparrow, R. (2007). Killer robots. *Journal of Applied Philosophy, 24*(1), 62–77.

Spearman, C. (1961). In J. J. Jenkins & D. G. Paterson (Eds.), *Studies in individual differences: The search for intelligence* (pp. 59–73). Appleton-Century-Crofts. https://doi.org/10.1037/11491-006

Stanovich, K. E. (2005). *The Robot's rebellion: Finding meaning in the age of Darwin.* University of Chicago Press.

Sternberg, R. J. (Ed.). (2000). *Handbook of intelligence* (1st ed.). Cambridge University Press.

Taşkın, M., & Sağsan, M. (2020). Social media intelligence from Turkish National Security Perspective: An exploration of the relationship between 'Detection Tactics' and 'Prevention Strategies' of PKK terrorist attacks. *Revista Argentina de Clínica Psicológica, 29*(5), 1948–1958.

Turchin, P., Hoyer, D., Korotayev, A., Kradin, N., Nefedov, S., Feinman, G., Levine, J., Reddish, J., Cioni, E., Thorpe, C., Bennett, J. S., Francois, P., & Whitehouse, H. (2021). Rise of the war machines: Charting the evolution of military technologies from the Neolithic to the Industrial Revolution. *PLOS ONE, 16*(10), e0258161. https://doi.org/10.1371/journal.pone.0258161

Turse, N. (2013). *Kill anything that moves: The real American war in Vietnam.* MacMillan.

Tversky, A., & Kahneman, D. (1992). Advances in Prospect theory: Cumulative representation of uncertainty. *Journal of Risk and Uncertainty, 5*(4), 297–323.

Urbina, F., Lentzos, F., Invernizzi, C., & Ekins, S. (2022). Dual use of artificial intelligence-powered drug discovery. *Nature Machine Intelligence, 4*(3), 189–191. https://doi.org/10.1038/s42256-022-00465-9

Weber, M. (1919). Politics as a vocation. In H. H. Gerth & C. W. Mills (Eds.), *From max weber: Essays in sociology* (pp. 77–128). Oxford University Press. https://perpus.wildanfauzy.com/Politik/Max%20Weber%20(1919)%20Politics%20as%20A%20Vocation.pdf

Wegner, D. M. (1987). Transactive memory: A contemporary analysis of the group mind. In B. Mullen & G. R. Goethals (Eds.), *Theories of group behavior* (pp. 185–208). Springer. https://doi.org/10.1007/978-1-4612-4634-3_9

Whalen, J. (2022, February 21). Russian drones shot down over Ukraine were full of Western parts. Can the U.S. cut them off? *The Washington Post.* https://www.washingtonpost.com/technology/2022/02/11/russian-military-drones-ukraine/

Wood, J. D., Ware, C. M., Correll, T., Heaton, J. E., McBride, T., & Haynes, J. T. (2018). Relationship between spiritual well-being and post-traumatic stress disorder symptoms in United States air force remotely piloted aircraft and intelligence personnel. *Military Medicine, 183*(9–10), e489–e493. https://doi.org/10.1093/milmed/usx032

Zaloga, S. J. (2011). *Unmanned aerial vehicles: Robotic air warfare 1917–2007.* Bloomsbury Publishing.

14 The Cosmopolitan Principle of Humanity in an Age of Autonomous Weapons

The Role of the Ethical Underpinnings

Berkant Akkuş

14.1 INTRODUCTION

Artificial intelligence (AI) applications have made relentless advances across a multitude of sectors in recent years (Stern, 2018). The controversies generated by large language model (LLM) platforms such as ChatGPT have proven to be effective public opinion lightning rods, triggering often fierce debates regarding both their utility and the ethical implications associated with humans and 'robot' interactions (Open AI, 2023). AI proponents argue that humanity's future prosperity will be driven by AI, especially as its technologies become increasingly sophisticated. A common pro-AI constituency refrain echoes this theme: AI will entirely resolve the mundane, time-consuming tasks across entire swaths of society, leaving humans with more time to create, problem solve, and enjoy the fruits of invention (Campos & Laurent, 2023).

AI detractors make equally vigorous assertions that draw their strength from the notion that AI technologies will inevitably evolve to the point where they operate beyond effective human control. In this grim dystopic future, human interests are not only subordinated to AI decision-makers (Selbst, 2019). AI systems will quickly evolve into freestanding entities that will lack the fundamental human ability to make complex ethical choices. The generalized fear expressed in many mainstream media reports is easily summarized: Machines will control humans, and not the other way around (Bagga, 2023).

These competing AI visions provide the following four-section critical discussion with its tension and intense intellectual appeal. There are few, if any more contentious AI merits' debate issues than those flowing from the selected discussion topic. A significant body of literature is emerging concerning AI-based weapon systems and frightening international humanitarian law (IHL) outcomes that may flow from their continued development and armed conflict use if military strategies and operational decisions are made by AI systems alone (The International Committee of the Red Cross [ICRC], 2015). IHL is a term that is used interchangeably across this scholarship with 'law of war', and 'law of armed conflict' labels (The International Committee of the Red Cross [ICRC], 2015).

Cosmopolitanism emerged during the European Enlightenment period (roughly 1700–1800) as an essential way to view human society (Bristow, 2023). Its central propositions are echoed in all leading 21st century international law and human rights frameworks: (1) individual is part of a single global community; and (2), this community and its values are anchored by the principle that every human life has inherent dignity, thus all community individuals have equal moral worth (Miller, 2021). The Universal Declaration on Human Rights 1948 (United Nations, 1948) linked dignity to fundamental rights guarantees: "All human beings are born free and equal in dignity and rights" (preamble).

DOI: 10.1201/9781003441700-14

The influential UDHR 1948 successor provisions captured within the European Convention on Human Rights 1950 (*European Convention on Human Rights*, 1950), Protocol 13 also uses 'dignity' in its literal meaning sense (Council of Europe, 2002), long with its 'inhuman' and 'degrading' companion concepts. The Convention Article 3 case law concerning the right to be free from torture, enslavement, and similar treatment references this right in terms including 'inhuman', and 'degrading' leaves little doubt concerning the influence that cosmopolitanism continues to assert across the present-day international community. These authorities include cases such as *M.S.S. v. Belgium and Greece* (*M.s.s. V. Belgium and Greece*, 2011), and *Khlaifia and Other v Italy* (*Khlaifia and Other v Italy*, 2016). These highlighted cosmopolitanism features (whose origins pre-date Enlightenment developments) are given fuller section two treatment.

It is also important to appreciate why the ECHR 1950 protections are highlighted in this project. Other international law instruments might seemingly serve present research purposes at least as well as the ECHR 1950 provisions its related ECtHR case law. The International Covenant on Civil and Political Rights 1966 (United Nations, 1966) Article 6 is a notable example (United Nations Human Rights Committee, 2018, [65]). It was decided that upon carefully reviewing which of the available international law instruments have consistently attracted the most scholarly and judicial attention concerning individual freedoms and personal dignity, ECHR 1950 and ECtHR decisions afforded the most attractive and accessible research options (Heselhaus & Hemsley, 2018 citing *Tyrer v. The United Kingdom*, 1978, [33]). Extensive database searches conducted using Westlaw, SSRN, and Google Scholar databases confirm that ECHR 1950-related research is more extensive and arguably more nuanced than what is available regarding ICCPR 1966 or other international law instruments cited in cosmopolitanism contexts (Niemi, 2021).

A philosophy of global ethics, cosmopolitanism acknowledges the emergence of an ethical community founded on transnational rights and obligations, transcending State interests and territorial bounds. A world community interest approach to norm-creation for new weapons technology is developed using cosmopolitan legal theory, specifically in the context of IHL (Ulgen, 2016). This method acknowledges issues of global concern that affect humanity, transcend the interests of particular States and the interstate dimension, and often call for transnational governance. The wealth of lethal autonomous weapon systems (LAWS) literature should benefit from the cosmopolitan perspective.

Lethal autonomous weapon systems (LAWs) are the AI technology subsets that bring the above-noted tensions into clearer focus (Pacholska, 2023). For present critical discussion purposes, LAWs are defined as any weapon systems that can advance critical military and battlefield operations without any direct human involvement. These operations include target acquisition, directing and coordinating attacks on enemy targets, and their destruction (Caron, 2020). While some LAWs are now deployed during combat to destroy or damage objects, the present discussion is directed at a more prominent LAWs – ethical debate element: the autonomous weapon systems designed to kill or injure humans (Amoroso & Tamburrini, 2020; Thurnher, 2014).

Their many detractors perceive LAWs as existential threats to many core IHL objectives (Amoroso & Tamburrini, 2018), namely: (1) Protecting anyone who is directly or actively participating in hostilities; (2) protecting civilian populations; (3) limiting (regulating) how warfare is conducted; and (4) protecting and respecting human dignity (The International Committee of the Red Cross [ICRC], 2015; Wagner, 2014). A succinct phrase summarizes the four noted IHL objectives while also reinforcing a contentious proposition that flows directly from the topic at hand: LAWs must be banned from the battlefield to ensure that fundamental human agency (and thus human accountability for military decision-making) remains intact (Egeland, 2016). The following sections set out the various arguments for and against prohibiting LAWs' use that take this proposition as its commencement point.

Viewed from this ethical perspective, the anti-LAWs constituency (those who endorse the above proposition) strongly resists the notion that life-and-death decisions (such as those made during virtually every military conflict) must not be 'ceded to machines' (Arms Control Association, 2018; Lobel, 2023). These arguments mirror the broader fears often expressed concerning AI systems generally as introduced above (Selbst, 2019).

These LAWs criticisms are not universally accepted as valid across the international law scholarship (Schmitt, 2013). The loosely aligned pro-LAWs academic cohort does not dismiss the human agency proposition outlined above, so much as these commentators see significant ethical value when autonomous weapon systems are understood using a results-oriented formulation (Arms Control Association, 2018). LAWs have greater potential precision and reliability than their conventional weapons counterparts, where AI equips these systems with safeguards against human error. When this argument is taken to its logical conclusion, LAWs might encourage heightened respect for IHL and human-ethical values, and thus reduce both the frequency and severity adverse humanitarian consequences that result from weapon system operational errors (Reeves et al., 2020).

There is an attractive realism that usually accompanies the pro-LAWs arguments. The ability to ensure that LAWs utilization does advance IHL, and ethics principles reflect the same problems that must be confronted by all other weapon system developers. The legality of any weapon use is inextricably linked to their design and how the weapons are used (Coyne & Alshamy, 2021). It is also acknowledged that a second argument adds further cogency to pro-LAWs positions. Military leaders and their governments have a positive duty to protect their own forces, and LAWs that are designed to eliminate human operation errors when deployed would conceivably promote these objectives (Arms Control Association, 2018). Further, in reasoning that tends to bring cosmopolitanism principles, IHL, and LAWs tensions as outlined above full circle, the following ICRC restatement acquires even greater critical discussion significance: "The principle of humanity forbids the infliction of all suffering, injury or destruction not necessary for achieving the legitimate purpose of a conflict" (The International Committee of the Red Cross [ICRC], 2015). One might assume that in an ideal military operations world, a perfectly functioning LAWs system and its precise targeting capabilities, is better aligned with how the ICRC conceives this 'principle of humanity'. This point receives more detailed section two and Three consideration.

The LAWs – IHL – ethics tensions receive a further accelerant when a related pro-LAWs position is fairly assessed. There is at least superficial merit in the notion that if these weapon systems are correctly utilized in military conflicts, better overall adherence to IHL because challenges that confront human soldiers such as post-traumatic stress disorder (PTSD) are likely reduced. LAWs critics offer an equally compelling rebuttal. They suggest that given the complex stew of legal and moral concerns that LAWs generate, only where human control over warfare conduct is guaranteed will LAWs 'autonomy' power potential for significant civilian harm and property destruction be avoided (Amoroso & Tamburrini, 2020). The book chapter research questions have been crafted in ways that ensure these tensions are carefully explored across sections two and three.

On this principled basis, if LAWs deployment directly or indirectly contributed to indiscriminate attacks and associated severe harm for human society, such weapons should be prohibited – the inherent values represented by human life must take priority over LAWs, irrespective of any benefits that these systems might promote within current IHL and human ethics frameworks (Amoroso & Tamburrini, 2020). The two research questions formulated for critical discussion purposes are designed to more fully probe the issues outlined above. These interconnected questions are summarized as:

1. What if any LAWs designs and their related operational features are compatible with cosmopolitanism and its 'principle of humanity'?
2. To what extent have international law scholars and military policymakers accepted that LAWs can be IHL compatible?

The sources cited in this discussion Introduction indirectly confirm why the project research has a strong secondary source focus. It is essential that LAWs – IHL – ethics issues are appropriately placed within a clear legal principles' framework. Debating the relative merits of LAWs must not occur in isolation from the law as defined by IHL and related human rights considerations (Creswell & Creswell, 2018). However, the two research questions also encourage LAWs discussions that extend beyond the relatively narrow confines of doctrine, normative standards, and 'black letter' legal rules (Tyler, 2017). Socio-legal research methods provide excellent ways to supplement the discussion as these techniques ensure that the cited sources include what might be broadly cast as the 'societal effects' that LAWs will have if their AI-based autonomy is given its fullest possible scope (Chynoweth, 2008).

The discussion points presented in the following sections are not based on any preconceived notions of how a fully LAWs-enabled IHL and human rights landscape might differ from current military operations (theory and practice). This approach represents a conscious effort to emphasize a balanced, objective view of a wide ranging, continually evolving scholarship that at times seems to struggle keeping up with LAWs technological developments (Kowalczewska, 2019; Vries, 2023). It is apparent that the weight and analytical directions taken in these commentaries tend to regard LAWs as fundamentally 'anti' IHL and thus systems that undermine human rights. The conclusions presented in section four tend to align with these prevailing views, but LAWs potentially positive international law effects cannot be entirely discounted. In the following section two legal and human rights concepts are presented with this research philosophy as guidance.

14.2 COSMOPOLITANISM, LAWs, AND ETHICS – AN INTRIGUING DYNAMIC

The book chapter research proposal included a question that also provides this section with a clear commencement point: Which IHL principles are traceable to the cosmopolitan heritage of international law? The answer to this question equips the reader with a sound legal platform from which they can fully engage with the discussion themes (Bellaby, 2021). When the Enlightenment era thinkers who gave cosmopolitanism theory its initial impetus described their work, Bristow suggests that these philosophers did not regard the Enlightenment as a historical period (Bristow, 2023). Instead, a group that included Grotius, Pufendorf, and Kant saw themselves as contributing to a process that included social, psychological, or spiritual development elements that were also unbounded by time or place (Bristow, 2023). Kant (1959) defined enlightenment in a well-known essay entitled ("An Answer to the Question: What is Enlightenment?"), in the following terms: (1) humankind's release from self-incurred immaturity; where (2), this immaturity reflects an inability to use one's personal understanding without having to either seek out, or then rely upon the guidance of another individual (Kant, 1959).

For the modern reader, it seems preferable to treat Kant's 'immaturity' descriptor as akin to the ongoing present-day societal evolution that has placed ever-greater emphasis on the individual, and their human rights-based ability to make autonomous decisions – irrespective of whether the 'community' standard means that others necessarily agree with them (Heselhaus & Hemsley, 2018). Kant provides an effective synthesis of how otherwise often widely divergent Enlightenment thinkers regarded their time and place in intellectual history. For Kant, true human enlightenment only begins when individuals undertake to "… think for themselves to employ and rely on one's own intellectual capacities in determining what to believe and how to act" (Bristow, 2023). The following discussion points taken from the current IHL literature sources illustrate the profound importance of Kant and his *Enlightenment* philosophy contemporaries' work in shaping modern IHL doctrines.

It is perhaps surprising – even unsettling – for newcomers to IHL studies that this international law discipline has its companion expressions 'law of warfare' and 'law of armed conflicts'. Humanitarian concepts are usually popularly associated with the notion that everyone assumes a positive moral duty to improve our human condition wherever reasonably possible (The International Committee of the Red Cross [ICRC], 2015). A useful technical humanitarianism definition is provided by Brewis et al. (2020): An active belief rooted in the principle that all human life has intrinsic value, whereby individuals engage in benevolent treatment including assisting others to reduce suffering and therefore improve the human condition motivated by moral, altruistic, and emotional considerations (p. 3).

The 'unsettling' description introduced above is linked to a seemingly self-evident warfare – humanitarianism philosophical contradiction. Throughout human history, countless wars have produced incalculable human suffering that has often resulted in humanitarian crises of remarkable magnitude. Ceaseless refugee crisis and related war-related displacement of communities are two notable examples (Bloch, 2019; The International Committee of the Red Cross [ICRC], 2015). IHL and humanitarianism would therefore seem to be a conceptually uneasy fit – the well-known aphorism 'War is hell' lends further strength to the argument that anything involving armed force and humanitarianism are strange bedfellows (Anderson, 2011).

When this reasoning is taken to a logical conclusion, conducting war in ways that are intended to either destroy, or at least neutralize an enemy, and humanitarianism concepts are seemingly a contradiction in terms. This logical inconsistency is at least partially resolved by considering how the international emergency aid organization 'Doctors without Borders/Médecins Sans Frontières' (MSF) explains how military conflict-related crisis emergency responses *should* be understood in humanitarianism terms (as quoted in the Harvard Law School Human Rights Journal, 2007). MSF accepts that humanitarianism's pivotal principles are humanity, neutrality, independence, and impartiality, ones that collectively support a "conviction that all people have equal dignity by virtue of their being human based solely on need, without discrimination among recipients" (Harvard Law School Human Rights Journal, 2007). For this reason, irrespective of a given military conflict and any horrific outcomes that might flow directly from it, humanitarian organizations are obligated to refrain from taking any active role in any events that might create an advantage for one side over another, or where the organizations' involvement might serve political, religious, or other agendas' interests (Harvard Law School Human Rights Journal, 2007).

These approaches to humanitarianism demonstrated in an armed conflict situations thus serve two vital purposes. The first is their embodiment of how human suffering can be alleviated without the organization having ulterior motive. Second, they create a backdrop against which operational tools can be developed, ones that assist in securing consent from the communities directly impacted by warfare to have humanitarian organizations operating nearby – especially in volatile, dangerous active combat situations (Harvard Law School Human Rights Journal, 2007).

As indicated above, 'successful' military operations usually involve killing enemy forces and destroying (or neutralizing) enemy property. If one accepts the admittedly general proposition that wars or other armed conflicts are part of our human condition, any international laws that make these conflicts more 'humane' are furthering worthy objectives (Tasioulas & Verdirame, 2022). The extensive Geneva Convention 1949 provisions have directly contributed to IHL, and the following selected Convention extracts also more fully inform the entire LAWs – IHL – ethics discussion (of numerous examples, ICRC, 2023; The International Committee of the Red Cross (ICRC), 2014, its Protocol I of 1977, Article 50). These early Convention deal primarily with the treatment of civilians, non-combatants, and prisoners of war under combat circumstances.

A lesser known but important Geneva Convention requires specific elaboration here – the 1980 United Nations Convention on Certain Conventional Weapons (Arms Control Association, 2018). It is remembered that as a core international law proposition, any nations that ratify a validly made international treaty like the CCWC are accepting a positive, non-derogable obligation to implement the relevant treaty terms into the State's domestic law regime (Vienna Convention on the Law of Treaties (1969), 1969, Articles 2, 23 through 30). The full Convention name also serves as an excellent summary of its international law scope and effect: "Convention on Prohibitions or Restrictions on the Use of Certain Conventional Weapons Which May Be Deemed to Be Excessively Injurious or to Have Indiscriminate Effects" (Arms Control Association, 2018 as amended in 1995, and 2003).

The further stated Convention purpose also embodies what 'Humanitarian' and the cosmopolitanism 'Principle of Humanity' would seemingly encourage across all societies: "… to ban or restrict the use of specific types of weapons that are considered to cause unnecessary or unjustifiable suffering to combatants or to affect civilians indiscriminately" (Arms Control Association, 2018, preamble). This Convention purpose also parallels the comment offered above regarding the ironies associated with concepts known as 'international humanitarian law' that are applied where human armed forces are intent on causing death and destruction where they deem such actions necessary to advance their strategic interests. The emphasis placed on 'unnecessary or unjustifiable suffering' is language plainly calculated to underscore a humanitarian interest.

The Convention purposes are furthered by the five Convention protocols defining the weapons that its provisions are devised to govern, or otherwise restrict the use of: (1) weapons that generate 'non-detectable fragments'; (2) landmines and booby trap devices; (3) incendiary weapons such as flame throwers; (4) blinding laser weapons; and (5), 'clearance of explosive remnants of war', including State obligations and best practices (Arms Control Association, 2018, Protocols I through V). These measures individually and collectively seem well-aligned with the 'humanitarian' element in IHL concepts as explained above – until one carefully reviews the weak, readily circumvented Convention monitoring and enforcement provisions (Arms Control Association, 2018). Among various examples suggesting that profound gaps exist between effective IHL control over how States use weapons of war, and actual practice, of the 126 States that have ratified the Convention, many have only endorsed two of the five Protocols (the legal minimum to be regarded as an international law State Party) (Arms Control Association, 2018; Vienna Convention on the Law of Treaties, 1969).

It is not surprising that the UN CCWC policymakers convened international talks in 2017 to deal with the new generation of LAWs and their potential for more lethal destruction than many currently restricted or prohibited weapons systems now defined with the Convention framework (The Economist, 2018). The article cited here references the history of nuclear weapons arms control, arguably the previously most discussed (and most worrisome) single weapons technological advance in human history. It is noted that notwithstanding the fears that have persisted across the entire international community since the two nuclear strikes the ended WWII in 1945 that another 'bomb' might end humankind, various nuclear weapons test bans have appeared to hold over time. It is noted that the Nuclear Non-Proliferation Treaty (*Treaty on the Non-Proliferation of Nuclear Weapons (NPT)*, 1968), along with the subsequent Threshold Test Ban Treaty (*Threshold Test Ban Treaty*, 1974) and the Comprehensive Nuclear Test Ban Treaty (CTBT, 1996) are prominent international law examples.

However, *The Economist* report also emphasizes a grim LAWs reality that was still emerging in 2017, one that has become further entrenched in the international community consciousness in the following six years: the 'fast-approaching' military robotics revolution has a different, and even more challenging character than nuclear weapons regulation, given the multidimensional "ethical, legal, policy and practical problems" that LAWs represent (The Economist, 2018). The article's general pessimism expressed regarding whether LAWs could be brought within the CCWC 1980, and its protocols legislative frameworks has proven prescient (Coyne & Alshamy, 2021). The following extract from another CCWC 1980 – LAWs research article effectively summarizes the current dilemmas that exist because these weapons systems remain unregulated under IHL or any other international legal instrument:

> … The world just blew a 'historic opportunity' to stop killer robots … Several states, including the US, Russia, the United Kingdom, India, and Israel, [remain] opposed to any legally binding restrictions… China has supported a binding legal agreement at the [CCWC 1980 talks], but has also sought to define autonomous weapons so narrowly that much of the A.I.-enabled military equipment it is currently developing would fall outside the scope of such a ban….

(Khan, 2021)

Khan (2021) also noted that eight years of efforts made by the UN 'Disarmament and International Security Committee' had resulted in little to no progress regarding whether or to what extent LAWs can be regulated or made the subject of outright prohibitions made against their deployment (*United*

Nations Disarmament Commission – UNODA, 2023). The current situation thus provides a prelimi-nary partial answer to the second project research question outlined in section one: International law scholars and military policymakers (at the UN level) have not yet taken a final position that LAWs can be IHL compatible, given that efforts to place LAWs under the existing CCWC 1980 framework have proven unsuccessful.

Some scholars also argue that the entire LAWs debate has stagnated because some international community stakeholders have not made the important conceptual differentiations between the two primary LAWs introduced in section one. Jenks (2016) observes that many international anti-LAWs campaigners have sought to ban all present and future LAWs – notwithstanding the fact that since 1980, over 30 nations have possessed weapons systems with the onboard capability to select and engage targets without further human intervention (see also the global Campaign to Stop Killer Robots, 2023; Jenks, 2016). Jenks (2016) contends that the UN is guilty of making the same concep-tual error, given that its Special Rapporteur to the Human Rights Council unequivocally argued that "autonomous weapons systems that require no meaningful human control should be prohibited" (Kiai & Heyns, 2016).

The Jenks (2016) proposal ensures that an 'autonomy focus' is directed at how LAWs make tar-get selection and then pursue engagement with the target. An attractive point is extracted from this approach. It is less important to define weapons systems as automated, automatic, or autonomous (three terms frequently encountered in the LAWs literature that are used interchangeably, notwith-standing their literal meaning distinctions), than devising an IHL regime that governs all weapons that have the built-in capability of selecting and engaging targets without any further human inter-vention (citing Heyns, 2016). The concluding section discussion points concern the most recent LAWs technological developments, one that provides a further factual linkage between cosmopoli-tanism – principle of humanity theory and the section three examples.

This Jenks (2016) analysis provides the foundation for a seemingly a sensible proposal to break the current international community impasse regarding LAWs, given the various stalemates in the UN committee negotiations process explained above. What Jenks (2016) has outlined has not been adopted – nor does it appear likely that a new CCWC 1980 protocol governing LAWs develop-ment and armed conflict deployment is on the immediate IHL reform horizon (*United Nations Disarmament Commission – UNODA*, 2023). However, for present discussion purposes the Jenks (2016) position assists in gaining a more rounded appreciation of the intricate section three LAWs and ethics interplays that give this IHL area its special interest. Jenks (2016) notes that as the UN has been unable to persuade its membership that a preemptive ban of future LAWs is necessary, a more limited moratorium on LAWs that are primarily designed to lethally target military personnel is a viable law reform option (contrary to the Campaign to Stop Killer Robots, 2023 reform focus).

The book chapter Introduction provides a sound, if generally framed basis for concluding that LAWs have profoundly altered prior, well-understood IHL, and ethics understandings. The follow-ing specific examples highlight why LAWs must be understood from a cosmopolitanism perspective if the international community can ever successfully regulate these weapon systems' use (the first research question restated).

Coyne and Alshamy (2021) trace the ongoing LAWs evolution to support a sobering weapons systems' aspect. The first-generation LAWs were primarily defensive systems. Since the mid-2010s, the world's major military powers (including China, Russia, and the US), have focused their weapons systems' research on LAWs that rely on machine learning (AI) to potentially operate offensively when searching out targets. It is this narrower, but compelling weapons research and development (R&D) direction that Coyne and Alshamy (2021) impacts the issues that are most central to this dis-cussion (Scharre, 2018). It is the notion that these targets will be determined and then attacked with-out any human control or oversight that prompts the most serious LAWs fears in the current societal mainstream and academic commentators alike (Coyne & Alshamy, 2021). The LAWs scholarship also confirms that the international community members who are engaged in ongoing LAWs R&D justify their work on the following bases: (1) lower monetary and human costs (autonomous systems simply require fewer personnel for their operation); (2) much faster data analysis (evaluating possible strategic options) and the reaction times associated with this decision-making; (3) targets are struck with higher accuracy; and (4) a perceived need to remain competitive with the LAWs that other (rival) States are developing (Trumbull, 2020). The following hypothetical scenario is based upon various reports published about the current Russia-Ukraine war (Dawes, 2023; Pacholska, 2023).

State X has devised and implemented a semi-autonomous weapons system (code named Vulture). It operates as a bomb and drone weapon hybrid. The device can hover in the air for extended periods while it waits for an appropriate target to come within its range. Vulture was originally designed to be operated with significant human control over its key decision-making maintained throughout its engagement with any enemy targets. In its ongoing armed conflict with State Y, X has suffered significant losses resulting from Y's missile bases located near the specific battle areas. The casualties resulting from Y's attacks have included military and civilian persons. In some instances, the Y bases are located near (but not in) Y's civilian districts. It is noted that Y is not using the civilian districts or other locations such as hospitals or schools as 'cover' for its missile bases (an IHL violation, as defined by ICRC [2023], Geneva Convention 1949, Optional Protocols 1977, Rules 5, 9 [ICRC customary IHL provisions].

Vulture has new AI-powered software equipped with algorithms that now allows Vulture to autonomously select targets. X's military leadership gives Vulture its operational approval. The Vulture hybrid is deployed over a Y installation. Its AI decision-making process determines that (1) if Vulture delivers lethal strikes into civilian areas, (2) Y's emergency response to treat the Vulture strike casualties will distract the Y forces and thus reduce the risk that Y will successfully launch missiles into X's territories. Vulture directs a series of offensive attacks against a Y school and civilian hospital, killing and wounding over 500 civilians – and taking no Y military casualties. The Vulture strategy was successful, to the extent that Y's emergency response did reduce Y's immediate ability to respond or otherwise counter what Vulture had achieved.

The X-Y scenario events are horrific when viewed through a cosmopolitanism – principle of humanity lens. Innocent civilians have been killed and wounded. Y state public property and essential infrastructure including hospitals, schools, its national energy grid, and other utilities have also been damaged in these Vulture attacks (Safronova, 2023). This X conduct cannot be justified under current IHL principles, but the complete lack of any meaningful LAWs regulation logically leaves the following counterargument available for consideration. On an initial reading, this position might seem remarkably amoral when places against the legal and ethical principles outlined in sections one and two. X's military and government leadership might argue that the dispassionate (and for many humans, cold-blooded) Vulture AI-based decision-making might appear to have achieved entirely immoral outcomes, the Y casualties contributed to fewer corresponding X losses – thus a positive saving of X lives. The following section three examples build on this controversial LAWs justification.

14.3   SELECTED LAWS AND ETHICS EXAMPLES

The X-Y scenario appears to engage LAWs concepts that are entirely antithetical to cosmopolitism and its humanity preservation objectives. The international community efforts made to seek comprehensive LAWs R&D bans would seem especially meritorious if a system like Vulture can autonomously make decisions that entirely overcome human morality and allow potential legal cover for the humans who ultimately decided to deploy Vulture against Y in any capacity. It is noted in passing that under international criminal law frameworks (including the International Criminal Court (ICC) statute (1998)), Part 2 'Jurisdiction', as established through Articles 5–8), the Y Vulture weapon attacks fall within the ICC 'grave crimes' definitions (especially 'crimes against humanity', Article 7, and 'war crimes', Article 8). The relevant Article 7 language includes the following qualification: Attacks directed against any civilian population "… means a course of conduct involving the multiple commission of acts referred to in paragraph 1 against any civilian population, pursuant to or in furtherance of a State or organizational policy to commit such attack" (ICC, Article 7(2) and Article 8).

The ICC statute language also includes this direct statement that also aligns with the principle of humanity: The Court will exercise its jurisdiction over "… the most serious crimes" that "deeply shock the conscience of humanity", where the ICC concludes that these 'grave crimes' prosecutions will ultimately contribute to their future prevention (ICC, 1998, preamble). What X's Vulture system perpetrated was an attack that was specifically devised and executed to target humans, as opposed to seeking out vehicles, infrastructure, and other Y weapons (Campaign to Stop Killer Robots, 2023). The more intriguing issue raised by the X-Y scenario is whether the X leadership decision to essentially give Vulture autonomous control over what became an armed conflict decision involving life and death does not absolve them of ultimate ICC statute criminal responsibility. If the X leadership can rely on its detachment from how Vulture decided to strike civilian targets as an ICC statute Article 7 defense, the concerns expressed regarding overall LAWs legitimacy as violating the principle of humanity would appear to gather even greater strength (its 'digital dehumanization', as explained by the Campaign to Stop Killer Robots, 2023; Human Rights Watch, 2023).

A final point anchors this section and its LAWs – ethics examples. The X-Y scenario illustrates how autonomous weapons systems do not possess the more sophisticated, nuanced human judgment that permits a combatant to lawfully distinguish between civilians and legitimate military targets as defined by the various Geneva Convention 1949 provisions. The dispassionate, calculated X attacks also arguably entirely dehumanize war by reducing the perceived risks presented to the participants, thus entirely eroding meaningful human control over all battlefield outcomes. This proposition may be accurate, but it also does little to reduce the warfare – humanitarian definition ironies highlighted in section two (Brewis et al., 2020; Caron, 2020).

Further, the relevant (broader) IHL literature has considered how the so-called combatants' privilege might be invoked by X's leadership and then applied by any adjudicative body (like the ICC) responsible for assessing individual and collective criminal liability for the Vulture attack. It is emphasized that no matter how broadly one might seek to interpret this privilege, it cannot be extended to justify or otherwise insulate a perpetrator from criminal liability where they direct attacks at a civilian population. The Inter-American Commission on Human Rights (IACHR) has accepted that in limited circumstances, the relevant IHL frameworks appear to sanction this privilege, but it can assume the appearance of a *de facto* "… license to kill or wound enemy combatants and destroy other enemy military objectives" (Inter-American Commission on Human Rights, 2002). Scholars such as Dormann (2003) observe that lawful combatants' conduct (i.e., where they have remained within all Geneva Convention 1949 and its protocol definitional boundaries), they cannot be prosecuted for any acts or omissions occurring during any military operations.

It is therefore irrelevant for these criminal liability assessment purposes that the relevant combatant conduct is otherwise a serious crime if committed in any other non-armed combat circumstances (Dormann, 2003). The current law regarding a 'combatants' privilege' is thus readily summarized. All military personnel engaged in any activities related to armed combat participation are insulated from ICC or other criminal prosecutions unless they have violated IHL standards such as war crimes (ICC Statute, Articles 5–8). This overview thus sets the stage for considering where LAWs R&D is likely to potentially influence renewed cosmopolitanism – principle of humanity thinking.

The section two X-Y scenario and the rapidly accelerating LAWs technological developments might create the appearance of 21st technology having rendered obsolete a philosophy rooted in the 18th century Enlightenment. The following discussion points confirm that while cosmopolitanism is not a term that appears very frequently in mainstream media – LAWs commentaries, its principles, continue to enjoy acceptance – albeit more indirectly than Kant might have preferred when he was crafting his theories (Bristow, 2023). The modern cosmopolitanism-themed literature gives this argument further positive stimulation.

Kowalczewska provides insights that readily reinforce why the principle of humanity retains its centrality in all LAWs – ethics debates. Kowalczewska suggests that there is at least some persuasive power in the proposition that all current IHL paradigms support the notion that IHL seeks to achieve a reasonable (therefore workable) practical balance between humanity and military necessity considerations (Kowalczewska, 2019; citing Wagner, 2014). If this view correctly captures contemporary IHL thinking, the law will never perfectly echo what Kant described in his original treatments (Kant, 1959) as a philosophical ideal. Conversely, Kowalczewska (2019) also accepts that the largely practical IHL applications set out in sections one and two can benefit from a fresh injection of theory and thus give IHL are better, more stable underpinning. Kowalczewska (2019) does not expressly include the following phrasing in her analysis, but the meaning extracted by her words is clear – the cosmopolitanism humanity emphasis is essential to make IHL more robust and relevant in crucial contemporary LAWs – ethics debates (Kolb, 2014; Kowalczewska, 2019).

If Kowalczewska (2019) is correct, cosmopolitanism theory has never been separated from IHL. She accurately contends that the accepted principles of humanity and military necessity doctrines do not expressly receive mention in the various cited sections one and two treaties, they nonetheless remain part of the IHL normative standards in terms of both their core justification and basic considerations. These combined humanity and military considerations are closely linked to what might be characterized as IHL's original norms that promote avoiding unnecessary suffering, useless property or infrastructure destruction, and the fundamental obligation assumed by all armed conflict stakeholders to ensure that everyone connected to combat receives humane treatment (Kowalczewska, 2019; Żeligowski, 2014, as translated).

At the same time, IHL represents an entirely utilitarian approach to how armed force is employed in disputes arising between States, given that all 'military necessity' concepts depend upon a fundamental international law recognition that every warring State is pursuing victory in the most cost-effective manner possible: use as few human and material resources as possible, with compromising on the State's ability to achieve victory (Pettit, 2016, and a 'globalized' principle of humanity ideal). For Kowalczewska (2019), and the other IHL commentators cited here, there is a prevailing and arguably attractive notion that connects their various IHL interpretations: as understood within all IHL contexts, military necessity must also include equity, where this term represents the various methods that might be employed to reduce human losses and military resources' expenditures.

When this IHL proposition is taken to its own logical conclusion, military decision-makers must remain responsible for the lawful conduct of all combat operations involving LAWs. This approach has another positive benefit for IHL at large. It tends to give a desired practical aspect to what Kleingeld and Brown described as the 'nebulous core shared by all cosmopolitan views' – all human beings can and should be citizens in a single community that shares common values that place humanity at their center (Kleingeld & Brown, 2019).

If these Kowalczewska (2019) analytical points are repurposed and applied to the section two X-Y scenario, an interesting idea regarding LAWs legality begins to take shape. It is noted across the IHL scholarship that the UN Disarmament Commission (2023) has enjoyed scant success at best in its efforts to encourage a LAWs-specific international treaty. However, the combined Kleingeld and Brown (2019), and Kowalczewska (2019) analyzes effect leads directly to the following potential outcome for X and its military leadership. The X leadership decision to give its now autonomous Vulture weapon full discretionary power over life and death (the AI-based decision to attack Y's civilian population) cannot ever be justified on principle of humanity bases. Only where military necessity outweighs humanity can X leadership avoid international criminal law liability for their decision (this approach reflects the further, patently obvious proposition that Vulture cannot be held criminally liable – only those with human agency can be prosecuted) (The International Committee of the Red Cross [ICRC], 2018).

This further agency observation also takes the entire discussion along a reasoning pathway that further underscores how IHL balances humanitarian and military necessity concepts in ways that appear to give the principle of humanity superiority in most instances where the two concepts conflict (The International Committee of the Red Cross [ICRC], 2015, 2018). The Martens Clause

(ICRC, 2023a; The International Committee of the Red Cross [ICRC], 2014) invites specific consideration in this context. The Clause (as restated in the 1977 Additional Protocols) provides that:

> In cases not covered by this Protocol or by any other international agreements, civilians and combatants remain under the protection and authority of the principles of international law derived from established custom, from the *principles of humanity* and from dictates of *public conscience ....*

(Protocols Article 1(2))

The direct linkages made between cosmopolitanism, humanity, public conscience, and the allied notion that these principles contribute to enforceable legal standards is further, likely irrefutable evidence that the original 18th century Enlightenment concepts have been preserved intact within modern IHL. The cited ICRC authorities (especially ICRC, 2018) give the Martens Clause an emphatic, pro-cosmopolitanism interpretation. The International Committee endorses the view that properly understood, the Clause acts as a strong safeguard against States relying on acting on the presumption that where any armed combat-related conduct is not explicitly prohibited by relevant international law instruments, States may proceed on the basis that the relevant conduct is lawful (International Court of Justice, 1996; The International Committee of the Red Cross [ICRC], 2018).

The Martens Clause is intended to function as an international law – humanity safety net – especially when new military technologies such as LAWs and their implications are understood (The International Committee of the Red Cross (ICRC), 2018). As explained throughout sections one and two, the international community has not been able to determine where LAWs necessarily 'fit' within accepted CCWC 1980 frameworks and the five Convention Protocols (Khan, 2021; Reeves et al., 2020). A principled argument can be made that unless, or until LAWs are given express international legal instrument approval, such weapons should not be deployed in combat. *In theory*, all individuals authorizing such LAWs use would therefore *potentially* face criminal prosecution under these Martens Clause effects. It is equally apparent that this theoretical position has numerous practical obstacles – most notably the reality that there is no persuasive case law authority that has defined the Clause scope and effect (The International Committee of the Red Cross (ICRC), 2018).

It is also likely that if such prosecution arguments were advanced in a criminal trial, prospective defendants could mount the following, equally principles defense. The Martens Clause was articulated at a time in the world's geopolitical history where new technologies of war weaponry were employed far more indiscriminately than is the case with LAWs like the X – Y Vulture hybrid bomb. Of numerous examples, landmines and cluster bombs are weapons that often strike blindly, and without any control asserted over their eventual triggering event. The ICRC (2015) observed that as these weapons indiscriminately affect civilians, they also generate a profound 'ethical revulsion' resulting from 'landmine carnage' and associated 'appalling humanitarian consequences' – outcomes that led to a 1997 prohibition of these weapon (The International Committee of the Red Cross (ICRC), 2015; United Nations, 1997). Similar ethical revulsion might flow from X's Vulture autonomously deciding to use missile strikes against Y targets that it knew through AI power would create civilian casualties.

The defense success or failure would necessarily turn on the fundamental precision associated with LAWs weaponry. The following admittedly counterintuitive reasoning might be offered by X. IHL instruments like the 1997 Convention rightly emphasized the humanitarian concerns that center on the weapon's random killing and wounding effects. For example, X might lay landmines near its Y border. Y military personnel crossing the border region might be killed or maimed by a mine explosion. Curious X schoolchildren or unknowing adults might unwittingly trigger a mine – X forces laying these mines have no precise foresight regarding who will suffer landmine harm. The well-known expression 'a callous disregard for human life' applies to why the 1997 Convention enactments have clear 'principle of humanity' value.

In the fuller X-Y scenario context, X leadership could legitimately point to a distinguishing LAWs – Vulture feature that might create criminal liability insulation from ICC statute Article 7 or similar IHL liability. By its nature, the Vulture AI is precise. The underlying algorithms process all available data that supports its autonomous decision to launch missile strikes against the noted Y civilian targets. Vulture AI reached this decision based on how *it* understood military necessity imperatives at that moment in the X – Y conflict. Unlike the 1997 Convention purposes (ones that seek to limit random, and thus presumptively unnecessary military actions), Vulture made a focused determination of what offensive measures it could take to minimize possible harm that would be inflicted on X military personnel and civilian populations. This was a decidedly unpleasant, and arguably ruthless application of cause-and-effect reasoning by Vulture, a process paraphrased as: (1) my AI generated data says that if I attack Y civilian targets 1, 2, and 3, (2) X lives will be saved, and X property damage will be reduced (Coyne & Alshamy, 2021).

This point is not entirely rejected in the cited ICRC materials, notwithstanding the International Committee's well-expressed concerns that LAWs generally do not conform to IHL values (The International Committee of the Red Cross (ICRC), 2015, 2018). The ICRC acknowledges that some types of advanced LAWs weaponry (notably remote-controlled armed drones with precision-guided munitions) carry the potential for greater offensive attack precision, with concurrently lesser risks that indiscriminate weapon use effects will result (The International Committee of the Red Cross (ICRC), 2015). However, if the LAWs are provided with, or otherwise accesses inaccurate information concerning the intended target, where the technology is not sufficiently sophisticated, or persons and objects protected under the Geneva Conventions 1949 provisions are deliberately or accidentally targeted, the fact that Vulture or any other LAWs has greater precision attack potential is not a logical IHL violation defense (Caron, 2020; Egeland, 2016). When these various sub-arguments are carefully appraised, it appears that in any IHL balancing of humanitarian versus military necessity interests, a defense mounted for LAWs deployment that results in civilian harm will have to overcome a prevailing, principle of humanity-based view that humanitarian interests will be given priority over military necessity – all other factors being equal.

De Vries's paper is perhaps an unlikely source for detailed cosmopolitanism – principle of humanity analysis, given the often-international scholarly reputations that many of the commentators cited in this discussion have fostered over their careers (Vries, 2023). This work has particular utility for present discussion purposes because De Vries uses both significant amounts of case law (the section on 'doctrinal' research method) and he supplements these sources with well-crafted socio-legal opinions (the second section on research avenue). Two points are extracted from this De Vries paper as especially relevant to determining where the IHL humanitarian – military necessity balance might lie in all current LAWs debates (citing Kayser & Beck, 2018; Vries, 2023). These points effectively supplement the Martens Clause elements highlighted above. De Vries also demonstrates how cosmopolitanism theories have effectively influenced all IHL instruments commencing with the Hague Conventions, as their preambles also reinforce. The Hague Convention drafters asserted that until more complete laws of war codifications are created,

> … populations and belligerents remain under the protection and empire of the principles of international law, as they result from the usages established between civilized nations, from the laws of humanity and the requirements of the public conscience.

(ICRC, 2023a, preamble)

It is interesting to observe that no more 'complete' codifications have been drafted in the intervening 116 years, in the crucial sense that a single 'Laws of War' instrument does not exist. These laws have remained amendable to largely piecemeal evolution, as evidenced by UDHR 1948, the 1949 Geneva Conventions, ECHR 1950, and the later CWCC 1980, and other subject-specific warfare regulation (Mégret, 2016). It is doubted that any intended complete warfare law codification would have achieved the stated Hague Convention drafters' future ambitions in any event. No matter where one may stand on the IHL – humanitarian – ethics questions generated by these weapons systems, LAWs are proof that military weapons technologies are being developed at rates faster than law reforms could ever match. The piece meal responses are thus more appropriate in these specific IHL contexts than codification (Egeland, 2016).

De Vries uses the well-known International Criminal Tribunal – Yugoslavia Tribunal proceedings (ICTY, as created under Statute of the International Criminal Tribunal for The Former Yugoslavia 1993 auspices) to indirectly make an excellent cosmopolitanism – principle of humanity point, one that also resonates loudly in this critical discussion. The ICTY declared that at its essence, IHL and related international human rights law are each concerned with protecting every individual's human dignity (irrespective of their gender (ICTY, 1993)). The Tribunal defined the general principles as ones promoting respect for human dignity as the "… basic underpinning and indeed the very *raison d'être* of international humanitarian law and human rights law…" (*Prosecutor v. Anto Furundzija (Trial Judgement)*, 1998).

One might conceivably stretch the X-Y scenario Vulture LAWs defense position in the following direction. Relying on the precision that Vulture brings to AI-based missile strike decision-making, X leadership could argue that Vulture promotes human dignity in two ways: (1) its precision ensures that when targets are struck, only the intended harms will result; and (2), the fact that Vulture attacks contribute to saving X citizens' lives is itself consistent with dignity, and not human life degradation. The 'stretch' description employed above leads directly to an unambiguous rebuttal argument directed against X. The sources cited throughout this work also equate human dignity to the notion that only humans, and not AI constructs like Vulture, should ever make the ultimate decisions regarding human life and death. In this crucial way, dignity is properly synonymous with human laws of war accountability – Vulture cannot assume a form that possesses true human agency, and thus dignity as De Vries explains it is a necessary cosmopolitanism element in 2023 and beyond (Vries, 2023). The section four discussion conclusions are also consistent with this approach.

14.4 CONCLUSION

The two research questions posed in section one now provides the basis for the entire critical discussion conclusions. Each question is addressed in order.

1. *What if any LAWs designs and their related operational features are compatible with cosmopolitanism and its 'principle of humanity'?*

   It is apparent from the assembled discussion sources that the weight of current scholarly opinion (as opposed to any universal consensus being supported) favors this first question being answered in the following fashion. It is evident that when considered from a technical, weapons design perspective, LAWs, and their ability to utilize AI-enabled features directly contribute to the impression that LAWs decision-making ability is limited by the weaponry's inability to rely upon human ethics and related 'principle of humanity' sensibilities when operating autonomously from human military leadership control. This fact, if taken alone, would suggest that LAWs are generally incompatible with cosmopolitanism, its associated notions of universal ethical standards, and the 'principle of humanity'.

   There is the possibility (one that is indicated by the section two LAWs sources) that as these remarkably sophisticated systems continue to evolve, there may be ways that LAWs designers can incorporate true 'human' ethical reasoning and discernment into their systems. Until such developments are possible, the fact remains that an improperly supervised or controlled LAWs (the X – Y Vulture scenario) will act in a manner incompatible with IHL and its cosmopolitanism- inspired ethics. For this reason, research question 1 also supports the conclusion that a moratorium on LAWs development such as the approach advocated by Jenks (2016) is a sound law reform position (notwithstanding the obvious resistance that such a measure will meet from US, China, and other nations now firmly committed to LAWs R&D).

2. *To what extent have international law scholars and military policymakers accepted that LAWs can be IHL compatible?*

Once again, it is doubted that there is any true broad international community acceptance of the notions that LAWs can be utilized in a manner that includes full IHL (and by extension cosmopolitanism – principle of humanity) compatibility. There are some tentative suggestions taken from the extensively cited scholarship that LAWs are IHL compatible because (among other claimed positive LAWs deployment outcomes) fewer military personnel are placed at risk in combat where LAWs are utilized. These arguments are less convincing than those that collectively support this second research question answer (paraphrased) – "when the various cosmopolitanism components that include the principle of humanity and dignity are fairly considered, LAWs are not compatible because by its nature, the systems' autonomous nature contradicts what humanity and dignity must mean in laws of war terms".

A similar comment made concerning research question 1 applies with equal force here. There may be future LAWs technological developments that swing the current IHL humanitarian – military necessity interests more favorably toward the LAWs side of this equation. That day has not yet arrived, and until it does LAWs cannot conform to IHL as constructed from its cosmopolitanism foundations. The X-Y scenario analysis also confirms that when LAWs are permitted full battlefield decision-making autonomy, their dispassionate, cold detachment from human ethics, is a recipe for IHL disaster.

REFERENCES

Amoroso, D., & Tamburrini, G. (2018). The ethical and legal case against autonomy in weapons systems. *Global Jurist, 18*(1), 5–31. https://doi.org/10.1515/gj-2017-0012

Amoroso, D., & Tamburrini, G. (2020). Autonomous weapons systems and meaningful human control: Ethical and legal issues. *Current Robotics Reports, 1*(4), 187–194. https://doi.org/10.1007/s43154-020-00024-3

Anderson, K. (2011). Targeted killing and drone warfare: How we came to debate whether there is a 'Legal geography of War'. In P. Berkowitz (Ed.), *Future challenges in national security and law* (pp. 1–17). Hoover Institution. https://digitalcommons.wcl.american.edu/facsch_legsrp/5

Arms Control Association. (2018). *Convention on certain conventional weapons (CCW) at a glance*. Arms Control Association. https://www.armscontrol.org/factsheets/CCW

Bagga, A. (2023, November 9). *SubscriberWrites: The AI safety debate: Will my AI rule the world?* ThePrint. https://theprint.in/yourturn/subscriberwrites-the-ai-safety-debate-will-my-ai-rule-the-world/1838936/

Bellaby, R. W. (2021). Can AI weapons make ethical decisions? *Criminal Justice Ethics, 40*(2), 86–107. https://doi.org/10.1080/0731129X.2021.1951459

Bloch, A. (2019). *Refugees in extended exile: Living on the edge*. Routledge.

Brewis, G., Götz, N., & Werther, S. (Eds.). (2020). *Humanitarianism in the modern world*. Cambridge University Press. https://www.cambridge.org/core/books/humanitarianism-in-the-modern-world/humanitarianism-in-the-modern-world/A7229B64C53CCF89CD10B57BE95191C0

Bristow, W. (2023). Enlightenment. In E. Zalta & U. Nodelman (Eds.), *The Stanford encyclopedia of philosophy*. https://plato.stanford.edu/archives/fall2023/entries/enlightenment/

Campaign to Stop Killer Robots. (2023). *Stop killer robots*. Stop Killer Robots. https://www.stopkillerrobots.org/

Campos, S., & Laurent, R. (2023). A Definition of General-Purpose AI Systems: Mitigating Risks from the Most Generally Capable Models. *SaferAI*. https://doi.org/10.2139/ssrn.4423706

Caron, J.-F. (2020). Defining semi-autonomous, automated and autonomous weapon systems in order to understand their ethical challenges. *Digital War, 1*(1), 173–177. https://doi.org/10.1057/s42984-020-00028-5

Chynoweth, P. (2008). *Legal research in the built environment: A methodological framework*. International Conference on Building Education and Research (BEAR) Content Language English Conference, University of Salford, UK. https://citeseerx.ist.psu.edu/document?repid=rep1&type=pdf&doi=3430f4cb45ab598979565208340e0c08fa1e9b97

Council of Europe. (2002). *Protocol no. 13 to the convention for the protection of human rights and fundamental freedoms, concerning the abolition of the death penalty in all circumstances*. Council of Europe. https://www.coe.int/en/web/conventions/-/council-of-europe-protocol-no-13-to-the-convention-for-the-protection-of-human-rights-and-fundamental-freedoms-concerning-the-abolition-of-the-death-1

Coyne, C., & Alshamy, Y. A. (2021). Perverse consequences of lethal autonomous weapons systems. *Peace Review, 33*(2), 190–198. https://doi.org/10.1080/10402659.2021.1998747

Creswell, J. W., & Creswell, D. (2018). *Research design: Qualitative, quantitative, and mixed methods approaches.* Sage.

CTBT. (1996). *The comprehensive nuclear-test-ban treaty.* CTBTO. https://www.ctbto.org/our-mission/the-treaty

Dawes, J. (2023, February 21). *War in Ukraine accelerates global drive toward killer robots.* The Conversation. http://theconversation.com/war-in-ukraine-accelerates-global-drive-toward-killer-robots-198725

Dormann, K. (2003). "The legal situation of "unlawful/unprivileged combatantsn". *International Review of the Red Cross, 85*(849), 45–74.

Egeland, K. (2016). Lethal autonomous weapon systems under international humanitarian law. *Nordic Journal of International Law, 85*(2), 89–118. https://doi.org/10.1163/15718107-08502001

European Convention on Human Rights. (1950). https://www.echr.coe.int/documents/d/echr/convention_ENG

Harvard Law School Human Rights Journal. (2007). *Harvard Law School Human Rights Journal.* https://journals.law.harvard.edu/hrj/

Heselhaus, S., & Hemsley, R. (2018). Human dignity and the European convention on human rights. In P. Becchi & K. Mathis (Eds.), *Handbook of human dignity in Europe* (pp. 1–24). Springer International Publishing. https://doi.org/10.1007/978-3-319-27830-8_47-1

Heyns, C. (2016). Human rights and the use of autonomous weapons systems (AWS) during domestic law enforcement. *Human Rights Quarterly, 38*(2), 350–378.

Human Rights Watch. (2023). *Killer robots.* Human Rights Watch. www.hrw.org/topic/arms/killer-robots

ICRC. (2023). *Hague conventions of 1899 and 1907.* https://casebook.icrc.org/a_to_z/glossary/hague-conventions

ICTY. (1993). *Statute of the International Criminal Tribunal for the former Yugoslavia, Security Council resolution 827 (1993).* https://legal.un.org/avl/ha/icty/icty.html#:~:text=Acting%20under%20Chapter%20VII%20of,violations%20of%20international%20humanitarian%20law

Inter-American Commission on Human Rights. (2002). *Report on terrorism and human rights. OEA/Ser.L/-V/ II.116 Doc. 5 rev. 1 corr.* http://hrlibrary.umn.edu/iachr/terrorism-index.html

International Court of Justice. (1996). *Legality of the Threat or Use of Nuclear Weapons, Advisory Opinion, ICJ Reports 1.* https://www.icj-cij.org/case/95

Jenks, C. (2016). The gathering swarm: The path to increasingly autonomous weapons systems. *Jurimetrics, 57*(3), 341–359.

Kant, I. (1959). What is Enlightenment? In L. W. Beck (Trans.), *Foundations of the metaphysics of morals and what is Enlightenment.* Liberal Arts Press.

Kayser, D., & Beck, A. (2018). *Crunch Time: European positions on lethal autonomous weapon systems (PAX/2018/07).* https://reliefweb.int/report/world/crunch-time-european-positions-lethal-autonomous-weapon-systems-update-2018

Khan, J. (2021, December 22). *The world just blew a historic opportunity to stop killer robots.* Fortune. https://fortune.com/2021/12/22/killer-robots-ban-fails-un-artificial-intelligence-laws/

Khlaifia and Other v Italy, Application no 16483/12 (Grand Chamber 15 December 2016). https://hudoc.echr.coe.int/fre#{%22itemid%22:[%22002-11454%22]}

Kiai, M., & Heyns, C. (2016). Joint report of the special rapporteur on the rights to freedom of peaceful assembly and of association and the special rapporteur on extrajudicial, summary or arbitrary executions on the proper management of assemblies ¶ 67(f), U.N. Doc A/HRC/31/66. *UN Human Rights Council.* https://digitallibrary.un.org/record/831673

Kleingeld, P., & Brown, E. (2019). Cosmopolitanism. In E. N. Zalta & U. Nodelman (Eds.), *The Stanford encyclopedia of philosophy* (2019th ed.). Metaphysics Research Lab, Stanford University. https://plato.stanford.edu/archives/win2019/entries/cosmopolitanism/

Kolb, R. (2014). *Advanced introduction to international humanitarian law.* Edward Elgar Publishing.

Kowalczewska, K. (2019). The role of the ethical underpinnings of international humanitarian law in the age of lethal autonomous weapons systems. *Polish Political Science Yearbook, 48*(3), 464–475. https://czasopisma.marszalek.com.pl/images/pliki/ppsy/48-3/ppsy2019305.pdf

Lobel, O. (2023). The law of AI for good. *Florida Law Review, 75*(6), 1073–1139. https://doi.org/10.2139/ssrn.4338862

M.s.s. v. Belgium and Greece, 30696/09 (ECtHR [GC] 21 January 2011). https://hudoc.echr.coe.int/fre?i=001-103050

Mégret, F. (2016). Part I Cross-Cutting Issues and Common Provisions, D The Geneva Conventions in Context, Ch.33 The Universality of the Geneva Conventions. In A. Clapham, P. Gaeta, & M. Sassòli (Eds.), *Oxford public international law.* Oxford University Press. https://opil.ouplaw.com/display/10.1093/law/9780199675449.001.0001/law-9780199675449-chapter-33?prd=OSAIL

Miller, J. (2021). Hugo Grotius. In E. N. Zalta & U. Nodelman (Eds.), *The Stanford encyclopedia of philosophy* (2021st ed.). Metaphysics Research Lab, Stanford University. https://plato.stanford.edu/archives/spr2021/entries/grotius/

Niemi, H.-M. (2021). The use of human dignity in legal argumentation: An analysis of the case law of the supreme courts of Finland. *Nordic Journal of Human Rights, 39*(3), 280–299. https://doi.org/10.1080/18918131.2021.1999576

Open AI. (2023). *Introducing ChatGPT.* https://openai.com/blog/chatgpt

Pacholska, M. (2023). Autonomous weapons. In B. Brożek, O. Kanevskaia, & P. Pałka (Eds.), *Research handbook on law and technology.* Edward Elgar. https://papers.ssrn.com/abstract=4388065

Pettit, P. (2016). The globalized republican ideal. *Global Justice: Theory, Practice, Rhetoric, 9*(1), Article 1. https://doi.org/10.21248/gjn.9.1.101

Prosecutor v. Anto Furundzija (Trial Judgement), IT-95-17/1 (International Criminal Tribunal for the Former Yugoslavia 10 December 1998). https://www.refworld.org/jurisprudence/caselaw/icty/1998/en/20418

Reeves, S. R., Alcala, R. T. P., & McCarthy, A. (2020). Challenges in regulating lethal autonomous weapons under international law. *Southwestern Journal of International Law, 27*(1), 101–118.

Safronova, O. (2023, November 18). *Russian Drone attack cut power supply to Ukrainian consumers.* Bloomberg. https://www.bloomberg.com/news/articles/2023-11-18/russian-drone

Scharre, P. (2018). *Army of none: Autonomous weapons and the future of war* (First ed.. W. W. Norton & Company.

Schmitt, M. N. (2013). Autonomous weapon systems and international humanitarian law: A reply to the critics. *Harvard National Security Journal, 4*(1), 1–38. https://doi.org/10.2139/ssrn.2184826

Selbst, A. D. (2019). Negligence and AI's human users. *Boston University Law Review, 100*(1), 1315–1376.

Stern, S. (2018). Introduction: Artificial intelligence, technology, and the law. *University of Toronto Law Journal, 68*(supplement 1), 1–11. https://doi.org/10.3138/utlj.2017-0102

Tasioulas, J., & Verdirame, G. (2022). Philosophy of international law. In E. N. Zalta (Ed.), *The Stanford encyclopedia of philosophy* (Summer 2022). Metaphysics Research Lab, Stanford University. https://plato.stanford.edu/archives/sum2022/entries/international-law/

The Economist. (2018, January 25). Autonomous weapons are a game-changer. *The Economist.* https://www.economist.com/special-report/2018/01/25/autonomous-weapons-are-a-game-changer

The International Committee of the Red Cross (ICRC). (2014). *The Geneva Conventions of 1949 and their Additional Protocols.* https://www.icrc.org/en/document/geneva-conventions-1949-additional-protocols

The International Committee of the Red Cross (ICRC). (2015). *What is IHL?* https://www.icrc.org/en/document/what-ihl

The International Committee of the Red Cross (ICRC). (2018). *Ethics and autonomous weapon systems: An ethical basis for human control?* https://www.icrc.org/en/document/ethics-and-autonomous-weapon-systems-ethical-basis-human-control

Threshold Test Ban Treaty. (1974). U.S. Department of State. //2009-2017.state.gov/t/isn/5204.htm

Thurnher, J. S. (2014). Examining autonomous weapon systems from a law of armed conflict perspective. In H. Nasu & R. McLaughlin (Eds.), *New technologies and the law of armed conflict* (pp. 213–228). T.M.C. Asser Press. https://doi.org/10.1007/978-90-6704-933-7_13

Treaty on the Non-Proliferation of Nuclear Weapons (NPT). (1968). https://disarmament.unoda.org/wmd/nuclear/npt/

Trumbull, C. (2020). Autonomous weapons: How existing law can regulate future weapons. *Emory International Law Review, 34*(2), 533–594.

Tyler, T. R. (2017). *Methodology in Legal Research* (3). *13*(3), Article 3. https://doi.org/10.18352/ulr.410

Tyrer v. the United Kingdom, 5856/72 (ECtHR 25 April 1978). https://hudoc.echr.coe.int/fre?i=001-57587

Ulgen, O. (2016). "World community interest" approach to interim measures on "Robot Weapons": Revisiting the nuclear test cases. *New Zealand Yearbook of International Law, 14*(1), 3–34. https://doi.org/10.1163/9789004345911_002

United Nations. (1948). *Universal declaration of human rights.* United Nations; United Nations. https://www.un.org/en/about-us/universal-declaration-of-human-rights

United Nations. (1966). *International Covenant on Civil and Political Rights 1966.* https://www.ohchr.org/en/instruments-mechanisms/instruments/international-covenant-civil-and-political-rights

United Nations. (1997). *Convention on the Prohibition of the Use, Stockpiling, Production and Transfer of Anti-Personnel Mines and on their Destruction of 1997.* https://legal.un.org/avl/ha/cpusptam/cpusptam.html

United Nations Disarmament Commission – UNODA. (2023). https://disarmament.unoda.org/institutions/disarmament-commission/

United Nations Human Rights Committee. (2018). *General Comment No. 36 (2018) on Article 6 of the International Covenant on Civil and Political Rights, on the Right to Life.* https://www.ohchr.org/en/calls-for-input/general-comment-no-36-article-6-right-life

Vienna Convention on the Law of Treaties (1969). https://legal.un.org/ilc/texts/instruments/english/conventions/1_1_1969.pdf

Vries, B. de. (2023). Individual criminal responsibility for autonomous weapons systems in international criminal law. In *Individual criminal responsibility for autonomous weapons systems in international criminal law.* Brill Nijhoff. https://brill.com/display/title/63377

Wagner, M. (2014). The dehumanization of international humanitarian law: Legal, ethical, and political implications of autonomous weapon systems. *Vanderbilt Law Review, 47*(5), 1371–1424.

Żeligowski, M. (2014). Zakazy i ograniczenia użycia środków prowadzenia działań zbrojnych w świetle międzynarodowego prawa humanitarnego konfliktów zbrojnych. In Z. Falkowski & M. Marcinko (Eds.), *Międzynarodowe prawo humanitarne konfliktów zbrojnych* (pp. 229–267). Wojskowe Centrum Edukacji Obywatelskiej.